Environmental Management

Concepts and Practical Skills

This contemporary textbook and manual for aspiring or new environmental managers provides the theory and practical examples needed to understand current environmental issues and trends. Each chapter explains the specific skills and concepts needed for today's successful environmental manager, and provides skill development exercises that allow students to relate theory to practice in the profession. Readers will obtain an understanding not only of the field, but also of how professional accountability, evolving science, social equity, and politics affect their work. This foundational textbook provides the scaffolds to allow students to understand the environmental regulatory infrastructure, and how to create partnerships to solve environmental problems ethically and implement successful environmental programs.

Marc Lame is Professor Emeritus in the O'Neill School of Public and Environmental Affairs at Indiana University. He is an entomologist and has taught courses on Environmental Management; Risk, Trust, Credibility, and Public Participation; Natural Resource Management and Policy; and Applied Ecology. He maintains a clinical practice diffusing environmental innovations. Marc is on the advisory committee to the US Environmental Protection Agency (EPA) and helped to develop the Integrated Pest Management (IPM) education program offered by the National Environmental Health Association and the Centers for Disease Control and Prevention.

Richard Marcantonio is a teaching assistant professor in the Department of Management and Organization at the Mendoza College of Business, and a fellow at the Kroc Institute for International Peace Studies at the University of Notre Dame. A scholar-practitioner of environmental management and peacebuilding, he has conducted environmental and social science research on five continents and with varied communities, partnering with governmental, private, and nongovernmental organizations to pursue positive human and environmental outcomes. He is author of *Environmental Violence: In the Earth System and the Human Niche* (Cambridge University Press, 2022).

"*Environmental Management: Concepts and Practical Skills* is an extremely timely book addressing the challenges that executives will face in the decades to come. It is useful to college professors, students, and practitioners in their careers."

Professor Jeff Anstine, North Central College

"*Environmental Management* offers sage advice, grounded in practical realities, for ethical and effective management of pollution and natural resource problems. Lame and Marcantonio have written a fantastic textbook, filled with real-world examples and concrete lessons, that instructors will find valuable for training future environmental leaders."

Dr. David Konisky, Indiana University

"In an era when environmental management is often clouded by partisan politics and rhetoric, this book is a breath of fresh air teaching the next generation how to manage *for* the environment."

Professor Rosemary O'Leary, University of Kansas

"The textbook is full of insightful details, from emphasizing that environmental management is managing both people and nature, to highlighting the importance of understanding the scale, effect, and history of an issue at hand, and using past knowledge to inform decisions while anticipating future conditions. It challenges prospective and seasoned environmental managers with tough but necessary questions, evaluating your effectiveness and inclusion of equitable practices."

Professor Brian Watts, Flood-Prepared Communities Initiative, The Pew Charitable Trusts

"As a natural resource manager and professional, the book, while environmental management focused, is still relevant, as many of the trends and discussions occur in my world the same as they appear in the environmental management sphere. It's a great book for being able to begin to understand the ever changing and evolving world of environmental management."

Ben Weise, Contra Costa Resources Conservation District

Environmental Management

Concepts and Practical Skills

Marc Lame
Indiana University

Richard Marcantonio
University of Notre Dame

CAMBRIDGE
UNIVERSITY PRESS

CAMBRIDGE
UNIVERSITY PRESS

University Printing House, Cambridge CB2 8BS, United Kingdom

One Liberty Plaza, 20th Floor, New York, NY 10006, USA

477 Williamstown Road, Port Melbourne, VIC 3207, Australia

314–321, 3rd Floor, Plot 3, Splendor Forum, Jasola District Centre, New Delhi – 110025, India

103 Penang Road, #05–06/07, Visioncrest Commercial, Singapore 238467

Cambridge University Press is part of the University of Cambridge.

It furthers the University's mission by disseminating knowledge in the pursuit of education, learning, and research at the highest international levels of excellence.

www.cambridge.org
Information on this title: www.cambridge.org/highereducation/isbn/9781009100243
DOI: 10.1017/9781009110068

First published 2023

Printed in the United Kingdom by TJ Books Limited, Padstow Cornwall 2023

A catalogue record for this publication is available from the British Library.

ISBN 978-1-009-10024-3 Hardback
ISBN 978-1-009-11206-2 Paperback

Additional resources for this publication at www.cambridge.org/lame-marcantonio.

Brief Contents

Contents

Figures and Map

FIGURES

MAP

Real-World Examples, Author's Notes, and Interviews from the Field

REAL-WORLD EXAMPLES

AUTHOR'S NOTES

INTERVIEWS FROM THE FIELD

Preface

The United States has been learning critical lessons on how leadership and management, or more importantly, poor leadership and mismanagement, are addressing two human-caused existential threats to our species and planet: the pandemics and earth-systems change, the "Anthropocene." In a speech recently (2020) delivered at the Indiana University O'Neill School of Public and Environmental Affairs, former administrator of the United States Environmental Protection Agency (US EPA) Gina McCarthy opined in a 50-year review of the agency, originally "the mission was fundamentally to protect public health and the natural resources upon which we all depend." While this book is not about epidemiology or the global scope of managing human impacts on the environment, it *is* about the primary (to protect human health) and secondary (to protect the environment) management goals of our national environmental protection establishment. In this book we focus on domestic environmental laws, management concepts and skills, with the knowledge they can impact or be a model for international environmental management.

Environmental Management: Concepts and Practical Skills is written for scholars of environmental management and the environmental management professional. It is primarily focused on (1) the public sector – those of "Service" (civilian and uniformed service) – who are accountable for regulating human impacts on the environment; but also (2) those professionals in the private sector who, as corporate citizens, strive to comply with environmental policies (laws, rules, regulations, and guidelines); and (3) those in the not-for-profit sector who wish to influence the implementation of environmental regulation and requisite policy formulation. In other words, this book is intended to be of "tri-sectoral" utility.

We wrote this book from a perspective of clinical practice as opposed to management theory, though we do include a healthy dose of both. We thus incorporate management theories and the practical, real-world application and considerations of those theories. Collectively we bring together the experience of a state scientist and regulator turned environmental management professor and federal advisor, and a military professional turned human–environmental systems scientist and public health practitioner. The combination of our academic training and expertise brings to bear a management practicum for what our former students and clinical colleagues have suggested to us as "professional requirements to understand science, policy and management." This book should be viewed as a text for teaching and training but also as a reference and practical manual that can be used throughout the career of the environmental management professional. Our teaching philosophy relies on intellectual stimulation accompanied by entertaining stories, mantras, irreverence, and humor. It is written by authors who are and have been "mission oriented," who have participated in the program, resource and political management necessary to attain objectives: one who is old and crusty and one who is fresh, tested, and enlightened.

Much like how one might address an essay exam question, we provide detailed solutions (often by way of explanatory lists) to environmental management concepts and skills backed up with scholarly evidence and expert opinion. We rely heavily on a number of books, seminal articles and examples which bring together commonalities encompassing applied ecology, the importance of leadership and administration, ethics, economic competitiveness, and communication such that the environmental management professional can develop a foundation for accomplishment.

Special features in this text include Real-World Examples, Author's Notes, Interviews from the Field, and end of chapter questions. The first three features are meant to provide the reader with a clinical perspective to aid in concept and skill clarification but might also be critical of current practices based on real-world experiences. The Real-World Examples provide mini-case studies from the firsthand experiences of environmental management professionals ranging from civil servants at the US Environmental Protection Agency (EPA) to environmental practice consultants to a farming-focused non-profit. These professionals provide keen insights from their area of expertise that offer lessons for environmental managers of any stripe. The Author's Notes are our own take on a range of issues from funding of environmental management research to firsthand accounts of the challenges of diffusing innovations. Environmental management is rarely easy, so we offer our experience in certain challenging circumstances to help others better navigate their way when faced with similar contexts. The Interviews from the Field are a further extension of firsthand accounts of environmental management, from two environmental managers with a wealth of experience in public, private, and non-profit environmental management practice. The end of chapter questions were developed to allow the scholar to organize text material for application. We urge the reader to take advantage of these features.

Environmental Management: Concepts and Practical Skills addresses:

- Why we manage the environment – history from Malthus through Carson to current trends of human impacts on the environment including contemporary threats to human health and biodiversity.
- What is environmental management? Defining environmental management, what causes environmental problems and the differences between environmental management and natural resource management.
- What are the fundamental elements required for the successful environmental management professional? The diversity of training, worldview, expertise and experience requisite for an environmental manager and groups of managers to be effective managers is greater today than it has ever been.
- The nexus of management fundamentals, service and environmental protection – including the functions and requirements of the public manager. Multiple perspectives and approaches are needed because each area contributes to and has an interest in the environmental issues we face today.

- Who manages the environment? The different professions involved in environmental protection. How do inside and outside participants influence environmental protection and how do courts manage the environment? From the courtroom to the landowner, decisions about environmental management plans are made every day with a multitude of preferences, outcomes and regulations involved in the process.
- The structure of environmental regulatory programs in the US: federalism – US EPA HQ/region/state dynamics, tensions and solutions; understanding specific environmental media, offices and how to address "the silo" and the importance of mission with regard to the actions of federal departments or agencies.
- Critical and basic issues for the environmental manager and their managerial impact. Special attention is paid to environmental justice, jurisdiction, funding, sustainability, and legal/professional accountability and how these factors can and should affect the program, resource and political management of the environmental management professional.
- What authorities – laws, rules, and regulations – are in place for the environmental management professional to protect human health and the environment? Providing a basic understanding of 13 US environmental laws, summarizing their intent and major provisions.
- The basic legal trends and how they impact the environmental manager in terms of public accountability, property, and civil rights.
- The ethics of management for the environmental professional to understand and ethically address the forces that influence environmental managers. As well, how to co-produce environmentally ethical solutions with stakeholders while considering the managerial implications of the rights revolution. What defines environmental leadership?
- Communication: Why do we communicate and what are the differences between public and private communication? What are the professional requirements and responsibilities (legal and ethical) to address strategic communication planning, barriers to communication, and how to maintain trust and credibility? What are the specific skills required for risk and crisis communication including public participation and environmental dispute resolution? An observation of the skills and concepts necessary for working with the public and media including using the public ombudsman and public information officer.
- Program implementation and the questions that must be asked and answered before implementation. What are the necessary strategic planning and personnel management skills, including task assignment and situation analysis? The skills needed to get communities to adopt environmental innovations? How best to conduct a program evaluation and report? What management systems work for the environmental manager?
- Contracting for environmental management: Why do we contract out work and what are the advantages and disadvantages of contracting?
- How the environmental manager uses and/or reformulates policies. Understanding the participants and process for policy formulation. Basic policy tools and rules for the implementation of public environmental programs.

Acknowledgments

We would like to acknowledge those who helped this book to come to fruition after some years in concept and development. In general, there are three groups who come to mind: those who believed this text should be published as a contribution to the profession and its professionals; those who contributed to the text with their assistance and expertise; and those who allowed for its publication.

The first group consists of our graduate assistants who believed our conception of a text would help students and instructors to understand environmental management better. More so, that we could actually come up with stimulating and authentic ways of explaining the concepts and skills needed for today's environmental manager. For at least five years (ten semesters on the Teaching Assistant timescale), this group helped to brainstorm ideas for its organization and provided updated outlines. More important was the fact that they made us better scholars and instructors of environmental management, assisting us to improve our teaching continuously. Finally, this first group included some particularly persistent folks – Leanna McKeon, Julia Savia, Rebecca Ciciretti, Taylor Michel, Elizabeth O'Brien, Ashley Scholl, Ben Weise, Ben Young, and Rachael Sargent, who, as they became professionals, would follow up with the question, "How is your book coming along?"

We could not begin to acknowledge the contributions made to this book without starting with Rosemary O'Leary, Daniel Fiorino, Robert Durant, and Paul Weiland, whose 1999 text *Managing for the Environment* provided so much of the foundation of our book. Of particular value were the three elements that allow an environmental manager to be successful: understanding complex and volatile issues and trends; co-production of solutions; and the implementation of those solutions. This foundation guided us as we built new "scaffolds" and skills to encompass the profession and its scholarship as they have progressed since the publication of their book.

From there, we thank those colleagues who taught us as we were teaching and while we occasionally needed better information to include in this text. There are many such colleagues, but a few standouts would include: Jim Barnes, who patiently explained environmental laws in terms of the authority they provided and the goals they aspired to. Jim was a large part of crafting many of those laws, and then implementing them with his general counsel and administration as the US Environmental Protection Agency began and evolved – truly an honorable public servant. Evan Ringquist, who explained environmental policy and required policy wonks to follow the science; and David Konisky, who has a deadpan read of what is environmental justice and how it might be addressed. I (M. L.) thank my colleagues at the Arizona Department of Environmental Quality: Ed Fox, Bill Wiley, Nancy Wrona, Ira Domsky, Bill Norman, Dot Roberson, and John Godec, who taught me many of the practical skills of environmental management. I (M. L.) also thank Leon Moore, one of the early cotton entomologists who implemented sustainable pest management practices (Integrated Pest Management) and always admonished, "If you are not getting your hands dirty, you are not doing your job."

We want to acknowledge our contributors, especially those who provided "Real-World Examples" from their perspectives as practitioners or researchers (or both) of environmental management: Tricia Balluff, Roy Fillyaw, Jenna Larkin, Marc Marcantonio, Stan Meiburg, Jessie Mroz, Stephanie Redick, Pat Regan, Jorgen Rose, and Mary Willett. We also owe thanks to the senior environmental managers who allowed us to interview them: Ed Fox, Roger Ferland, and Tom Neltner. And thanks to Victoria Andersen, Becca Haussin, and Emily Szwiec, former students and TAs who allowed us to use their homework as examples, read our draft chapters, manufactured figures, and provided feedback.

There are two essential entities that allowed us to publish this book. First, our editors and publisher. Our readers should be aware that finding a world-class publisher and convincing them to publish your book is HARD. Cambridge University Press is a world-class academic publisher, and we want to acknowledge their confidence in us. Finally, and most importantly, we must acknowledge our families who pretended to believe our excuse, "We have to work on our book!" – thereby allowing us to skip out on certain family functions. But, more than that, they listened to us, guided us, and allowed us to really dig into this project and make it happen. Without them it would have been a non-starter. Thank you.

1 Introduction to Environmental Management

Science is the key foundation of everything EPA does. Science has defined the challenges, pushed the discoveries, it has operated as the foundation to design new solutions … it has been EPA's professor, our prosecutor and our protector.

Gina McCarthy,
Environmental Protection Agency administrator, 2012–16

1.1 Introduction

Every year tens of thousands of Americans and millions of people worldwide die from pollution (Fuller, Sandilya, and Hanrahan 2019). Pollution is the single largest source of human-caused death, killing 15 times more people than all violent crime and warfare combined (Landrigan et al. 2017). While there are many professions that contribute to the monitoring of these deaths and to the development of technical solutions that address pollution of the water, air, built environment and our land, it is the understanding of risks and the application of these solutions that will save lives and protect our environment. The genesis of the environmental movement and the subsequent demand for environmental management derives, in part, from Rachel Carson's (1962) book *Silent Spring* in which she illustrated what follows when humans disassociate themselves from the ecosystem and are denied "the right to know" regarding their health and the health of the environment.

An ecosystem is "A unit of nature in which living and non-living substances interact, with an exchange of materials between living and non-living parts" (Odum 1971). This unit could be our body, where living microorganisms interact with the non-living molecule, oxygen (O_2). If these substances are not in balance our bodies fail. Thus, in many ways, what our doctors are always admonishing us to do is to manage *our* environment. Other professionals look at ecosystems within local, state, or national jurisdictions and, of course, at the planetary level.

Environmental management in the broadest sense addresses how to keep these units of nature in balance in terms of what humans put into the ecosystem and take out. From a curricular perspective, **environmental management** is most often taught in terms of pollution, how humans unbalance the ecosystem by what they put into it, whereas **natural resource management** is taught in terms of how humans unbalance the ecosystem by what they take out of it. In general, the issues, trends, and human management in each are the same and there is much overlap. Probably the reason for these two "tracks" is that pollution laws regulating air, waste and water are most often enacted by standalone state and federal agencies, i.e., they are more centralized, and natural resource laws are regulated by a multitude of agencies associated with a natural resource: the Fish and Wildlife Service, the Conservation Service, the Bureau of Land Management, the Forest Service, etc.

Different cultures throughout history, from Indigenous peoples' reverence for an integrated human–ecological kinship, to pagan naturalism, to Talmudic protections of trees, to the American Evangelical embrace of efforts to curb global climate change, have been and are concerned with environmental protection. Famed conservation biologist Aldo Leopold, considered the progenitor of the modern sustainability movement, argued that it is only "when we see land as a community to which we belong, we may begin to use it with love and respect" (Leopold 1949). This idea of nature and human well-being as being intertwined is represented in the cosmologies and worldviews of myriad human cultures throughout human history and still today (Francis 2015; Fuentes 2017; Grim 1997).

The major theme of *Silent Spring* is that rather than maintaining that "mankind" dominates the natural world, humans must understand that they are an integral part of the ecosystem. Pre-Carson in the US there were three dominant "environmental ethics" that categorized how humans view their "place" in environmental protection: John Muir's *preservation ethic* – that we should protect the natural environment in a pristine, unaltered state; Gifford Pinchot's *conservation ethic* – humans should put natural resources to use but also we have a responsibility to manage them wisely; and Aldo Leopold's *land ethic* – humans should view themselves and "the land" as members of the same community and people are obliged to treat the land in an ethical manner (Westover 2016).

These ethics were originally applied to natural resource management. But, as the "ecology movement" grew from *Silent Spring*, it can be argued that Leopold's land ethic was the one that most aligned with the realization that human health is inextricably connected to a balanced ecosystem, complemented by Carson's recognition that human-produced pollution posed a new and grave assault on human health. The preservation and conservation ethics of Muir and Pinchot gave rise to organizations such as the US Forest Service and the National Park Service; and Leopold's land ethic motivated the development of natural resource management organizations such as the United States Fish and Wildlife Service (USFWS). It was Carson's *Silent Spring* combined with subsequent environmental disasters (e.g., the burning of Cuyahoga River and the Santa Barbara oil spill, both in 1969) that led to the development of the first national environmental management policy and agency in the world: the National Environmental Policy Act (US EPA 1969) and the US Environmental Protection Agency (EPA; Barnes, Graham, and Konisky 2021), respectively.

1.2 Environmental Management Is People Management

Often environmental scientists mix up environmental management with applied ecology which can be defined as a scientific discipline which uses ecological concepts to solve environmental problems. To be clear, humans can practice population ecology, study the effects of pollution on the ecosystem, and develop tools to manage the environment, but environmental management is people management. For example, whether or not we yet have an understanding or even a consensus that global climate change exists, what its causes are, and how to mitigate them, it will not be solved unless and until humans change their current behavior. Further, the mission of environmental managers, like that of the US EPA, is often "to protect human health and the environment," and in that order, i.e., human health takes primacy over ecological balance, though they are usually interdependent.

These two disciplines, environmental management and applied ecology, are integral and in fact are part of a "scientific continuum" (Figure 1.1) which ultimately can allow for a balanced ecosystem.

1.3 The Successful Environmental Manager

Managing for the Environment (O'Leary, Durant, Fiorino, and Weiland 1999) is, to date, perhaps the most foundational text for our work as environmental managers. It defines environmental management as "an interactive process wherein we learn how social institutions can best reconcile humankind's needs and aspirations with the limits that the natural world imposes" (O'Leary et al. 1999). It outlines and explains the three criteria for environmental managers to be successful:

1. Understanding volatile and complex issues and trends
2. Co-producing with the community methods for dealing with those issues and trends
3. Delivering those methods effectively in a dynamic, politically charged, and legally contentious environment populated by interorganizational networks of actors with often competing interests. (O'Leary et al. 1999: xxiv)

1.4 We Don't Own Environmental Solutions, But We Are Accountable for Their Implementation

In short, the environmental manager must realize how laws, legal and social trends along with public sentiment affect their everyday management decisions and that often solutions

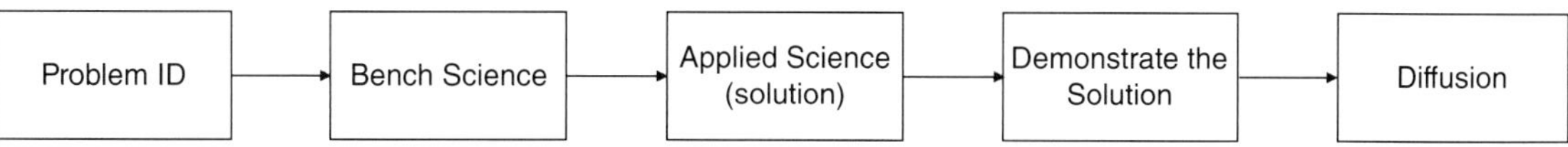

Figure 1.1 The scientific continuum.

or interventions meant to address these issues as specific environmental problems must be produced with others, especially non-scientists. Just as a pilot does not fly a plane without a co-pilot or ground control, the public manager is accountable for good management but requires assistance. Allowing stakeholders to "co-produce" ethical environmental solutions requires leadership by the environmental manager and the mindset that the manager does not "own," in the sense of controlling the decision as it applies to the solution or situation. As the reader will learn, many management disasters have been blamed on "bureaucrats" and "technocrats," who are often depicted as people disengaged with the people affected by the environmental problem at hand and found to be at fault for not addressing their needs which usually expand past the "science" of the issue; and too often the accusers are right.

Whereas there are those who since the publication of *Silent Spring* believe that the basic missions of environmental regulatory agencies have not been properly focused on human health, in fact the mission has always been "protecting human health and the environment." And at its inception this mission has had the primary goal of protecting human health, with a secondary goal of protecting the environment. Indeed, Carson herself pointed out that unless we address both we will fail as a species in the global ecosystem and are doomed.

1.5 Why Do We Need to Manage the Environment? From the Tragedy of the Commons to the 90/10 Rule

While our global ecosystem has been unbalanced by natural forces like epidemics, earthquakes, volcanoes, droughts, and other geologic events, today it is often human activities that cause this imbalance (Rockström et al. 2009; Steffen, Richardson et al. 2015; Steffen et al. 2018). Of course, one can argue that humans as an indigenous species of our planet are also natural. Nonetheless, Garrett Hardin (1968) in his seminal article "The Tragedy of the Commons" illustrates a simple but critical point regarding humans: without regulation there are humans that will take advantage of "the commons" and exploit them to ruin. The commons from our perspective are those resources such as air, water, and land that are held by all such that all can benefit. Common law such as the "Public Trust Doctrine" was first addressed in Roman law for navigable waters to be preserved for the benefit of all (WEF 2020). There are even contemporary movements to establish common law for air (Nevitt and Percival 2018).

From my (M.L.) clinical perspective – that of one who has conducted, studied, and taught environmental law enforcement for more than three decades – this comes down to the realization that 90 percent of people will do the right thing if they know what it is and they have or are given the means to do it, while 10 percent will not because they are either greedy, crazy, or both.[1] Whether it is individuals, communities, companies, or nations, there is an

[1] The 90/10 figure, a general rule of thumb for various organizations, regulatory schemes, etc., does have evidence to support it, even though the exact ratio is variable in every context. For example, the number of polluters included in the Toxic Release Inventory, managed by the EPA, that exceed their permitted amount usually hovers roughly around 10 percent (US EPA 2020a, 2020b). Obviously, despite being a small percentage of the whole, they have a big impact on ecosystems.

unequal distribution of pollution responsibility, with a small percentage of polluters responsible for most of the accumulated pollution in the US and the pollution shaping the global ecosystem today (EPI 2018; Fuller, Sandilya, and Hanrahan 2019; Hickel 2020b). For example, the top 1 percent of wealthy individuals in the world today emit twice the amount of greenhouse gases that the bottom 50 percent of people combined emit (UNEP 2020).

The consequences of this "tragedy" as witnessed by John Muir were the steady degradation of public lands by the late 1800s, the linkage of septic pollution to cholera in the mid-1850s (Snow 1849), and subsequent US public health water regulations in the early 1900s. After World War II, when belching smokestacks signified returning prosperity and coal heating was common in most homes in the West, cities in America and Europe experienced their first air quality emergencies killing tens of thousands. A single pollution event in the UK between December 5 and 9, 1952, now called the Great Smog of London, caused by a combination of frigid temperatures driving increased rates of coal use, high atmospheric pressure, and windless conditions, led to the death of an estimated 10,000–12,000 people (Stone 2002). A similar event, though causing fewer deaths, occurred in New York City in 1966, just before the advent of the Air Quality Act of 1967 and subsequently the Clean Air Act of 1970 (Barnes et al. 2021).

During the same period, with the advent of synthetic agrochemicals and an attitude of "better living through chemistry," industrial agriculture produced what Carson described as a "silent spring" (Carson 1962). Carson illustrated that the public's environment was so degraded by the indiscriminate use of pesticides that nature (particularly birds) was silenced. And until it happened, no US citizen would have believed a river would catch fire as the Cuyahoga River did in Cleveland in 1969. The list of historical examples goes on and on, impacting our "common" reliance on air, water, and biodiversity for our well-being and our desire to appreciate nature. As we will address in this text, the causes of these tragedies are basically twofold: the 90/10 rule described above and disregarding the public's right to know.

Environmental management professionals of today and of the future hopefully will be ready, willing and able to correct past administrative abuse and neglect with an increased acknowledgment of management concepts and skills that can result in success. They will have to deal with more contemporary tragedies including the facts that at least 7 million of the world's population die annually from ambient air pollution (Burnett et al. 2018), while at least 9 million die of some form of toxic pollution (Fuller, Sandilya, and Hanrahan 2019; Landrigan et al. 2017); 80 percent of all human wastewater is discharged back into the ecosystem untreated, causing cholera, dysentery, and a host of other enteric diseases, especially to children (UNESCO 2017); and the frequency and cost of "natural" disasters, from heatwaves to tropical cyclones, have been increasing since 2000, in the US causing a record US$300 billion worth of damage in 2017 (IPCC 2018; NOAA 2020b). And disproportionately it is disenfranchised communities that are exposed to these risks (EPI 2018; Konisky 2015; Landrigan et al. 2018; ND-GAIN 2019; Watts et al. 2019). One can continue listing these tragedies, whether due to mining operations, agrochemical abuse, or the myriad other ways humans are reshaping the global ecosystem.

Why do we manage the environment? Because the policy and implementation decisions of environmental managers are life-and-death decisions.

1.6 Who Manages the Environment?

Theoretically, we all manage the environment, as a matter of societal good and maintaining commonwealth. At the individual level we manage the environment by voting, running for office, and knowing about and participating in public processes at the federal, state, and local levels. There are professions that environmental agencies, the private sector and non-profit organizations value whether they be policy- or science-based: the law, politicians (yes, even them), financial management, accounting, communication, policy analysis, medicine, natural sciences (the whole gamut from agriculture through zoology), and engineering, to name a few. At this point, advanced thought related to understanding "complex and volatile issues and trends," co-production and implementation could also include historians, economists, sociologists, and anthropologists, to understand the how and why of human impact and behavior. In the end, all humans are environmental managers, albeit at radically varying scales, because we all put things into the environment: from cooking fires, to vehicle exhaust, to the water and chemicals that leave our body when we excrete, all humans contribute substances to the ecosystem. But in regulating these actions certain credentials are favored in the environmental management arena, with a healthy mix of environmental and social science and policy training usually being the best bet for success.

In the US, our three branches of government (executive, legislature, and judiciary) manage the environment at all levels. Executive agencies meant to enact laws, rules, and regulations have different jurisdictional authorities and responsibilities. Each agency acts in accordance with its mission. Some, like the EPA, regulate to protect human health and the environment, whereas the US Department of Agriculture (USDA) is tasked with the mission to protect and promote American agribusiness. Both missions can overlap whereas, depending on the contexts of the issue at hand, the EPA would typically have "primacy" and USDA might act in a "consultant" capacity in an environmental management problem; though agencies do not always play nice with each other despite often having similar goals.

Agencies that have primacy with regard to natural resource management, such as the US Fish and Wildlife Service in the Department of the Interior (DoI), would take the lead in a natural resource management issue while the EPA would then act as a consultant. At the same time many of the executive agencies also manage the environment in terms of their own pollution emissions, as required by the National Environmental Policy Act of 1970 (which we will discuss further in Chapter 4). One such is the Department of Defense (DoD), which is the largest source of pollution in the federal government and the single largest consumer of oil in the world (Crawford 2019). These are just examples of how complex the management of the environment is at the executive level.

Of course, it is the legislative bodies (e.g., Congress) that produce the laws under which the executive branch operates. As well, it is this branch of government that generally determines the fiscal resources required for executive enactment, i.e., they make the budget. Finally, it is the judicial branch (the courts) that interprets if the laws, rules, and regulations being enacted by the executive branch are constitutional. Further, one can argue that this branch can also conduct enforcement when an executive agency is not fulfilling its mandate per the law. For example, in 2007 Massachusetts and 11 other US states sued the EPA for not regulating carbon dioxide and other greenhouse gases, claiming that the Clean Air Act required the EPA to regulate any air pollutant from a motor vehicle that could endanger public health and welfare. Often it is the courts when forced by environmental activist organizations and sometimes by government (states suing the feds or other states) that ultimately manage the environment. Interestingly, while the US Supreme Court ruled in favor of Massachusetts et al. and declared that the EPA must regulate greenhouse gases, the EPA has yet to develop and implement such regulations.

1.7 Private and Non-profit Sector Management of the Environment

Aside from government (the public sector), management of the environment is also conducted by the private sector in terms of how lawful they are but also in terms of their participation in the policymaking process (lobbying) and their ability to serve as models and "peer enforcers" for achieving or degrading environmental standards. As mentioned above, often it is the not-for-profit (sometimes called non-governmental organizations – NGOs) sector's environmental activist organizations that practice environmental management by monitoring how well the public and private sectors are protecting human health and the environment. Perhaps most important it is these organizations that manage the environment by suing both government and private-sector entities to fulfill their mandates and holding them accountable for their actions (or sometimes inaction). We will discuss this management of the legal impacts of enforcement and public participation more throughout this text.

1.8 What Are Some Special Skills Required for the Professional Environmental Manager? Or What Skills Can Get You Hired and Promoted?

While the professional environmental manager must practice the three objectives O'Leary et al. (1999) outline for success (understanding, co-producing, and delivering), they themselves must develop certain skills or strategies for this "adaptive" or "high-level" management. This section provides a short introduction to each skill and the foundational knowledge required of the successful environmental manager. Importantly, none stands alone, and each can be considered a discipline of its own. Skills unlike concepts take practice. Whether it be in heart surgery, karate, tennis, cooking, or management, one only obtains expertise through repetition.

The following skills (in bold text below) and knowledge, and corresponding short summary, are skills that are critical to the practice of environmental management. These skills are laid out and evaluated in detail in different chapters throughout the book but collectively they are the essential tools in the environmental manager's toolkit for operating in the tri-sectoral environmental management arena.

Situational Analysis – The professional environmental manager must be able to "paint a picture" for any environmental situation, for two basic reasons: to properly address the situation, and to make it understandable to superiors, peers, and subordinates. Executive memos or briefing papers are sometimes used as terms for this situational analysis; however, brevity is the key. Situations in terms of issues, legal trends, authorities (laws), jurisdiction, communication barriers, and policy impact with the recommended specific management actions should be included in this analysis. Another skill closely associated with situational analysis is being able to prioritize such that the environmental manager **operates with their "ducks in a row" and can identify threats and opportunities** for efficient and effective environmental management.

Navigating Government or Co-producing Solutions with Your Stakeholders – The professional environmental manager must learn how to navigate who has responsibility for achieving goals related to environmental management, who determines the goals and resources in terms of laws, rules, and regulations and who interprets and ultimately enforces environmental laws and their mandates. Critical to co-production of ethical environmental solutions is the implementation of **communication skills** particularly those associated with **public participation and crisis management**.

Skills critical to being able to deliver or implement co-produced solutions would be **leadership skills** to provide reality, expectations and give inspiration. **Strategic Planning** – The professional environmental manager must be able to develop a logical plan to solve problems, otherwise they will fail to achieve their goal or, at best, squander valuable resources. Strategic planning is a skill which when obtained allows the manager to "see the route and produce a roadmap." Other skills critical to mission-oriented management are those associated with **compliance assurance**, **quality control and assurance**, and **continuous improvement**.

Diffusion of Innovations – The professional environmental manager must be able to affect the behavior of communities needing to adopt innovations for protecting human health and the environment. Having established that it is human behavior and subsequent activities that unbalance the ecosystem, how can those behaviors be changed? There are management techniques related to communication science that can do so.

Policy Formulation – The professional environmental manager must be able to understand the process and participants that are critical to policy formulation, for two basic reasons: first, so they may wrap their head around what is being asked of them (their mission) and, second, because they often know what needs to be done and how to do it, so must influence the process such that the mission is achievable.

1.9 Conclusion

Environmental management is the practice of a range of skills, from strategic planning to program implementation, in a tri-sectoral world that is bound and shaped by multiscalar environmental laws, policies, and norms. In short, it is complex and ever-evolving. While the science of environmental hazards is a critical component of any environmental management issue, in this text we focus on the production, integration, communication, and full-spectrum management of people, programs, resources, and politics for effective environmental management practice at the local to national scale and in the public, private, and non-profit arenas.

1.10 End of Chapter Questions

1. Why were environmental regulatory agencies created?
2. What is the stated mission of the US EPA?
3. Name the author of each of, and describe, the three dominant American environmental ethics.
4. In terms of what you are managing, how is environmental management different than natural resource management?
5. Define ecosystem.
6. Diagram the "scientific continuum."
7. In what two ways are the "courts" responsible for managing the environment?
8. Describe the three elements O'Leary thinks are required to successfully deal with issues of environmental management.
9. What is Carson's overriding consideration regarding humans and the environment?

INTERVIEW FROM THE FIELD 1.1 The tri-sectoral landscape

Tom Neltner, a tri-sectoral environmental manager

Tom, the reason I asked to interview you is because you are the only person that I've worked with that I believe has worked in all three sectors. As a chemical engineer you did some environmental management with Lilly, and after you earned your law degree you were in senior management with IDEM and then you advocated for environmental health with a number of different NGOs. So, what do you see are the major differences in environmental management between the three sectors?

The state agency regulates, but they are effectively subsidiary to the federal government. So everybody's got somebody else in charge. Here at the state you've got the

INTERVIEW FROM THE FIELD 1.1 (cont.)

federal EPA in charge. And if you're at EPA, you've got Congress involved and the White House involved. And even at the state, you've got the governor and EPA involved.

In industry I answered to somebody at the company who's doing the actual environmental management, running the wastewater treatment plant, overseeing the air pollution control device, or making sure the permits are in order and the paperwork is filed. Or you may be the one that's trying to do the higher-level corporate compliance instead of just the facility compliance – the internal regulator if you will. So everywhere there's a hierarchy.

In the advocacy groups there is also a hierarchy, and you're accountable often to your funders who want to see results, whatever that may be. And while you are also independent, just like the state of Indiana is independent from EPA, you're often looking to the national groups for guidance and support to help understand what to do and understand the details of issues. So everybody's got somebody overseeing them. And that it's important to realize.

And, what do you find the three sectors have in common regarding their management?

First, I've always found that no matter where you're at, there are dedicated people who care deeply about the environment and health. Some of the most strident advocates, the best advocates, work inside of companies because they felt that they could have the biggest impact from within a company. They're not necessarily seen as environmental advocates.

As an NGO, the key is finding those people in the agency and in the companies that are really focused on getting the job done right. There are a lot of people that are doing their job, but they define the job very narrowly and aren't necessarily focused on the outcomes. The goal is to find the people that are focused on outcomes. They're the ones that are really trying to manage and protect the environment. Don't ignore the others. In an NGO, you have some of the same differences. I try to recognize that there are outstanding people in any of those sectors and to work with them and leverage them. That's a big one.

Second, while we all may feel like we're the smartest person in the room, there's usually a lot of smart people there. So I tried to work from the assumption that there were always smarter people in the room. As a result, my approach to environmental management was it's OK to have an opinion but share the idea before you go public with it. Share ideas with people who are more knowledgeable. I still do that when I'm at EDF and I'm working on a blog about packaging and chemicals in packaging. I will try to share it with the agency to get their feedback and with the packaging companies. In essence, I share it with the people who don't like it and say give me your feedback and then listen carefully to it. They all know it's my work, but they often make it better because they have more knowledge

INTERVIEW FROM THE FIELD 1.1 (cont.)

about the process. They have insights I don't have. Don't assume that you know the most about a problem. Everybody sees a little piece of it. The analogy I always like is the elephant, right? Everybody knows a little part of the elephant. And it's only when we all get together and communicate can we really define how big the elephant is.

As a longtime environmental advocate who is also a lawyer, what are some basic problems with environmental advocates?

Well, you don't have to be a lawyer to do advocacy, but you have to do advocacy to be an advocate.

I often find environmental groups and advocacy groups use a strategy, especially at the state level where they're really resource-limited. As a result, they find themselves in "ready, fire, aim" mode. The challenge is how to shift to "ready, aim, fire" with resource limits, time crunches, and limited ability to control the agenda. Timing matters for success, and that timing may be different at the federal and state level.

So along with that, why have you chosen to stay in the not-for-profit sector for all these years?

Well, I started off with a very clear intent that I wanted to work for industry. That happened when I was 15 years old, I decided I wanted to become a chemical engineer, become a lawyer, work for industry for six years. And I wanted to learn everything I could about how the industry operates. Then I wanted to go and work for government for six years. By age 37, I wanted to be a full-time environmental advocate. It was a very conscious decision to learn as much as I could about a sector, but with the non-profit sector I wanted to change the world, that is the altruistic part of me. But I also recognize that I needed to have that experience. It also helps to have income complement that sound experience. Eli Lilly paid for my law degree. I got experience and that helped me be more effective.

I took a big gamble when I went to the non profit side and didn't have any dedicated source of revenue. And it was difficult for the first year. But I had the confidence from working at Eli Lilly and the state and sufficient resources. Within the NGOs, I have moved around based on my interests and the fit. I am now at EDF for six years and three months – three months longer than I've been anywhere so far. I like it. I may stick around because the job keeps growing within me.

So, what should students or newly minted environmental managers know about working in the not-for-profit sector?

If you're working in the advocacy sector, it's about funding. A lot of your decisions are shaped by funding. And it's the liberating moment when you have flexibility that isn't driven by funding. When I left the state and started Improving Kids Environment, I made a

INTERVIEW FROM THE FIELD 1.1 (cont.)

decision that I was going to do what I needed to and then find the funding. It's a little along the lines of find your path and know your passion, and then find a way to pay you to pursue it.

And I was lucky. I found agencies, donors, people who are willing to fund it, but rarely did it shape what I was going to do. I was going to follow the priorities that were out there. One thing I did find is that in children's environmental health advocacy, I was progressively pulled from focusing on the companies to housing to our neighborhoods. So, as I was trying to set priorities, I found myself moving away from what I think of as industrial facilities, partly because there were strong regulations on it. The gap was children's health in homes and neighborhoods if I wanted to protect children. So, that became a major focus for 11 years – 6 at state level and 5 at national level. In essence, I went with where the data took me. Scientists want to do that but get pulled in different directions.

I talk to the students about the three C's, co-optation, collusion, and corruption, and that as environmental managers, they have an obligation to understand what those are and be able to recognize them in terms of adversaries, but also in terms of keeping their own center. Do you have any opinion about that?
Haven't thought much of it in those terms but I have seen it. I have watched me lose battles that I should have won simply because there was an arrangement. I tend to take people at their word and not question their motives and go from there.

As an advocate, you have the bully pulpit. It's a little harder if you're in a company where you are making compromises. You have to make some compromises, but to set some principles. A friend of mine told me when I was going to volunteer to serve on an EPA advisory committee, "You don't join an advisory committee until you have clearly defined up front what are the conditions under which you're going to quit" – when you won't compromise anymore because you defined it up front. Otherwise, it'll become incremental, and you could be seen as corrupt or co-opted or whatever. So, I've tried to do that even at all my current jobs and it's helped me.

I've been big on transparency, but there are real limits to it. We've pushed the issue for many years. Transparency in labeling, for instance. The reality is most people are too busy for that. So, you have to recognize the limits of transparency. That said, when it comes to a government operation, transparency is essential. You need to be transparent because you're making decisions with government money on behalf of the taxpayer. That's when transparencies are at a premium. It's also when you realize that you can do a public records request or a Freedom of Information Act request on virtually everything out there. So that serves as a real tool, and I've leveraged the public records requests and the agency responses to requests almost every day of the job.

INTERVIEW FROM THE FIELD 1.1 (cont.)

What are the most pressing of the immediate environmental problems facing today's environmental manager?

Well, I've been focused on environmental health. Because I'm a chemical engineer, that means a chemical focus. That's the path I have taken. And I love it because I understand chemicals, I understand toxicology, I understand those issues.

The reality is, the most pressing challenge is climate change. And everything we do needs to be shaped around how can I address with climate change – asking how do I make sure we're advancing climate change. It is an existential issue. So, increasingly I'm framing the work around that because we have to.

One area I am increasingly focused on is environmental justice. It's asking, what about the communities that lack some of the benefits I've had as an old white guy? Then recognizing that and how do we lift up the people that haven't had those benefits. Especially in the underserved communities that have historically struggled with red-lining, with segregation, with underinvestment.

So, the last one, what concepts and skills should a competitive environmental management candidate have?

I look at people differently. I don't focus as much on the degree because I found that experience can be better than a degree in many cases. A couple of things that I really do look for: One is that ability to listen. In an interview, if the person doesn't listen or listens to a question and then misses it, or doesn't ask clarifying questions, that's a showstopper for me. It really is a barrier because if you're not able to listen in an interview to a question, you're not going to be very effective in any job.

The other one is an ability to write. Increasingly that's hard, but I find groups using a master's degree as a surrogate for being able to write. I think that's a lousy surrogate. I was a lousy writer when I graduated as an engineer. Law school helped me think but not necessarily write. Getting edits and understanding the red ink – or track changes today – helped me learn. Over the past six years, an amazing woman in her 20s taught me more than anyone else. If you cannot write clearly and efficiently, everything else is harder. We almost always give candidates a writing assignment to see can they think and write clearly.

That said, there's talent out there all around.

Another issue is not being afraid to learn about chemicals. I run into a lot of people that as soon as you say chemistry, they run away. For me, I was excited because the periodic table was just beautiful. But not everybody feels that way or needs to. The key is, are they willing to get into the details? So that's the other thing I'm looking for is somebody who's willing to sweat the details or is so good at communication that they can translate those details into something really powerful.

INTERVIEW FROM THE FIELD 1.1 (cont.)

One of the things I found that's been really valuable is framing things around the economics. In our society, it really does matter. We worked on the question of whether replacing a lead drinking-water pipe has sufficient societal value to replace it. We took several documents from EPA, put them together, and were able to say that for every pipe replaced there's $22,000 in long-term social and economic benefit – more than four times the cost. The benefits were primarily in reduced cardiovascular disease deaths from lead. And that estimate does not include the benefits to kids' brains.

Social and economic analysis, while it's harsh and it's frustrating, it is how a lot of people who don't necessarily care about chemical risks are thinking about it. Therefore, I found it really valuable to dig into socioeconomic impacts and raise questions. It is interesting how it affects decisionmaking. It's not direct but makes a subtle difference. It is one of the reasons why we were able to get Congress to provide significant funding for lead pipe replacement (assuming the bipartisan infrastructure bill passes).

2 Roles of the Environmental Manager in a Tri-sectoral World

Skill: *Situational analysis – i.e., understanding the situation.* The successful environmental manager must be able to effectively analyze complex and often conflicting information, directions, and stakeholders when managing the environment. This skill we call situational analysis. This is a skill that requires the environmental manager to analyze and translate the management factors in terms of a situation summary – which issues, legal trends, specific laws, jurisdictions, policy participants, policy processes, barriers to communication are applicable – *and* make management recommendations to address the situation. This is an "everyday" or field skill that managers should use; however, as you will note in subsequent chapters, it is also closely related to strategic planning. It is a skill equally important to public, private, and not-for-profit environmental managers in terms of regulating pollution, complying with regulations and understanding those you wish to influence.

2.1 Introduction

The environmental manager must be able to "paint a picture" for stakeholders and other managers for any environmental situation for two basic reasons: to properly address the situation; and to make it understandable to superiors, peers, and subordinates.

REAL-WORLD EXAMPLE 2.1 Issue reports

Issue Report: This report is an actual example of the final of five reports required for the graduate level environmental management student to develop situational analysis skills. Issue reports 1 and 5 may use the same news or other article (to show comparative progress), but reports 2, 3, and 4 require different press articles.

The student should assign themselves a fictitious but related, regulatory environmental manager's position allowing them that perspective for their analysis and recommendations

REAL-WORLD EXAMPLE 2.1 (cont.)

for each issue report. We recommend the articles chosen by the student pertain to pollution (what humans are putting into the ecosystem), and are current and domestic (unless pertaining to international environmental management).

Issue report 1 – situation summary, possible policy issues, jurisdiction, legal authorities and trends, with a short description of the program, resource and political management required.

Issue report 2 – same requirements as issue report 1, with an added section on barriers to communication.

Issue report 3 – same requirements as reports 1 and 2, with an added section on co-production.

Issue report 4 – same requirements as reports 1, 2, and 3, with an added section on strategic planning.

Issue report 5 – same requirements as reports 1, 2, 3, and 4, with an added section on policy formulation.

Article: "Environmental racism is here: Indiana city 'bombarded by lead' confronts EPA chief," www.washingtonpost.com/news/morning-mix/wp/2017/04/20/environmental-racism-is-here-ind-city-bombarded-by-lead-confronts-epa-chief.

Summary: The town of East Chicago, Indiana, home of one of the USS Lead Superfund Sites, is facing the effect of regulation rollbacks set by Scott Pruitt. This USS Lead Superfund Site was considered in 2009 one of the nation's most contaminated sites. The small town of East Chicago is a largely black and Hispanic town. The residents of East Chicago are concerned and frustrated that the regulation rollbacks set by Scott Pruitt will impede any progress on remediating the contamination. One area of particular concern is the West Calumet housing complex, which was built on the land of a former smelting facility. Since 2009 when the EPA designated the site as a priority cleanup, very little has been done to treat the lead contamination due to stalled plans.

My role – regulator for Office of Waste, the Indiana Department of Environmental Management

- Issues:
  - Accountability – There was a lack of responsibility from the EPA officials and the former smelting facility in regard to the proper disposal of the waste and the cleanup efforts afterwards. The plans for cleanup should not have been pushed back for more than five years following the designation as a USS Lead Superfund Site.
  - Sustainable *Development – One* issue regarding the hazardous contamination polluting the water systems and the ambient air is the long-term effects it has on

REAL-WORLD EXAMPLE 2.1 (cont.)

future generations. Lead is a toxin that has severe negative effects on pregnant women and children. If this problem is not solved and there is little regulation to enforce cleanup, this waste will continue to affect the lives and health of citizens as it has been for several years.

- Environmental Justice – The city of East Chicago consists predominately of black and Hispanic communities, who are most affected by the contamination and rollbacks on regulations. These communities' health has been put at risk and is the target for Scott Pruitt's political agenda in the process.

- Jurisdiction:
  - EPA Region 5
  - Local: City of East Chicago
  - State: Indiana Department of Environmental Management
  - EPA
- Legal Authority:
  - Clean Water Act (CWA) – Parties involved could have also followed guidelines to eliminate lead contamination and make the water "swimmable and fishable" again. They failed to follow these guidelines and allowed for the quality of water to continue to decline in nearby bodies of water such as Lake Michigan.
  - Safe Drinking Water Act (SDWA) – The water supply of East Chicago was contaminated with lead and other pollutants, causing the public to have no access to clean drinking water. The drinking water for both areas was considered not safe to consume due to high levels of contaminants, indicating contamination in the groundwater.
  - Comprehensive Environmental Response, Compensation, and Liability Act (CERCLA) – The site is listed as a USS Lead Superfund Site, and the article discusses that the plans for cleanup by EPA officials were delayed and now possibly eliminated with the rollbacks on regulations. Since this is a Superfund site, there should have been more thorough testing and a stronger effort, or an effort altogether, to remediate the lead contamination.
  - Superfund Amendments and Reauthorization Act (SARA) – The public affected by the contamination of their water supply and ambient air should have been made aware of the scope and specifics of these contaminants well before they were. The information about the severity and health effects of the hazardous contamination should have been clear and timely for the public. Also, the specifics about the cleanup efforts and why they were continually pushed back should have been clear to the public.

REAL-WORLD EXAMPLE 2.1 (cont.)

- Clean Air Act (CAA) – With lead being one of the criteria air pollutants, the amount of contaminates along with lead that were released into the air exceed the National Ambient Air Quality Standards (NAAQS), causing concern and the eventual listing as a Superfund site.

- Legal Trends:
  - The municipality as a mini water pollution control agency – The contamination in the water supply of East Chicago should have been detected and treated earlier by the municipality. There was a lack of accountability from the municipalities to ensure safe drinking water.
  - Increased reporting requirements – More documents should be available to the public and made to ensure proper cooperation and knowledge of issues and programs. This would help prevent events like this happening because it requires businesses and other entities to report their impacts and actions in a timely manner.
  - Erosion of government immunity – The lack of action from the regional and federal employees of EPA and the city of East Chicago to clean up the Superfund site for several years should be accounted for. The public was knowingly put at risk when plans to clean up the contamination were stalled.
- Management:
  - Program – As an environmental manager:
    - I would develop new permits to prevent further and future lead contamination of water systems and ambient air.
    - Also, within this plan, I would implement a program to monitor for compliance with the permits and health effects of the lead that was leaked into the drinking water, soil, and air.
    - Enforcement is necessary to ensure safe drinking water and clean ambient air.
    - Technical assistance is needed to provide accurate and updated lead testing and monitoring for the city of East Chicago.
    - Barriers to Communication
      - Difference in frame of reference: government officials may have different motives and interests, such as the economy, when communicating to the public and other officials about their plans for East Chicago. In order to overcome this barrier, there must be an understanding of all viewpoints, including environmental and economic.
      - Prejudice: When communicating to the public, other officials along with myself must be aware of the differing racial and economic backgrounds of the public. We must not have any pre-existing notions about our audience and must treat them all equally.

REAL-WORLD EXAMPLE 2.1 (cont.)

        - Language: As an environmental manager, I must communicate clearly to others the issues at hand without using an abundance of technical jargon. That would only confuse my audience and cause uncertainty and frustration among the public.
    - Resource – Resources that are needed for this program include:
        - writers for the new, stricter lead permits
        - lawyers for potential lawsuits from enforcement
        - employees for compliance and monitoring programs
        - money for technology for technical assistance
        - time to create and issue permits
        - offices for employees
        - new technology for technical assistance
        - proper training for employees
    - Political – In order to receive the resources:
        - the public needs to be informed and involved with a "public participation" process by the respective jurisdictions using press releases and public meetings/hearings
        - there needs to be cross-jurisdictional communication with governmental employees about their involvement with the program
        - relationships need to be formed with politicians for support with the program
        - there needs to be formal communication with legal authority to ensure enforcement of the program
        - a policy analysis should be conducted regarding the effectiveness of the current policy streams and if a more desirable policy alternative should be available.
- Co-production:
    - As part of my program and political management, I will need to co-produce with the community and officials methods for remediating the site and preventing the occurrence of a similar event in the future. I will hold a town meeting with the citizens of East Chicago, where they can provide their concerns and ideas for a program that the city and the EPA could implement in order to resolve the issue. EPA officials will also need to co-produce with the community ways they can improve their methods of reporting and remediation.
- Strategic Plan:
    - In order to efficiently implement a program, a strategic plan must be made to ensure the success of the program. As an environmental manager, the mission of this program is to protect human health and the environment. The vision would be to provide a safe and clean environment for the community of East Chicago, Indiana.

REAL-WORLD EXAMPLE 2.1 (cont.)

In order to achieve the vision and mission, goals and strategies must be established. The goal of the program is to remediate the site and to prevent future similar rollbacks of regulation. Objectives would be similar to the program management, which includes permitting, monitoring for compliance, conducting enforcement, and technical assistance. In order to reach the goals and objectives, strategies will be put in place. For example, to permit for the sites within the West Calumet housing complex, relationships need to be formed with politicians for support and employees will have to create the permit. In order to conduct enforcement, you will need employees and lawyers to take action if entities are found to be out of compliance. These are a few examples of strategies that would ensure a successful and efficient program.

- Policy Formulation Process:
  - Inside participants – the President and Governor of Indiana may be pro-industry and anti-regulatory, additionally, the EPA has been unable to take appropriate action due to lack of funding on Superfund cleanups in disenfranchised areas for decades and Indiana US Senators, Congressmen and Governor not pushing for cleanup.
  - Outside participants – the mayor and citizens of East Chicago, the Natural Resource Defense Council (NRDC). Pollution industries who do not want to be accountable as "responsible parties" (RPs).
  - Problem stream – remediation of Superfund sites is often delayed due to legal actions by RPs and/or EPA Regional Offices' lack of resources. Disenfranchised communities are often disproportionately affected, creating an environmental justice situation.
  - Policy stream – current policy is to decrease funding for remediation in R5 and to continue to deregulate pollution industries.
  - Political stream – citizens and local/state politicians feel that they are being discriminated against and have a right to clean water, air and land. Potential RPs (PRPs) view the President's Administration as industry-friendly and are lobbying for deregulation. The President and EPA Administrator are trying to address Superfund with a "more with less" attitude.
  - Window of opportunity – newly tested levels of lead are higher than ever before and families with children were being exposed for years without their knowledge. The community is now outraged and has more scientific evidence showing inaction and probable discrimination. There is abundant press coverage.
  - Possible reformulated policy – accelerate actions for these types of sites on the NPL and make sure they are funded so as to decrease exposure to vulnerable populations. Require more transparency to affected communities.

2.2 The Management Profession

As an entomologist I (M.L.) was not trained to be a manager and it was my experience that many scientific and/or technically educated practitioners were not subjected to the principles of management. In fact, most of us have had jobs, prior to our formal training and afterward, where we had to suffer from mismanagement. Conversely, initially trained as a military officer prior to an environmental manager, I (R.M.) was explicitly trained in leadership and management as it is the primary skill needed to accomplish a mission with a team. You could be sound on tactics, techniques, and other knowledge, but without leadership and management to organize and enable people, you will never succeed. The same is true for the environmental manager, especially as they progress through their career.

Management is defined as "the act or art of managing: the conducting or supervising of something: the judicious use of means to accomplish an end" (Merriam-Webster 2021e). This can apply to an individual or group's efforts and implies that the management conducted is to be efficient in terms of resource use and effective in terms of goal achievement (Starling 2005).

Unfortunately, many science and technically oriented professionals do not consider management to be a profession. Like me (M.L.), many folks believe that information, science, facts are enough to implement programs. My initiation into that degree of ignorance began when I was hired by the University of Arizona to be an agricultural Extension specialist (USDA, Agricultural Extension Service) with the idea that I could extend the knowledge of the University such that farmers would reduce the risks from pests and pesticides by adopting Integrated Pest Management (IPM). This was a mistake that environmental managers make, and it can squander precious resources and cost lives.

2.3 Framework and Functions of the Environmental Manager

In subsequent chapters we will address a number of important management concepts related to implementation, but for now we will concentrate on what we consider to be the framework for addressing environmental issues and situations with some management basics. Like many scientists we have found that using categories and processes helps to make sense of how the world works whether learning phyla and biological cycles or more abstract social constructs, or whether managing a public sector agency, a company, or a not-for-profit advocacy group. Reading Grover Starling's *Managing the Public Sector* helped me envision the basic roles of the manager to be *program management*, *resource management*, and *political management* (Starling 2005). We often refer to this as the PRP framework that can be applied to most, if not all, environmental management issues. Each component of PRP is intertwined and overlaps with action from another role, i.e., each is interdependent but distinct. For the program and resources components of PRP, we can further apply the Permitting, Monitoring, Enforcement, and Technical Assistance (PMET) framework to parse complex issues, to serve as a checklist or reference to ensure we are accounting for all of our primary function as managers, and to consider and weigh different mixtures of available tools and resources to solve the environmental issue at hand most effectively and efficiently. To ground these frameworks and show their utility, we will now apply them.

2.4 Program Management

Program management requires knowing what you are trying to achieve and what actions and strategies are required to effectively achieve those objectives or goals. To best understand this concept one can apply it to a personal and well-understood goal – earning an A grade in a specific course.

To achieve your objective of earning an A your program management must be designed such that you perform in an exceptional manner on all assignments and evaluations (quizzes and exams). In general, to do this, the functions you will build into your program and then conduct are to attend class/office hours, participate in class, read required readings, complete assignments on time and with excellence, and study for evaluations. This is your Program Management for earning an A grade.

2.4.1 Applying Strategies to the Functions of the Environmental Manager

The environmental manager might have a specific goal such as implementing a new rule regarding decreasing effluents into the waters of the US under the National Pollutant Discharge Elimination System (NPDES) (US EPA 2014f). Strategies to achieve this goal could include planning, research, environmental assessment, education, public relations, and establishing and enforcing environmental laws, rules, and regulations. However, we find it more helpful to categorize these strategies in terms of the operations an environmental agency would employ. In general, managers in environmental regulatory agencies, and conversely those in the regulated community regarding compliance, in their program, resource and political management roles have or support **four basic functions/strategies**:

1. *Permitting* – is based on permitting entities to pollute within the standards a specific law, rule or regulation allows. Generally, there is a permit to produce pollutants (application) and a permit to operate. Like all of the regulatory functions, just to entertain issuing a permit involves staff and equipment to develop standards and determine a toxicological profile of the pollutants, and modelers to anticipate their environmental fate. Once an application is applied for, it then is reviewed by engineers and permit writers. Note: permit writing is a complex process and this text applies it in terms of management and is not a textbook on permitting. Resources for permitting will be provided at the end of this chapter.
2. *Monitoring for compliance* – upon issuance of a permit, polluting entities must be monitored on a regular basis by internal and external (regulatory) experts. Each inspection requires a report. The findings must be in compliance with the permit and are reviewed by agency compliance officers.

3. *Enforcement* – most environmental agencies are law enforcement agencies. If a regulated entity is operating outside of its permit it is subject to enforcement. At this point, the regulatory agency and its attorneys would issue a notice of violation which ultimately results in a compliance agreement (a civil violation) that could range from technical assistance and notification to fines, to a cessation of operations. Incarceration can happen in rare circumstances (criminal violation).
4. *Technical assistance or education* – as we will discuss in future chapters, technical assistance and education is often the preferred option as enforcement is expensive in comparison, particularly when pollution can be mitigated to acceptable levels or prevented. This function requires "change agents" for the diffusion of environmental innovations, and environmental regulatory agencies use partnerships, workshops, education credits for professional certification, and professional improvement programs, as just some examples, to provide technical assistance to regulated communities such that they can comply with regulations and/or as part of a compliance agreement.

This is your basic Program Management component of the tri-sectoral framework to address an environmental situation.

2.5 Resource Management

(Note: not to be confused with Natural Resource Management, it happens!)

Resource management requires knowing what resources are necessary to manage an environmental situation and how to use those resources for maximum efficiency and effectiveness. For the most part, management resources fall into one of three categories: (1) funding (which is used to acquire staff and materials); (2) time; (3) political will. Political will is "the firm intention or commitment of a government to carry through a policy, especially one that is not immediately successful or popular" (Oxford 2021b). We expand that definition to include the manipulation of a superior to commit support for your management objectives. Political will often comes in the form of a new policy, program initiative, law, rule or regulation. In a very real and ethical sense these are *your* resources. It requires prioritizing or reprioritizing your budget, how many hours per week you can actually work, and your professional credibility.

Going back to your goal of earning an A grade, what resources are necessary such that you can conduct a successful program to achieve that goal?

Your program management consisted of attending class/office hours, participating in class, reading required readings, completing assignments on time and with excellence, and studying for evaluations. Thus, the resources you will need will be funding for tuition (and probably room and board), books, and computers. You are the staff required to conduct the required activities and will need to prioritize your time (time management) such that you will properly attend class (on time and fully attentive), turn in assignments on time, and put in the time for the amount of studying necessary. "Political will" allows you to find the best professor get the most out of office hours, have meaningful class participation, and obtain attendance, test administration, and other classroom policies that might benefit you. Once these resources have been obtained they must be managed correctly as we all remember running out of money, time and goodwill and the effect it had on our grades. This is your Resource Management for earning an A grade.

Again, environmental managers in their resource management role, just like in their program management role, have or support **four basic functions/strategies**:

1. *Permitting resources* – you will need funding (appropriations) to pay for personnel that understand the goals of the applicant and the legal constraints dictated by the regulations that apply to the applicant. Staff may include scientists, engineers, permit/rule writers, and public participation specialists and they require administration support (office space, information technology, etc.). Your staff also require time to draft permit language and manage for efficiency and effectiveness in terms of training, professional improvement, and performance evaluations. Further, it takes time to implement the permitting process such as public notification or possibly going to court. Political will can impact the issuance or denial of a permit, and policies regarding social justice, regulatory takings or the economy may come into play.
2. *Monitoring for compliance resources* – you will need funding for personnel ranging from chemists to accountants specifically trained to inspect or audit permitted facilities, non-permitted sites or contractors, and the surrounding environment which might be impacted by the permitted or non-permitted pollution. Monitoring equipment and sample analysis, whether in the field or laboratory, is often expensive to purchase and maintain. Monitoring is labor-intensive and requires time for travel, equipment, analysis, etc. You will need political will to achieve or maintain trust and credibility regarding risk assessments and monitoring sites.
3. *Enforcement resources* – you will need funding for personnel such as attorneys to prosecute those who are out of compliance, along with some of the same technical personnel used in permitting to craft compliance agreements. Court time can be onerous and is often added to your other duties. Improving or creating legal authorities allowing for enforcement such as legislative action also takes time and often a "legislative liaison." Again, enforcement is labor-intensive and requires time for travel, equipment, analysis, etc. You will need political will to achieve or maintain trust and credibility regarding protecting the public or killing jobs and the subsequent impact on legislation.
4. *Technical assistance/education resources* – resources for this strategy or function often include those in the other functions but will also require a "change agent" or a "change agent corps" in order to get a community to adopt an environmental innovation (diffusion – see Author's Note 7.1 and Section 7.4 in Chapter 7). One of the best-known examples of this is what used to be known as the Agricultural Extension Service of the USDA where they extended the knowledge of Land Grand Universities to the farmer. The manager will either add or contract these change agents or they will have to make time for existing qualified personnel to act as change agents. At a minimum this change

agent corps must be maintained such that they can respond to any community with any environmental innovation.

This is your basic Resource Management component of the tri-sectoral framework to address an environmental situation.

2.6 Political Management

Political management is basically how you will obtain your resources: where will your funding come from; how will you justify your time; will the higher administration give their support? Thus, the manager first and foremost must establish an organizational need or mandate for initiating and maintaining a program and utilizing its resources. Strategies for your political management could involve reprioritizing, appeals based on data, compliance with mandates, or "scrounging." **Scrounging** is defined as "Seek[ing] to obtain (something, typically food or money) at the expense or through the generosity of others or by stealth" (Oxford 2021a). This time-honored, competitive management technique has been used by effective managers time and again to achieve their objectives in resource-scarce environments.

Returning to our example of seeking an A grade in class, leveraging your experience and knowledge, how will you obtain and maintain the required resources such that you can conduct a successful program to achieve a goal of academic excellence?

Of course, this depends on whether we funded ourselves through our higher education, had help from others (e.g., our parents) or a hybrid of both. While we are sure all of our readers would never use stealth or manipulation, we might have – in the most respectful and good-intentioned desire to survive and succeed, of course. We needed tuition, books, computers, time to attend class and study, room and board, even the goodwill of our professors – everything that it takes to be able to be an excellent student. We competed for scholarships and grants, if we had a job, we had to prioritize our schedule, and if we "came up short" we either had to drop out (permanently or temporarily) or ask for help. And what was our "ask"? One we remember is, "If I don't have a computer, I won't be able to pass this class!" Another was, "My professor says if I don't get this required text, there is no way I can get an A!" To our employers, "If I take this shift I will have to drop out of school!" and to our professors . . . well you get the point. And yes, there are good reasons for using exclamation points. This is your Political Management for earning an A grade.

REAL-WORLD EXAMPLE 2.2 Political management in environmental management

In my (M.L.) early days of teaching environmental management (after practicing as an environmental manager), I heard a firsthand story from a state environmental agency administrator. For some years, their agency had been underfunded by a state legislature

REAL-WORLD EXAMPLE 2.2 (cont.)

who was heavily influenced by a manufacturers' association (polluters). After continually providing evidence of poor environmental management due to lack of resources (specifically regarding water quality monitoring and enforcement) the administrator confided to the sympathetic governor that they could no longer protect human health and the environment in their state.

They devised a scheme where they would turn over "primacy" on the Clean Water Act to the EPA Region. The consequences of this action would be that water permits would have to be processed by regional personnel, resulting in a drastic increase in application time. Further, monitoring and enforcement would also be conducted by regional personnel known for rigorous monitoring and vigorous enforcement. When this possible action was announced, the manufacturers' association quickly decided their members would be better off under state management and convinced the legislature to increase funding for resources.

This is a somewhat rare but clear example of political management. It was my impression that the head of the state environmental agency and the governor had hoped for this outcome but were bluffing to some extent, as giving back authority for any federal law is not easy. Further, both individuals had been correct in understanding that certain stars had aligned regarding the public's attitude in favor of water quality protection and manufacturers' fear of federal oversight. Nonetheless, this was a very risky move by both individuals that required political will. Could this happen today? Yes, but only if it was conducted by astute leaders willing to take risks.

As I have listened to numerous state environmental agency heads over the years it is clear that most are extensions of their governors. Some governors want their executives to strictly follow the law while others would rather laws are "bent" in favor of jobs. Many years ago, Indiana Governor Mitch Daniels stated that the Indiana Department of Environmental Management's "job one" was creating jobs. Obviously, he would not lend his political will to strictly enforce environmental laws and preferred his environmental agency to simply be a "permit shop." No doubt, today's environmental manager will constantly be enterprising when it comes to creative political management.

2.6.1 Basic Strategies for Political Management

The environmental managers in their political management role must justify their program management and resource management to conduct or support the **four basic functions/ strategies**.

Justifications can and do overlap with regard to what is mandated, what is the "right thing to do" and what function or program is most pressing at that moment. Below are some examples of how managers can garner more support.

1. Certainly, the best "ask" or reason for obtaining support is that you have a mandate to meet a goal. These goals as you will read in subsequent chapters can be a specific goal such as implementing a new rule regarding decreasing effluents into the waters of the US. Often a mandate is not being implemented for a variety of reasons, such as lack of funding, time or political will. However, when the courts find that the responsible organization is not fulfilling its mandate, often because they are sued by a not-for-profit environmental advocacy organization, then at that point the regulated or regulating entity must comply. It is the environmental manager's responsibility to inform their administration of this situation and recommend what should be done to address it.
2. Another reason to obtain support is "optics" – how will it look? Basically, is your organization being accountable to its mission of protecting human health and the environment or is it being sensitive to the economic and/or cultural concerns of the community? These are issues we will discuss in depth in Chapter 4 that pertain to all environmental situations. They can supersede specific laws, rules, and regulations and can, in fact, stimulate the formulation of environmental policies. In general, having good "optics" is something most administrations want to do if they can because often the cost of not considering and addressing these issues will offset the resources anticipated.
3. A verifiable crisis is another way to conduct political management. Webster defines crisis as "an unstable or crucial time or state of affairs in which a decisive change is impending" (Merriam-Webster 2021d). The environmental manager must be able to document the instability in operations and make credible recommendations regarding what changes can or need to occur. This is not business as usual but is common. Natural disasters like hurricanes, which release hazardous materials into residential neighborhoods and contaminate drinking water quality or air quality, often are met with emergency response but when that is inadequate a crisis occurs (see for example Marcantonio, Field, and Regan 2020).
4. Managers often have planning in place to allocate or reallocate resources including obtaining the political will to support new policies under the circumstances described above. However, some of the best opportunities for political management come in a "postmortem" review of what could have been done better with the proper resources once an administration deals with forced mandates, incurring negative press and public sentiment and not being prepared for a particular crisis. The environmental manager must always be able to provide an objective evaluation of what the situation is and how it should be addressed.

This is the basic political management that can be used to provide the funding, time, and political will to address an environmental situation.

You will learn more about the implementation of environmental programs throughout this text. This chapter on management allows you to consider the three most important questions (Starling 2005: 402) that must be asked and answered in order to successfully address implementation:

(1) What action has to be taken?
(2) Who is to take it?
(3) Do these people have the capacity to do it?

2.7 Conclusion

While the management approach and general focus areas will differ between the public, private, and non-profit sectors, Program, Resource and Political Management factors are key functions of all environmental management practitioners no matter the arena. The PRP framework helps the environmental manager parse complex situations into a simple and user-friendly form, while ensuring these critical components are always considered and addressed.

REAL-WORLD EXAMPLE 2.3 Military and environmental management: a comparative look

USAF Colonel John Boyd, a fighter pilot ace and military strategist, developed a decisionmaking paradigm called the "OODA Loop," which stands for Observe, Orient, Decide, and Act. This cycle to observe and collect information in an actor's environment, orient on the decisive or greatest influencing factors, deciding on which and how to counter them, enact or implement a strategy, and then observe the results and re-implement a refined strategy accordingly, has been employed by military leadership and business strategists alike, and has been a preeminent concept in training leaders to execute a rapid decisionmaking cycle. The idea is to understand and employ this process in a circular manner that outpaces your adversary. For the military this is the opposing enemy force; for an environmental manager this can be represented by attempting to actively counter any negative influencing environmental stimuli, from non-point-source water pollution draining into the Mississippi River to greenhouse gases catalyzing global climate change.

Having recently completed my military service as a Marine Corps Infantry Officer, I (R.M.) am now in the inchoate stages of retraining as an environmental manager. Little did I know that while pursuing this new endeavor I would find so much transitivity between military leadership and environmental management. With new perspective, this really should not have come as a surprise. Both military leaders and environmental managers work in an environment characterized by resource finitude, and extremely dynamic systems, against adversaries that threaten the livelihood and health of the impacted party, a complex web of differing values and interests from all parties involved, and with a mission-oriented focus that compels the manager to deliver a maximally effective – if imperfect – outcome.

REAL-WORLD EXAMPLE 2.3 (cont.)

Program, Resource and Political Management (PRP)

Environmental managers and military managers alike have standard analytical processes and frameworks with which to frame, investigate, and assess issues. These tools allow managers from differing backgrounds to find common ground from which to begin orienting on an issue. In the remainder of this example I will seek to illuminate correlations and connections between military and environmental management through the environmental planning framework of PRP.

Program Management

Program generation and development, both for military and environmental managers, requires traversing functional and jurisdictional boundaries. For example, a Marine infantry platoon commander must integrate into his planning any adjacent units, supporting units, and higher headquarters units ensuring synchronization and mutual understanding. A failure to appreciate and account for all of these variables can result in mission failure and lives lost. An environmental manager must bridge jurisdictional lines when an issue impacts several regulatory bodies' areas of responsibility (which is the case more often than not), and engage and include political and public leadership as well as the populace affected by the issue.

Program management for environmental managers includes four key responsibilities: permitting, monitoring, enforcement, and technical assistance, all of which have direct parallels to military management. In permitting, an environmental manager must screen and certify polluting parties to conduct their actions in accordance with established environmental standards, ensuring proper documentation of actions and fulfillment of associated mandates. Military managers must permit and empower their subordinates to conduct training, equipment maintenance, administrative upkeep, and various other tasks that require meeting established unit standards and the published Commander's Mission Essential Task List. Once a permit has been issued or person empowered, the manager must monitor the execution of the co-produced plan and supervise those enacting it. Managers on both sides must supervise to ensure critical standards are maintained and that the personnel enacting the intent are making progress toward the desired endstate. The major difference between the two is merely in the event itself (i.e., environmental control vs. infantry maneuvers) but the methods of monitoring and quality assurance are very similar.

Another illustration of parallels is in the enforcement of standards. Military managers conduct back-briefs prior to execution and after-action reviews after the fact, as well as occasional direct intervention during execution to maintain standards amid changing

REAL-WORLD EXAMPLE 2.3 (cont.)

circumstances while maintaining effectiveness. Environmental managers conduct inspections of permitted field work and can audit the paperwork of polluting entities, along with a host of other active and passive enforcement efforts. Finally, both environmental and military managers provide technical assistance to their regulated entities or supervised personnel, respectively, in order to enable them to overcome potential or identified shortfalls and maximize efficiency. Efficiency, then, is a virtue, whether in the optimal distribution and employment of military personnel and firepower, or in the human, financial, and technical resources employed by an environmental manager. Managers bring to bear their own abilities, as well as their discretion to employ outside subject matter experts to train and educate for proficiency, updated standards and technology, and stay ahead of their adversaries' capabilities.

Resource Management

The next step in our PRP framework is Resource Management. In both worlds, resources are often scarce with many competitors vying for them, each believing their own need to be preeminent in importance and immediacy. Effective military managers are required to identify the assets on-hand and those that are potentially available elsewhere in order to begin developing a plan to gain and effectively employ these assets. Whether radio batteries, ammunition, or time in a training area, it is the crafty manager who can best justify his need and plan for efficient, effective usage that is usually granted those resources. The environmental manager follows the same process of asset identification and allocation. They face some of the exact same general considerations, such as time and money, and others of a different sort, driven by a difference of mission, such as preventing cyanide leaching in a mining retention pond vs. conducting a 10-kilometer movement-to-contact operation. Environmental and military managers face the same challenges in analyzing their problem, resource matching and utilization for optimal performance, and acquiring resources in a competitive environment, and thus must conduct a structured yet flexible process to develop a resource management plan.

Political Management

Political Management is the final step in the PRP framework. "Political" can be big P or little p as it applies to persons of differing titles and positions. For virtually any manager working in or with a human organization, managing people, with their varying personalities, strengths, weaknesses, biases, and potential, is an inherent and essential function that must be primary in priority. Military managers must manage relationships with personnel of various ranks, personalities, and priorities, inside and out of their unit,

REAL-WORLD EXAMPLE 2.3 (cont.)

coupled with issues caused by relatively frequent turnover of those personnel. Significantly, they must build and manage relations at their peer level, across supporting and adjacent units, and with their superiors in order to garner support for their plans. Without effective relationship management inefficiency and idle time will reign. Similarly, environmental managers must maintain relationships with varied regulatory agencies in their assigned region ranging from their representatives to the national legislature (big P) to representatives from the state Council on Environmental Quality (little p). Environmental managers deal with internal personnel rotational issues (revolving door: personnel rotating between public and private sector) and leadership rotation (elected officials changing), facing the same issues military managers do. Both types of managers must shape conditions and leverage relationships to ensure their programs gain the support and resources necessary to attain their desired endstate.

Whether the issue at hand is a combatant enemy force or a pollution source negatively impacting human health and the environment, the processes to orient on these issues are very similar in kind and require managers with the same attributes: a critical mind to analyze the situation; an understanding of logistics and asset-to-issue pairing; cross-boundary communication and co-production; ability to create buy-in from all stakeholders; and the ability to generate a clearly defined endstate with an adaptable plan to achieve it.

2.8 End of Chapter Questions

1. Why is the skill of situational analysis important?
2. What are the basic components of a situational analysis?
3. Define and give an example of Program Management.
4. Define and give an example of Resource Management.
5. What is "scrounging"?
6. Define and give an example of Political Management.
7. What are the four basic program management functions inherent to US environmental regulatory agencies?

3 Issues and Legal Trends That Impact Your Environmental Management

Skill: *Identify, analyze and prioritize* how environmental issues and legal trends impact your program, resource and political management. More specifically, how the environmental manager can protect their program by getting their "ducks in a row." It is critical that the environmental manager know the key issues and legal trends in the profession to effectively accomplish their, and their institution's, mission effectively and efficiently. Understanding these issues will allow the environmental manager to ensure they are not only complying with the law but also with the expectations and needs of the citizenry they serve.

3.1 Introduction

There are some basic issues and legal trends that will *always* impact your environmental management performance and program. Some are relevant in every situation, and often more than one or even all are pertinent. These "volatile and complex" issues and trends were foundational in O'Leary et al.'s (1999) *Managing for the Environment*, and are no less relevant today. As you will notice they overlap yet if categorized and used correctly each concept will allow the environmental manager to understand just what their management situation is and how to navigate it. For each issue and trend (listed below) we provide Windows or mini–case studies to contextualize the issue, to show how it can be applied, and to demonstrate what it means for the environmental manager.

The effective environmental manager needs to understand how the issues apply to *their* environmental situation and how they impact their program, resource and political management in terms of permitting, monitoring for compliance, enforcement, and technical assistance or education, i.e., the key frameworks for the environmental manager that we presented in the previous chapter. In this chapter we will address five issues and seven legal trends the environmental manager will encounter in the day-to-day environmental issues they will face. The frequency with which the issues emerge, and the direction of the trends, are always in flux and require the environmental manager to actively keep a pulse on them to maintain their proficiency in their profession. We thus provide the general contexts of each over the last decade or so, and then examples either ongoing or in the recent past to highlight the status of the issues and trends today.

We cannot emphasize enough that the skill of understanding and applying these complex and volatile issues and trends is *difficult* and must be practiced. Thus, each issue and trend lends itself to an exam question as a forum developing this skill. Each of the five issues and seven trends may be questioned this way:

Q – "How does a specific issue or legal trend (listed below) impact your environmental management in terms of program, resource and political management?" ... or "what are three managerial implications to each issue or legal trend?"

How to answer – The three managerial implications would be the program, resource and political management in terms of the environmental manager's basic functions (permitting, monitoring, enforcement, and technical assistance). Students need to coin those implications in terms of how those functions relate to the meaning of each issue or trend. – Understanding and applying!

Below we list and describe each issue and legal trend. We have found it helpful to assign a diagnostic word or phrase to each to aid in understanding where they might fit into the situational analysis.

3.2 The Issues

The environmental manager has to operate within and navigate an exceedingly complex world of policies, practices, science, funding, and myriad other components of the profession. So, to boil all of this complexity down is no simple task and will inherently overlook or oversimplify certain aspects of the job. However, we have identified **five key issues** (see Figure 3.1) that are critical and omnipresent in the world and daily functioning of the environmental manager.

The five issues for the environmental manager:

1. Accountability
2. Environmental Justice
3. Ecosystem Management
4. Sustainable Development
5. Unfunded (underfunded) Mandates

3.2.1 Accountability

Diagnostic: You are accountable – to your organizational mission and the applicable environmental laws.

If an environmental situation falls under your jurisdiction or job description, you are accountable for how you address it. The six core values of administrative responsibility are:

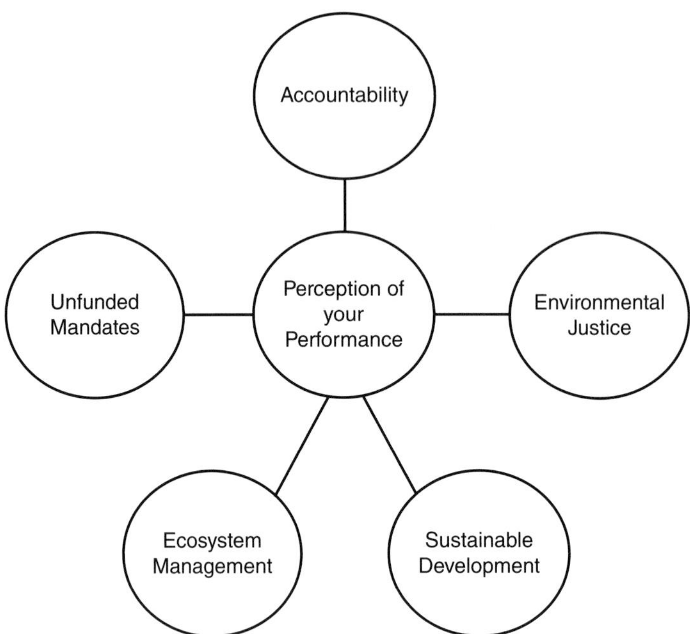

Figure 3.1 The five issues for the environmental manager.

responsiveness, fairness, flexibility, honesty, accountability, and competence (Starling 2005). It is along these core values that the environmental manager and their performance is assessed by the organization they work for, the institutions they are regulated by, and the public they are charged with protecting. They should be ever present in the mind of the environmental manager from the onset of an environmental issue until its resolution or transition to another manager or responsible authority. When that environmental problem or situation comes to light and demands management, first understanding why and to what statute or standard you are accountable is critical as it will help you to develop a strategic management plan to address the problem. Most environmental managers will be accountable according to the mandate for their office or manager position to enact or comply with an environmental law, rule or regulation relevant to the scale at which they work, i.e., municipal, state, or federal level of environmental management.

As well, professional codes of ethics might apply. As O'Leary (2020) makes clear in *The Ethics of Dissent*, we all have "ethical obligations" to country, family, the law, or humanity as we will discuss in Chapter 6. Beyond your legal accountability, two additional principles further highlight the importance and salience of your accountability: your actions can impact the lives and welfare of others; and "ethics is more than just thinking about right and wrong ... it is doing right, not wrong" (O'Leary 2020: 21). While we have mentioned ethics, the concept of competency such that your actions are effective and efficient is a measure of your management and it is the ethical responsibility of the manager to be competent in all requisite aspects of their assigned position. Conversely, **misfeasance** (the improper performance of lawful duties), **malfeasance** (the performance of duties that are forbidden by law or commonly accepted moral standards), and **nonfeasance** (the failure to perform required duties) would be considered a lack of competent management, and unethical (Starling 2005).

Examples of environmental managers not being accountable can include a bureaucratic response resulting in your failing to address or deferring a problem; disregarding the law; lack of transparency; becoming corrupted or co-opted; collusion; or not being to perform your duties in a competent manner.

3.2.2 Environmental Justice

The results of your environmental management should not be socially unjust – diagnostic: disenfranchised population.

Environmental justice is increasingly an issue that many managers will deal with. It is *not* about justice for the environment but rather the equal distribution of the environmental risk to human health and equal participation in the processes and policies that determine the amount and distribution of risk (Agyeman et al. 2016; Kramar et al. 2018; Shrader-Frechette 2002, 2007). Konisky (2015) defines environmental justice as "the disproportionate environmental risks experienced by low-income and racial and ethnic minorities, particularly African Americans, Hispanics, and Native Americans" in the US (p. 2). Similarly, O'Leary et al. (1999) note that the environmental justice movement "is concerned with unequal distribution of environmental risks based on income, race, or both" (p. 11). The EPA defines environmental justice as "the fair treatment and meaningful involvement of all people regardless of race, color, national origin, or income with respect to the development, implementation, and enforcement of environmental laws, regulations, and policies" (US EPA 2011). Our concept of environmental justice includes the factors addressed in these definitions as well as the unequal distribution of environmental benefits, such as the environmental services made available from a previously contaminated site, like a Superfund site, that has been remediated through an environmental management program.

In framing what environmental justice is and includes, three distinct but overlapping basic principles are often utilized:

- **Distributive justice** is the idea that all members of society have the right to equal treatment, and that outcomes should be distributed fairly;
- **Procedural justice** is the notion that people have the right to be treated equally in political decisionmaking regarding the distribution of goods and services; and
- **Corrective justice** also involves fairness, but specifically in how instances of lawbreaking are resolved. (Konisky 2015: 15)

The US federal government first addressed the issue of environmental justice in 1994, during the Clinton administration, with Executive Order 12898 – Federal Actions to

Address Environmental Justice in Minority Populations and Low-Income Populations. The order mandated all federal agencies to *consider* equity in environmental and public health policymaking and management (importantly, note that it calls for agencies to consider equity as opposed to enforcing equity) and to develop a plan for implementing environmental justice programs and practices (US EPA 2013e). While varied progress to this end has certainly been made since the order was issued, many scholars argue that, to date, the federal government has failed to fully implement the intent of the order (Konisky 2015). For example, minority communities are still substantially overrepresented in the neighborhoods around contaminated sites such as Superfund and brownfield locations (US EPA 2017).

We see the issue of environmental justice first and foremost as an issue of who is empowered to have fewer risks and more benefits, and how that distribution is determined. Environmental managers need to consider, in their area of responsibility, what are the demographics and community composition near those places where most hazardous and solid waste sites are located. Environmental managers must ask: How are disenfranchised communities considered regarding establishing health-based or risk-based standards? What is the priority ranking and implemented response to imminent environmental threats regarding who is exposed? And do the affected communities have the ability to participate in and substantively contribute to the decisionmaking processes that impact their health and their environment? To execute their program, resource and political management functions effectively and fairly, environmental managers must consider these questions and others related to environmental justice for the area and citizenry they serve.

These questions allow the environmental manager to address what researchers believe are the three reasons that environmental inequities exist:

1. intentional discrimination, where the proportional risk to disenfranchised communities is simply and knowingly discounted;
2. "neighborhood transition" or "minority move-in," where the character of the community changes with the siting and operation of an environmentally objectionable facility. One of the most common arguments or challenges to environmental managers in all sectors is, what if a community of workers grows up adjacent to a potentially dangerous environmental area or endeavor? We often hear, "This facility is the number one employer in our area!"; and
3. "political capacity," where disenfranchised communities lack the capacity to participate in decisions that affect their health and their environment. (Konisky 2015)

Thus, the environmental manager must consider and act when situations of environmental (in)justice are possible. Is the permit and permitting process unjustly subjecting disenfranchised communities to environmental risk? Are those responsible for monitoring for compliance doing so equally in disenfranchised communities as they are in empowered communities? Is enforcement being conducted such that disenfranchised communities as well as empowered communities are being protected, *and* are the industries which might be out of compliance being held accountable? Are these communities being communicated with in an ethical manner? And of course, are these programmatic functions being provided the same resources as other communities where successful environmental management is occurring?

If the environmental manager is not able to answer these questions with regard to their program, resource and political management they must ask why. At the very least, the manager must assure that the regulated facilities or pollution producers are operating according to whatever environmental laws and rules they are subject to and within the confines of their operating permits. They must also assure that the rights of the communities involved are being communicated (the "right to know") and acted upon. The environmental manager's ability to effectively communicate environmental justice issues is a critical skill only growing in importance (Polk and Diver 2020), a subject addressed more thoroughly in Chapter 6.

As many of the considerations we pose highlight, there are many overlaps between environmental justice and accountability. Environmental justice is not simply an issue that the environmental manager must consider, but also a principle with binding laws and policy that the manager is accountable to. The exact statutes are dynamic – that is, ever-changing – and likely will see increasing change as environmental issues increase in effect locally and nationally, with environmental justice likely to only grow in importance, salience, and in public scrutiny (Agyeman et al. 2016; Lee 2020).

REAL-WORLD EXAMPLE 3.1 Accountability and environmental justice

Environmental justice is about the unequal distribution of, and ability to affect, environmental risk. Accountability is about who is responsible both for causing an environmental hazard and for permitting, monitoring, and enforcing the regulations applicable to that hazard. Environmental justice and accountability issues tend to transcend scales and authorities such that who is accountable and responsible for a community being unequally exposed to an environmental risk can be difficult to determine or it can be many entities: the environmental regulators failing to act on their mandates; a polluter emitting more than their permit allows; or a city planner siting and permitting a pollution facility in a low-income area. Many people are often ultimately accountable even if not held responsible.

The Flint Michigan Water Crisis of 2014 is an example of failures at multiple levels of environmental management regarding accountability and environmental justice. The US Office of the Inspector General determined the following (OIG 2018):

> The circumstances and response to Flint's drinking water contamination involved implementation and oversight lapses at the EPA, the state of Michigan, the Michigan Department of Environmental Quality (MDEQ), and the city of Flint. Since January 21, 2016, the EPA has overseen the implementation of its emergency administrative order and amendment issued in response to the drinking water contamination.

REAL-WORLD EXAMPLE 3.1 (cont.)

EPA Region 5 and EPA headquarters officials have worked with the MDEQ and Flint personnel to help improve the city's water system. As of May 2018, the state of Michigan and city of Flint have completed some actions and are working on remaining actions.

Michigan: Under the MDEQ's supervision, the Flint water system did not adhere to two Lead and Copper Rule requirements: (1) develop and maintain an inventory of lead service lines needed for sampling, and (2) maintain corrosion control treatment after the water source switch in April 2014. The rule requires utilities to minimize consumers' exposure to lead in drinking water. As the primacy agency, the MDEQ is responsible for enforcing this rule for Michigan water systems. The MDEQ did not issue a notice of violation or take other formal enforcement action regarding either requirement until August 2015. Instead, the MDEQ advised Flint public water system staff to conduct additional tests and to delay corrosion control treatment installation. The decision to delay corrosion control treatment prolonged residents' exposure to lead.

The EPA: The agency retains oversight and enforcement authorities to provide assurance that states with primacy comply with Safe Drinking Water Act requirements, such as those in the Lead and Copper Rule. However, Region 5 did not implement management controls that could have facilitated more informed and proactive decision-making when Flint and the MDEQ did not properly implement the Lead and Copper Rule. While Flint residents were being exposed to lead in drinking water, the federal response was delayed, in part, because the EPA did not establish clear roles and responsibilities, risk assessment procedures, effective communication and proactive oversight tools.

Accountability and the seeking of environmental justice is still being processed and pursued in the case of Flint, with eight current or former state officials, including the former state governor, facing criminal charges (Booker 2021). The EPA and Michigan DEQ have spent over $400 million to rectify the issue. This is a step in the right direction, but accountability and the reduction and ultimate removal of environmental injustice still remains a critical issue for environmental managers as many instances go unaddressed and thus persist. The pursuit of accountability and environmental justice are essential to protecting human health and the environment.

3.2.3 Ecosystem Management

Ecosystems do not pay attention to human boundaries – diagnostic: pollution crosses jurisdictions within ecosystems.

The air blows and the water flows, to paraphrase Psalm 147:18. **Ecosystem management**, as an issue for the environmental manager, is *not* about managing an ecosystem but rather how we manage environmental situations within a shared ecosystem, for instance a "watershed" or an "airshed." This issue is about jurisdiction and jurisdictional boundaries. Rivers flow through different and often many jurisdictions with whatever pollution is discharged into their waters, just as the air stream moves our atmosphere and its airborne pollutants across our planet. Recent studies exemplify this fact, for example demonstrating that about half of air-quality-related premature mortality that occurs within a state is from toxic air emissions produced in a neighboring or more distant state (Dedoussi et al. 2020). Downwind and downstream states are unequally affected by their upwind or upstream counterparts, posing complex cross-state jurisdictional challenges for environmental managers.

Program and resource management for this issue involves permitting such that pollution from entities in your jurisdiction does not impact other jurisdictions. Monitoring for compliance is critical to this issue to make sure pollution from your jurisdiction is not impacting another jurisdiction or determine whether pollution impacting your jurisdiction is coming from outside of your area of responsibility and, if so, how much and from where. Enforcement becomes a function of program and resource management with regard to not allowing the pollution of your neighbors, but also relies heavily on your political management in terms of how you work with a neighboring environmental manager who is allowing their jurisdiction to pollute yours. As well, technical assistance and education involves your ability to effectively assist the tracking and measuring of the pollution produced in your jurisdiction and the pollution arriving in your jurisdiction from external sources, and then educating the external producers on their effects.

Other than assuring you are conducting your program and resource management functions correctly, ecosystem management can be done "the easy way" or the "hard way" – that is, voluntary cooperation or legal action taken with neighboring jurisdictions. Arguably one of the best examples of how this ecosystem management should but often does not work is the many jurisdictions abutting the Chesapeake Bay. This body of water has been out of compliance for decades, particularly regarding non-point-source runoff pollution from agriculture and municipalities. This has forced a mandate (a consent decree initially issued in 1999 but still in force today) for the EPA to coordinate permitting, monitoring, enforcement, and technical assistance efforts on a regional basis as the municipal and state-level environmental managers have been unable to coordinate an effective response between them (Baliles 2006). Another, more contentious example is when the states of the Northeastern US sued the Midwestern states who, through their power generation, were emitting pollution which caused acid rain, thereby damaging the forests and lakes of New England (Figura 2020).

Ecosystem management as an issue for the environmental manager is expected to increase in importance as the cumulative build-up and corresponding effect of human-produced pollution in the environment grows (Alexeeff et al. 2012; Alves et al. 2012; Solomon et al. 2016; Steffen, Richardson et al. 2015; US EPA 1999). For example, it is becoming clear that due to the future conditions expected from anthropogenic climate change, thousands of contaminated sites around the US are at high risk of flooding, with the toxic materials at those sites becoming redistributed to the surrounding areas and potentially areas well downstream (Kiaghadi and Rifai 2019; Marcantonio, Field, and Regan 2019, 2020). Mapping and understanding the complex interactions of the area the environmental manager is responsible for, and all of the externally produced materials that make their way into that area, is a growing issue but also a critical skill for the effective environmental manager.

REAL-WORLD EXAMPLE 3.2 Ecosystem management: toxic exposure and climate change

Increasingly environmental managers are being faced with interactive risks. That is, many of the pollution types and sources that environmental managers work against interact and create new or additional risk to human health and the environment. To effectively address these risks environmental managers must approach environmental problems using ecosystem management: a holistic approach to assessing and addressing various pollution sources and types to comprehensively deal with them.

For example, there are hundreds of thousands of contaminated sites throughout the US (US EPA 2021b). The US has also experienced an increased rate of extreme weather events, especially flooding events both inland and along the coasts (Garner et al. 2017; NOAA *2020b*; Taherkhani et al. 2020). Taken separately these phenomena already pose substantial risk to human health. But flooding also has the potential to redistribute the toxins present at contaminated sites, increasing the risk that they pose to human health (Marcantonio, Field, and Regan 2019, 2020). What does this mean for environmental managers? It means that a single source of pollution or environmental risk cannot be assessed in isolation. So, a city environmental manager who has multiple Superfund sites within their area of jurisdiction needs to consider what the risks to those sites are, given future expected climatic conditions; what are the potential risks to human health given the population or land use surrounding those sites; and then how to mitigate risks before they are realized. It also means that for effective ecosystem-scale management, environmental managers must work across jurisdictions to effectively reduce or mitigate the risks they face. While it may be a municipal environmental manager charged with the oversight of a contaminated site that is enrolled in the federal Superfund program, if their municipality is in an area with an increasing risk of flooding, they cannot directly reduce that risk because they cannot regulate the greenhouse gas emissions driving it. As a result, their policies and practices locally will focus on adaptation or site protection while working with national-level partners, namely the EPA, to shape new, efficient regulatory processes to reduce further contributions to climate change. Similarly, the municipal environmental manager might work with the Army Corps of Engineers to effect change on the water body that imposes the risk of flooding, managing together an adaptation to the risk. There are myriad partnerships to be made, but part of ecosystem management is working across jurisdictional boundaries to form a comprehensive approach to management.

Human modification of landscapes and land use intensity continues to increase (Ellis et al. 2021), and correspondingly these modifications affect each other so that one tweak here impacts another tweak previously made and so on. This is how systems work. Ecosystem management is recognizing this complexity and producing permitting,

REAL-WORLD EXAMPLE 3.2 (cont.)

monitoring, enforcement, and technical assistance programs that are not blind to it but rather embrace it by addressing multiple pollution sources and streams at once and mapping out how they interact. Ecosystem management as an approach is essential to protecting human health and the environment.

3.2.4 Sustainable Development

Sustainable development is about the future – diagnostic: affects the future of the ecosystem.

A person who understands sustainable development is a person "who knows that the world is not given by [their] fathers, but borrowed from [their] children" (Berry 1971: 3). This quote goes to the nub of sustainable development. According to the EPA, "Sustainability is based on a simple principle: Everything that we need for our survival and well-being depends, either directly or indirectly, on our natural environment. To pursue sustainability is to create and maintain the conditions under which humans and nature can exist in productive harmony to support present and future generations" (US EPA 2021b). Persistent and/or prolonged environmental impacts is what this issue is about for the professional who must be forward-looking in their management. Some environmental problems can resolve over time if we stop contributing to them, while others will get worse even though we have ceased to pollute.

One of the oft-used phrases in an older and now outdated form of environmental management is "the solution to pollution is dilution," employed mostly in reference to air and water pollution. Indeed, there are many instances where once pollution was better managed, i.e., reduced or stopped, the affected air and waters "healed" and became more sustainable. On the other hand, certain pollutants are not subject to dilution, reasonable degradation, or bioremediation. These instances are mostly associated with solid and hazardous waste ranging from plastics to dioxins (Allen et al. 2019; Smith et al. 2018; US EPA 2014d; Wright and Kelly 2017). For example, there is increasing concern over what have been termed "forever pollutants" or "forever chemicals," such as per- and polyfluoroalkyl substances (PFAS) – a group of about 4,700 chemicals used to make fluoropolymer coatings and products that resist heat, oil, stains, grease, and water (Reade, Quinn, and Schreiber 2019) – that have been determined to be in the blood of nearly every American alive today (US CDC 2019). While the toxicity and other risks to human health are still debated and being actively investigated, it is clear that they can cause adverse human health outcomes and they do not breakdown in the environment thus requiring remediation for removal (US EPA 2016b).

What are the implications of sustainable development for the environmental manager? Your permitting, monitoring for compliance, and enforcement will be dependent on standards involving the "environmental fate" of pollutants and are applicable to air, waste, and water. For example, the EPA's Environmental Fate and Effects Division (EFED), a sub-department within the Office of Pesticide Programs, "is responsible for evaluating and validating environmental data submitted on environmental properties and ecological effects of pesticides" (US EPA 2014e). The mandate for this division requires that they ask: Is the permit application addressing an emission that is even permittable regarding health- and risk-based standards both today *and* in the future? The environmental manager will have to consider what kind of monitors, and for how long, should be in place (for example, regarding monitoring of "leaching" from landfills) for a permitted action, what current and future uses this permitted use will restrict or constrain, and how this permitted use will interact with the emissions from adjacent or otherwise contributing emissions sources to the area under management. In other words, environmental managers must think in a complex and systemic way, taking the long view in all permitting assessments they conduct.

The environmental manager will encounter many laws and statutes that address or contain aspects of sustainable development concerns, several of which we will discuss in Chapter 4. For example, the Comprehensive Environmental Response, Compensation, and Liability Act, aka the Superfund Act, relies heavily on the enforcement of remediating hazardous waste, many varieties of which have exposed several generations of people and will continue to expose future generations. In fact, environmental managers from the municipal to the federal level are currently engaged in monitoring over 500,000 active and inactive hazardous waste sites throughout the US, all of which present implications for sustainable development (US EPA 2014c, 2017, 2021b). This large legacy of contaminated sites exemplifies why environmental managers must consider the future implications and effects of the permits they approve today as any new sites of emissions production that they allow to be created today will likely exist well into the future.

3.2.4.1 Greenwashing

Greenwashing is a term that is applied to activities or branding that, despite the alleged and often marketed intent of the program, are the opposite of sustainability in that they are not *actually* safe, sustainable, or, on net, better for human health and the environment; for example, they often favor profit today (and tomorrow) over human health and the environment. This concept is derived from the old and time-proven strategies of unscrupulous individuals that "paint or wash" their activities and brands white so as to "purify" them. Sometimes this practice is used directly by polluting industries; other times greenwashing efforts come from not-for-profit groups, trade associations and think tanks who are funded by those industries. Many of us were brought up to believe not-for-profit organizations were on the side of the angels because they were philanthropic or, given their name, not seeking profit. Unfortunately, like anything else, these entities can be co-opted for nefarious purposes, under the guise of producing "neutral science" or third-party assessments of industry.

Just as there is much debate about what is truly "sustainable" – as evidenced by the myriad definitions one can find – so too is there often controversy over whether a particular corporate program or a "green" certifying NGO is really just practicing greenwashing. The lines here are often blurred and rarely is there a conclusive answer. There are many concepts and frameworks that can potentially fall under the umbrella of greenwashing: **triple bottom line** (i.e., social, environmental, and financial bottom lines or outcomes); **ESG** (environmental, social, and corporate governance); and other similar frames that outline the purported social, environmental, and financial interests of a corporation or other organization. The general argument is that the most often practiced order, or order of concern, is first financial (profit), and then the social (people), and environmental (planet) implications. Again, it is often not easy to parse these complex contexts and the potential motivations and realized effects of such programs, making the job of the environmental manager that much more necessary but also difficult.

Ways that citizens and environmental managers can identify greenwashing are by following the branding and the money. As all of us have heard before, "if it sounds too good to be true, it probably is." Whether sham recycling, unrealistic plastic life stream, or certain schemes for carbon sequestration and storage, they all tend to play to what we *want* to hear: "consuming plastic without environmental cost," "saving the planet with pesticides," "better living through chemistry." Or, they are groups whose names evoke our common values, such as heritage, competitiveness, and construction, but are driven by underlying motives that in fact co-opt and abuse those values (Mayer 2017). *Always* investigate what individual, group, or think tank is funding the programs or groups that are promoting a "greener way." Finally, investigate the claims in terms of real cost–benefit and base on your own audit. As it turns out, some of our most revered universities have centers and initiatives that are funded to create the illusion of benefit to the environment by practices that in fact can harm human health and the environment (Mayer 2017). As the sayings go, never take anything at face value, or don't judge a book by its cover. Whichever way you want to characterize it, the point is that the environmental manager must always dig into the motivations and positioning of the person or organization providing them with information or guidance, to avoid the pitfalls of the false promises of greenwashing.

REAL-WORLD EXAMPLE 3.3 What is sustainability?

Sustainable development, and "sustainability" more generally, has become in recent years a topic of great focus in research and practice. Partly due to its wide use and interest, it has been given many different definitions. The EPA defines sustainability as "the conditions under which humans and nature can exist in productive harmony to support present and future generations" (US EPA 2014g). While every organization, whether it is an environmental organization or not, has its own definition, most tend toward a concern for continued human use of the environment for future generations.

Due to these various definitions, each of which represents the preferences of the people who put the definition together (i.e., the definitions are different because of what people think should count or not), different measures are used to determine if an action is

REAL-WORLD EXAMPLE 3.3 (cont.)

sustainable. For example, the United Nations Sustainable Development Goals (SDGs) contain 17 indicators of sustainability ranging from human poverty to terrestrial and aquatic functioning (UN 2021). Despite having a long list of indicators, there are many people who argue that the SDGs still do not actually measure sustainability because they do not, for example, account for material consumption rates (Hickel 2020a; O'Neill et al. 2018). The point being, how sustainability is measured and defined matters and there seems to not be, as yet, a consensus around a single definition.

So, what does this mean for the environmental manager? Generally, it means that there are no easy decisions and that one of the big challenges of sustainability is that it may require changing practices that allow people to survive today. As an example, from my (R.M.) own work with mining in Sierra Leone, many of the communities I work with are choosing to make tradeoffs to sustain their livelihoods today through practices that will not allow them to utilize the land they are using in the future. Local artisanal miners have few other options other than mining because the land they traditionally farmed has been or is being mined by large foreign mining operations. So to make a living they too must mine, but it renders the land unproductive for future use and emits substantial amounts of toxic and non-toxic materials into the river that is their primary source of water, and that of many communities downstream (Marcantonio and Fuentes 2020; Marcantonio, Field et al. 2021). I have been working with the Conservation Society of Sierra Leone and the Sierra Leone Environmental Protection Agency to address these issues and develop sustainable solutions, and we (like many others pursuing similar interventions) have found that addressing the pollution not only requires permitting, monitoring, enforcement, and technical assistance for the process underway – that is, the mining – but also the support and development of alternative livelihood options so that people have an alternative. Otherwise, any imposed regulatory intervention will likely be subverted because of the few other options available and the need for survival.

This example highlights how the environmental manager is again about people and relationship management more often than not, and how the role of the environmental manager can take many different forms just as "sustainability" does. Ultimately, the environmental manager often has to expand outside of their traditional bounds to make an intervention sustainable and to achieve the aims of the different definitions of sustainability to protect human health and the environment.

3.2.5 Unfunded or Underfunded Mandates

You must do this, but we won't pay for it. Or do more, with less – diagnostic: requires prioritization or reprioritization of your resources.

The last of the five policy issues that impact all environmental managers is budgetary: you are told or required to do something but not given the necessary funding to effectively do it (O'Leary et al. 1999). Unfunded or underfunded mandates impact every environmental manager we are aware of, but certainly not all environmental situations. All environmental managers often must deal and operate with scarce resources. It is then their job to prioritize program and resource functions, with the preference of the governing body for which they work often being motivated by political considerations and optics, i.e., for elected leaders and political appointees it is not always the science that matters but the salience to the public (Pralle 2019). To be clear, politicians and/or administrators catering to politics are loath to increase revenue (taxes, fees, prices, etc.) but are keen to give people what they want, maybe with the best of intentions. Unfortunately, the result is a mandate to increase services without funding or with insufficient funding. Or, when there are situations when budgets are cut or sequestered, there is often a demand that services remain uninterrupted. Either way, the manager must prioritize and allocate their resources balancing the demands of meeting their mandates and the expectations of the governmental leadership and the public that they serve. Meeting all requirements may not always be possible, requiring the environmental manager to evidence and, sometimes, defend why they made the decisions that they did.

Again, management under these conditions involves tradeoffs, balancing resources – money, time, political will, etc. – to meet mandates. A useful first step is to determine whether or not the lack of resources is perceived or real, and if there are any additional resources that can be tapped. For instance, under the Superfund Act (CERCLA) some managers make the mistake of believing it is their own funding that is required to finance a management program whereas it is sometimes the obligation of the responsible party (RP) to provide the necessary resources, though the enactment and requisition of these resources will require time and political will from the environmental manager.

The environmental manager must always consider: Which of your programs is most important and which function of each of these deserves allocation? Can any of your resources serve more than one mandate? Can you do more with less and still provide effective, efficient service? And, if not, what are you doing about it in terms of documenting or demonstrating impending non compliance with the law or failure to protect human health and the environment? These are the questions that will allow you to determine your program, resource and political management as operational resources become less available or as your mandates increase.

3.3 The Legal Trends

The legal trends are those trends or "commonalities" that apply to your environmental management decisions which impact the legality of your actions (see Figure 3.2). Considering each trend, the two general questions you must ask are, "am I liable?" and "what is the cost of this decision?" At the very least, each trend can cost you time, if not money, either in pursuit of avoiding liability or by spending time in court. As O'Leary et al.

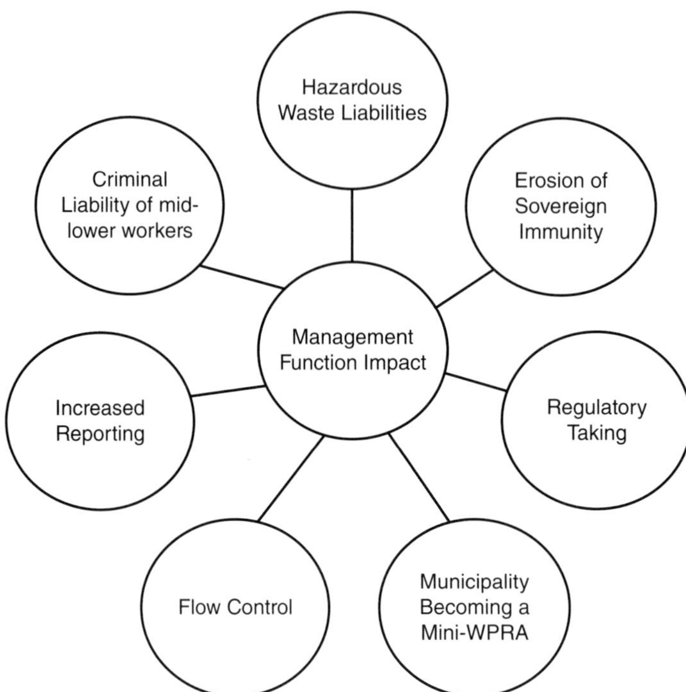

Figure 3.2 The legal trends that impact environmental managers.

(1999) point out, these trends "cut across the rights and responsibilities enunciated in the US environmental laws," complicating efforts to protect human health and the environment (p. 45). Over time the effect of these trends across different management situations has varied based on legal, policy, and political factors, but in general they have continued to be important directions in environmental management such that the environmental manager should be aware of them and their implications for their profession.

The legal trends:

- Erosion of government immunity
- Hazardous waste liabilities for government
- Criminal liability of lower and mid-level government employees
- Increased reporting requirements
- Liability for regulatory takings
- The municipality as a miniature water pollution control agency
- Flow control

3.3.1 Erosion of Government Immunity

Yes, you can sue city hall – diagnostic: a community can hold government (regulators) accountable.

The principle of sovereign immunity is that our government cannot be sued without its consent. The erosion of government immunity as a legal trend demonstrates how government is increasingly waiving its sovereign rights under a number of environmental laws, thereby inviting citizens and their advocacy groups to challenge the government and its corresponding environmental management practices and policies in the courts. To put this in numbers, in the 10-year period between 2009 and 2018 there were about 13,000 environmental cases filed in federal courts either against environmental regulatory institutions directly or challenging the implementation of the statutes they are mandated to enforce (Lazarus 2019). The largest percentage of these cases involved the Clean Water Act, but other federal environmental laws such as the National Environmental Policy Act and CERCLA were also included in numerous cases (see Chapter 4 for descriptions and key components of these laws).

What does this mean to the environmental manager? It means that if they are guilty, or perceived as guilty, of misfeasance, malfeasance, or nonfeasance their program management could suffer in terms of public trust (a type of political will) and other resources, such as personnel time and funding having to be diverted into mounting a defense against the allegations from individuals or groups. Thus, the manager must assure their program management is responsive, fair, flexible, honest, accountable, and competent in terms of their permitting, monitoring for compliance, enforcement, and technical assistance.

3.3.2 Hazardous Waste Liabilities for Government

Government pollutes too – diagnostic: when a public entity may be violating environmental law.

It turns out that local, state, and federal government agencies produce hazardous waste pollution. For instance, the waste production can occur at facilities owned (or once owned) by them, such as landfills, storage facilities, or military installations. In others it has to do with the transportation and treatment of waste conducted by government entities such as sewage treatment at a publicly owned treatment works, aka a sewage treatment plant. Hazardous waste liabilities incurred by government are increasing, are extensive, and cost billions of dollars to clean up (US EPA 2014c, 2015e, 2021b). For instance, approximately 101 of the 1,327 current Superfund sites on the National Priorities List are on current or former military installations (US EPA 2021i).

What does this mean to the environmental manager? It means that public sector environmental managers can become part of the "regulated" community and must comply with the permitted operations/allowances dictated by environmental waste laws, in particular the Resource Conservation and Recovery Act. We will discuss the environmental Acts in our next chapter, and compliance in subsequent chapters. Importantly, it also means that environmental managers will essentially have to manage laterally – that is, organizations or entities that are part of the same government structure as they are – and manage upwardly, meaning their bosses. For example, a municipal environmental manager often must regulate the municipal utilities that service their area, requiring that they hold accountable the governmental body that they work for.

Obviously, this can make the environmental manager's job precarious, as they are expected and legally required to attain the mandates in their charge but may have to navigate applying the standard to an entity that has power over them. Here again, political management is a key function of the environmental manager, with open and transparent discussion and documentation of your findings and actions an important part of your process and protection. The next two trends further affirm the need for open and transparent environmental management.

3.3.3 Criminal Liability of Government Employees

The "Nuremberg Defense" doesn't work for public servants – diagnostic: the environmental manager is asked/ordered to violate environmental law.

The trend for the criminal liability of lower- and mid-level government employees is costly in terms of the actual liability but even more so with regard to public trust. Know that most infractions of environmental regulations that are enforced are a matter of civil liability, resulting in "damages," fines, and compliance agreements. However, when a government employee knowingly commits an environmental crime, covers up, or does not disclose wrongdoing, that becomes a criminal offense and is punishable with incarceration. This punishment applies to employees who were "just following orders" from superiors who are breaking the law. Probably the most contemporary example of this is with the city of Flint, Michigan, where criminal charges were brought from mid-level city employees all the way up to the governor (OIG 2018; US EPA 2021f).

What does this mean to the environmental manager? First and foremost, one must know what is legal and what is not legal. Relying on one's superior to know the law and obey the law can have consequences. I (M.L.) remember being heavily pressured by a politician to do something I knew was illegal and my response was "Senator, are you asking me to break the law?" In short, it's *your* welfare or freedom which is at stake and not necessarily your superior's.

3.3.4 CYA: Increased Reporting Requirements

Diagnostic: requires documentation to CYA.

CYA, or "cover your ass," is a time-honored skill for all professionals. We bring it up, not to deviate from the seriousness of our profession, but because this phrase is well known in most professions. It refers to professional survival and consists of forward thinking for possible trouble, keeping up to date on your legal and professional requirements, and always leaving a paper trail (assuming your activities are legal and ethical).

The environmental manager has always had to document their actions regardless of the sector. They increasingly are faced with stiffer reporting requirements so as to meet legal and regulatory standards and provide a defense for possible lawsuits. These requirements not only protect you and your organization but also are critical to your resource and political management as they are a demonstration of program performance; they provide evidence to decisionmakers that is important to program maintenance or enhancement even if the program has reached its goals and should be discontinued such that the manager can move on to other priorities.

We will discuss program evaluation and reporting in Chapter 9 and note here the fact that there are laws which require documentation and reporting. The Government Performance and Results Act (GPRA) holds federal agencies accountable for achieving program results and abiding by the Federal Managers Financial Integrity Act where:

1. obligations and costs are in compliance with applicable law
2. funds, property, and other assets are safeguarded against waste, loss, unauthorized use, or misappropriation; and
3. revenues and expenditures applicable to agency operations are properly recorded and accounted for to permit the preparation of accounts and reliable financial and statistical reports and to maintain accountability over the assets.

3.3.5 Liability for Regulatory Takings

The king must compensate you – diagnostic: when regulation might impact private property.

Environmental agencies have a *liability for regulatory takings* and increasingly are being challenged. This trend goes back to the English Magna Carta which heavily influenced the US Constitution (fifth amendment) and Bill of Rights. It basically means that the king cannot take your property without just compensation. Make no mistake, the king could take your property, but the Magna Carta guaranteed the citizen compensation, just as our government can take (seize) property under the power of "eminent domain." An often misunderstood point of the legal trend is that there is really no actual taking of property but rather a regulatory taking of the value of property.

Environmental regulations can and do prohibit property owners from conducting certain activities such as developing the land, agricultural production, or industrial use. Implementing those regulations which might deprive the owner of economic benefits depends on whether that implementation "substantially fails to advance a legitimate governmental interest" and "whether implementing the regulation deprives the owner of all economically viable uses of the property" (O'Leary et al. 1999: 63). Further, the environmental manager must be able to answer the question, "do state environmental regulations merely make explicit what already was implicit in any property title (that is the right to regulate its use) or are they decisions that come after a person acquires title that were not originally implied? If it is the latter, they are takings that governments must compensate" (O'Leary et al. 1999: 65).

The environmental manager often will encounter this trend with regard to endangered or threatened species or protecting water quality and whether an activity is permittable. Their program must first determine if it is a taking in terms of its legitimacy regarding the environmental impact of the activity, how the regulation is crafted and its impact on the economic fate of the property. Second, if it is a regulatory taking, how much compensation is due to the property owner. Resources required would be: funding for rule writers, attorneys, scientists, economists and possible compensation; extra time when or if a challenge ends up in court; and political will as takings by the government of any sort can be unpopular.

3.3.6 Municipality As a Miniature Water Pollution Control Agency

Your municipality just became an environmental regulatory agency – diagnostic: when a municipality must increase water quality protection with regulatory programs.

Municipality as a miniature water pollution control agency is a trend that requires a city or county's publicly owned treatment works (POTW) to conduct the functions of a regulatory agency. Where these larger municipalities (munnies) historically were regulated with regard to their discharges into the waters of the US, they now are responsible for implementing regulations to identify and permit industrial users, monitor those users for permit compliance, conduct enforcement, and provide education or technical assistance (O'Leary et al. 1999). Thus, the environmental manager will incur the expenses of these program functions in terms of: funding for technicians, inspectors, enforcers, educators, and necessary equipment; their time, as now the public works official takes on the responsibilities of an environmental manager; and political will. Imagine what your mayor or county council would say if you shut down the number one employer in town?

3.3.7 Flow Control

Yes, you have to take our sh . . . waste even when transported across state lines – diagnostic: waste transported across state lines.

Flow control is a legal trend resulting from the answer to the question: "can states or local governments keep trash out of their districts?" The answer is no, as it would violate the Commerce Clause of the Constitution. Flow control has to do with the flow of commerce across jurisdictional boundaries. Thus, states and local governments must accept waste from other districts. The impact of this trend on the environmental manager is that they must accept pollution from entities in other jurisdictions. However, they can require those entities to obtain and comply with local permits and pay applicable fees and taxes. Perhaps the most impacted environmental management function would be that of monitoring the content of what is "flowing" into your district. For example, recent evidence shows that most of the health impacts from environmental pollutants in a given state are from pollution produced in a neighboring state (Dedoussi et al. 2020).

The managerial impact of flow control has to do with the question, "what do we do with this waste?" Thus, the political and fiscal realities of dealing with solid waste in landfills and transfer stations or other hazardous waste–processing operations such as incineration, sludge composing, and recycling can be onerous. Further, local government environmental managers, in particular, are under constant pressure to vet technologies to deal with incoming waste, and "sham recycling" continues to be a challenge. Finally, as with plastic waste, the environmental manager is dealing with a "moving target" involving the rapidly changing technical landscape. Additional resources might include consulting scientists or engineers.

3.4 Conclusion

Each one of the issues and trends described in this chapter impacts the environmental manager's program, resource and political management as they affect permitting, monitoring for compliance, enforcement, and technical assistance. The environmental issues and trends discussed are ever-evolving and thus the environmental manager must keep a pulse on what is happening, both in their area of responsibility and nationally, to most effectively and efficiently plan for and carry out their mission to protect human health and the environment.

3.5 End of Chapter Questions

1. Why is it important for the environmental manager to be able to *identify, analyze and prioritize* environmental issues and legal trends?
2. How does a specific issue or legal trend (listed below) impact your environmental management in terms of program, resource and political management? . . . Or what are three managerial implications to each issue or legal trend?
   (Hint: The three managerial implications would be the program, resource and political management in terms of the environmental manager's basic functions (permitting, monitoring, enforcement, and technical assistance). What you need to do is coin those implications in terms of how those functions relate to the meaning of each issue or trend. – Understanding and applying!)
3. How does the issue of accountability impact your environmental management in terms of program, resource and political management?
4. How does the issue of environmental justice impact your environmental management in terms of program, resource and political management?
5. How does the issue of ecosystem management impact your environmental management in terms of program, resource and political management?
6. How does the issue of sustainable development impact your environmental management in terms of program, resource and political management?
7. How can you and the professional identify greenwashing?
8. How does the issue of unfunded or underfunded mandates impact your environmental management in terms of program, resource and political management?
9. How does the legal trend regarding liabilities associated with the erosion of sovereign immunity impact your environmental management in terms of program, resource and political management?

10. How does the legal trend regarding liabilities for increased reporting impact your environmental management in terms of program, resource and political management?
11. How does the legal trend regarding flow control impact your environmental management in terms of program, resource and political management?
12. How does the legal trend regarding liabilities for regulatory takings impact your environmental management in terms of program, resource and political management?
13. How does the legal trend regarding hazardous waste liabilities impact your environmental management in terms of program, resource and political management?
14. How does the legal trend regarding municipalities becoming a water pollution control agency impact your environmental management in terms of program, resource and political management?
15. How does the legal trend regarding criminal liabilities impact your environmental management in terms of program, resource and political management?
16. What are the three basic reasons environmental inequities exist?
17. What are the three basic principles that frame environmental justice?
18. What are *three* implications for you as an environmental manager *specific* to the legal trend regarding liabilities for becoming a mini-water pollution control agency? How does each trend impact your environmental management in terms of program, resource and political management?

4 Environmental Regulation

Be still and the earth will speak to you . . .

Carl Moon – Navajo Nation, 1904

Skill: *Matching legal authority to environmental goals.* To develop and learn the language of the environmental management profession, with a specific focus on the terms and concepts of environmental law, to memorize the intent and major provisions of each environmental law, and to be able to compare and contrast the laws and their provisions for more effective and efficient diagnosis. Mastering this skill will enable the professional to better understand their legal accountability and the "arena" – i.e., the areas of operation and expectations of their work delimited by legal statutes – in which they are managing. In other words, to be able to respond to opposition and take advantage of opportunities.

4.1 Introduction

Environmental regulations such as laws and rules are a subset of the volatile and complex issues and trends that must be understood by the environmental manager. As well, the issues and trends discussed in Chapter 3 apply to all of these laws in terms of our accountability to the law, and how its enactment and enforcement affects equity, the distribution of pollution across jurisdictions, our future sustainability, and the ability to implement environmental management programs.

As both citizens and professionals we have witnessed that creating and implementing regulations is both volatile and complex, as politics is considered a "blood sport" and lawmaking is likened to "making sausage." We are not legal scholars or lawyers and do not intend or feign to teach the reader of this text to become one. Rather, we write this chapter to help environmental managers understand the basic US federal environmental laws, specifically their primary intent and the provisions specified in the laws designed to meet ("provide" for) that intent. Environmental laws are both proactive and reactive; that is, they can be enforced when they are broken (reactive) but they are also designed to be proactive by reducing or monitoring pollution at its source through permitting and other methods. We consider and discuss both the proactive and reactive characteristics of the environmental laws presented.

4.2 Why Do We Have Regulations?

From the perspective of environmental protection, regulations provide two critical benefits: They can prevent or reduce the "Tragedy of the Commons" (Hardin 1968); and, when well crafted, they can increase industry competitiveness (Porter 2011). As we discussed in Section 1.5 in Chapter 1, regulations are necessary to protect common goods or common pool resources such as public lands and waterways. Whether one applies the Tragedy of the Commons, the 90/10 rule, or the historical reasoning and precedence for common laws or doctrines, there is an understanding that without well-crafted regulations the common good will suffer (Barrett 2003; Kraft 2017; Ostrom 1990). There are widely differing opinions on what constitutes a well-crafted regulation, but as environmental managers it is our responsibility to carry out and enforce whatever the regulations are, but also to "manage up" – i.e., work with legislators and environmental policy decisionmakers who craft the laws we implement and enforce – by helping shape environmental regulations to be effective and efficient. This is the balance between policy (including environmental laws and regulations) and management (designing and implementing plans to comply with environmental laws and regulations). But whatever the regulations are, it is up to the environmental manager to develop a program (remember the three components: program, resource and political management or PRP) to implement, monitor, and enforce regulations knowing that there will always be some entity that, whether intentionally or unintentionally, will pollute afoul of the law and need to be corrected to protect human health and the environment.

We all remember that there was always some nasty kid in second grade that did bad things such that we all missed recess or snacks. Of course, that child knew the rules but just believed he would not have to follow them and seemingly did not care if we suffered the consequences.

Thus, regulations are not much good if they are not followed and that is where our strategic management of regulatory enforcement comes in: permitting, monitoring for compliance, enforcement, and technical assistance (PMET). The PMET process can be found in many practices of everyday life. All motorists, for instance, have had management regarding motor vehicle laws as a requirement for gaining a driver's license, their speed is monitored and sanctioned by the police, and they may occasionally be subjected to technical assistance for retesting or as an alternative to paying a ticket fine. Similarly, PMET for environmental management is an iterative and adaptive process for managing the practices of pollution producers that must take place day in and day out.

The idea that regulations can increase competitiveness can be controversial, but we contend, based on the data, that regulations when well-crafted can and do increase the competitive advantage of managed polluters (Ambec et al. 2013; Dechezleprêtre and Sato 2017; Porter 2011). There are two scientific concepts that help illustrate why this is the case, both having to do with forcing change and the impact of that force. First, there is the second law of thermodynamics, which is a law of physics but for our purposes is best explained by the psychologist Steven Pinker: "The ultimate purpose of life, mind, and human striving: to deploy energy and information to fight back the tide of entropy and carve out refuges of beneficial order" (Pinker 2017). In other words, without the input of energy, things tend to get disorganized, i.e., inefficient.

The disorder of a closed system is always increasing unless there is an external force acting on the system to produce order and thus direct the system to a more productive end. In our case, environmental regulations are the force acting on the system of production to force it to limit pollution, i.e., waste byproduct from inefficient production processes, and to become more efficient with inputs used in the production process. For example, the Lawrence Livermore National Laboratory of the US Department of Energy (DoE) produces a report every year recording the efficiency of energy use in the US. In 2019 the estimated efficiency of US energy use was just over 30 percent, meaning that almost 70 percent of all energy produced – with corresponding pollution emissions – was wasted (LLNL 2020). The well-crafted regulations of policymakers and their implementation by environmental managers can be a force put to this system to reduce inefficiency and create better order.

The other illustrative concept, a biological one, is that mutation or adaptation of an organism or system is often the result of external forces. Porter (1991) makes the argument in his "America's Green Strategy" that well-crafted regulations force or incentivize innovation by industry that in the end – or on net – increases efficiency, resulting in increased competitiveness. Often such change requires upfront investments, and these tend to be the source of angst for many industries and other producers, but the payoffs more often than not are positive, especially when accounting for the total human health and environmental impacts of the processes (Ambec et al. 2013; Dechezleprêtre and Sato 2017; Esty and Porter 2002; Porter 2011), i.e., the things that environmental managers are charged with protecting.

4.3 Environmental Laws of the United States

The following laws, while not perfect, are considered "well-crafted" regulations and they are, and shape, a large part of the environmental management professional working environment. Most are enacted or implemented by the EPA, though some, such as the Endangered Species Act, are implemented by other regulatory agencies – but all within the executive branch of the US government – such as the US Fish and Wildlife Service of the Department of Interior. Federal regulations are codified annually in the *US Code of Federal Regulations* (CFR). Title 40: Protection of Environment is the section of the CFR that deals with the EPA's mission of protecting human health and the environment (US EPA 2021k).

As noted in earlier chapters, states have "primacy" or have been authorized by the EPA to enact these laws by way of their state statutes that approximate the CFR 40 regulations.

Each federal environmental law has a primary intent and a number of major provisions. For the purpose of this textbook, we will use language that is not legalese but more practical, so as to be more understandable and memorable. We try to relate the intention of each law to the mission of the EPA, and each provision is a statement of what needs to be "provided" to achieve that intent. As a baseline of foundational knowledge, we believe it critical for the environmental manager to know the intent and major provisions of each law and to be able to compare and contrast them, and also to leverage each appropriately in pursuit of their mission. This skill will let the professional better understand the "arena" they are managing in.

4.4 The "Environmental" Laws

The first three "environmental" laws listed below, which were the first three national environmental Acts passed by the US Congress, are general in nature and in our opinion stem directly from and are related to the problems and solutions Carson discusses in *Silent Spring* (Carson 1962). They apply to the actions of government, landowning citizens, and industry and the effect they each have on both human health and the environment. Further, they provide a non-media-specific process for preventing pollution. The second set of "environmental media" laws, related to "byproducts of process as pollution" laws, pertain to specific "media" – air, waste, and water.

There is one air law, two water laws, four waste laws (two for active waste production and two for inactive waste sites), two "materials" laws, and one oil law. There are other federal laws that pertain to these media, but these 10 laws (in addition to the three non-media-specific laws discussed above) are the primary federal environmental laws. The media laws are designed around primary, health-based standards for human health and secondary, welfare-based standards for what humans consider important beyond public health such as the environment. Usually, their major provisions are a result of technology-forced legislation, with health-based standards requiring upgrades in design and technology. As well, the media laws require some kind of state planning.

The "materials" laws, for pollution that is not a byproduct but a result of direct emission of material into the environment, pertain to toxic materials (toxics and pesticides) and are "economically balanced laws" in their regulatory design. These toxic materials themselves – as opposed to the byproduct of their transformation or use as an input – directly affect human health and the environment with or without becoming waste. Rather than being health-based, they are "risk"-based and require a risk–benefit or cost–benefit analysis to be conducted in the impact assessment and permitting decisionmaking process. In other words, they regulate materials that are intentionally put into the environment for a particular process – be it killing weeds or treating a tree disease – rather than materials or emissions put into an ecosystem as an incidental outcome of a production process.

4.4.1 National Environmental Policy Act: NEPA 1969

This law was a direct, national response to Carson's (1962) *Silent Spring* acknowledging that human activity often has a negative and systemic effect on the environment. Lynton Caldwell operationalized Carson's book while crafting NEPA by explicitly including a requirement that there should be an Environmental Impact Statement (EIS), based on applied ecology, connected to all governmental actions. Thus, NEPA forced governmental entities, for the first time, to account for and assess the environmental impacts of their actions. Examples of these governmental actions could be building or enhancing freeways, airports, military installations, dams, or other major actions as defined in 40 CFR 1508.1 Definitions of NEPA (US EPA 2021c).

Intent: To protect human health and the environment regarding government actions by requiring that all federal entities consider and assess the expected environmental impacts of their actions.

Major Provisions

- The declaration of national Environmental Policies and Goals
- The establishment of action-forcing provisions for federal agencies to implement those policies and goals, including the production of an Environmental Impact Statement for all major federal actions
- The establishment of a Council on Environmental Quality in the executive office of the president

REAL-WORLD EXAMPLE 4.1 NEPA and the US Department of Defense (DoD)

The passage of NEPA substantially affected how the US military conducted business. As one might expect, the US military puts lots of inputs into the environment in the process of training, maintenance, and other quotidian tasks required to sustain force readiness, deterrence, and other tasks assigned to the military by Congress and the commander-in-chief, i.e., the president. From impact areas filled with exploded and unexploded ordnance to massive modifications of coastlines, the US military has a substantial potential impact on human health and the environment, with most of its actions constituting a "major federal action" that triggers the application of NEPA. The DoD developed their own DoD-wide regulations and NEPA review process, the 9-FD-f regulations, and further pushed guidance to each of its subordinate components to develop environmental management plans.

In complying with the DoD directive to conform to the intent and provisions of DoD 9-FD-f, each of the major military branches – the Departments of the Army, Air Force, and the Navy (which includes the Marine Corps) – established their own policies to meet the intent and requirements of NEPA. For example, the Army passed the 32 CFR 651

REAL-WORLD EXAMPLE 4.1 (cont.)

Environmental Analysis of Army Actions policy and established the US Army Environmental Command at the Aberdeen Proving Ground in Maryland, though it is now located at Fort Sam Houston in Texas. The environmental managers within this command have even put together a yearly planning tool for base and high-level unit commanders to consider the environmental impacts of their actions and how they can reduce them without diminishing training and unit effectiveness (see link for the 2019 version in our suggested resources) (USA 2020).

The DoD is responsible for more emissions than the combined total of the bottom 140 countries ranked by total annual emissions, leaving much work to be done and a continuing substantial problem for environmental managers (Belcher et al. 2020). However, it is hard to overstate the effect of NEPA on the DoD and the US federal government, as well as the environmental management plans of nations worldwide that, in response to the passage of NEPA in the US, developed their own but very similar national policies.

4.4.2 The Endangered Species Act: ESA 1973

The Endangered Species Act is another Act that can be linked to Carson's *Silent Spring*. It has two lead agencies, the Department of the Interior US Fish and Wildlife Service (US FWS) and the National Oceanic and Atmospheric Administration (NOAA) Fisheries Service.

Intent: The Act aims to provide a framework to conserve and protect endangered and threatened species and their habitats (ESA 1973). Initially the stated intent was to "protect and enhance biodiversity" as passed in 1973 (and amended in 1978, 1982, and 1988). Later, the US Supreme Court found the intent of the ESA "was to halt and reverse the trend toward species extinction, whatever the cost" ("Tennessee Valley Authority v. Hill," 437 U.S. 153, 1978).

The US EPA, Office of Pesticide Programs (OPP) acts as a consultant to the lead agencies regarding the environmental fate of pesticides and their ecological effect on listed species.

Biologist E. O. Wilson considers the decline of biodiversity to be an existential threat to humankind and to the entire global ecosystem, not just the species close to the brink of extinction, because it

> . . . is an unintended consequence of multiple factors that have been enhanced by human activity. They can be summarized by the acronym HIPPO, with the order of letters corresponding to their rank in destructiveness.
>
> **H** – habitat loss, including that caused by human-induced climate change
>
> **I** – invasive species (harmful aliens, including predators, disease organisms, and dominant competitors that displace natives)

P – pollution
P – human overpopulation, a root cause of the other four factors
O – overharvesting (hunting, fishing, gathering). (Wilson 2010: 75)

It goes without saying that Wilson, an entomologist by training, is a proponent of including invertebrates as possible endangered species, and indeed they are (Wilson 2010). However, there is resistance to this inclusion which flies in the face of preserving biodiversity. Just one example of the critical need for including insects is the idea of "pollinator protection," as without pollinators humankind would pay a huge price in terms of food costs and eventually its own possible extinction (Christmann 2019; Ratto et al. 2018). In other words, the effects of an input into the environment – an action regulated by environmental managers – are having a direct impact on non-human species that risks the health of the environment and subsequently also poses risk to human health. So, while non-human species may be considered a natural resource and therefore part of the natural resource manager's domain, the environmental manager is responsible for regulating many relevant and consequential inputs into the environment that pose risk to endangered species. As such, ESA can be a powerful tool for the environmental manager.

Major Provisions

- The federal government must determine whether species are endangered or threatened. If so, they must list the species for protection under the ESA on the appropriate endangered or threatened species list (section 4).
- If determinable, critical habitat must be designated for listed species (section 4).
- Absent certain limited situations (section 10), it is illegal to "take" an endangered species (section 9). "Take" can mean kill, harm, or harass (section 3).
- Federal agencies will use their authorities to conserve endangered species and threatened species (section 7).
- Federal agencies cannot jeopardize listed species' existence or destroy critical habitat.
- Any import, export, interstate, and foreign commerce of listed species is generally prohibited (section 9).
- Endangered fish or wildlife cannot be taken without a take permit. This also applies to certain threatened animals with section 4(d) rules (section 10).

REAL-WORLD EXAMPLE 4.2 ESA, the EPA, and pesticides (US EPA 2014h)

EPA Responsibilities

When registering a pesticide or reassessing the potential ecological risks from use of a currently registered pesticide, EPA evaluates extensive environmental fate and toxicity data

REAL-WORLD EXAMPLE 4.2 (cont.)

to determine how a pesticide will move through and break down in the environment and whether potential exposure to the pesticide will result in adverse effects to wildlife and vegetation. The EPA routinely assesses risks to birds, fish, invertebrates, mammals, and plants to determine whether a pesticide may be licensed for use in the United States.

EPA's pesticide risk assessment and regulatory processes ensure that protections are in place for all populations of non-target species.

Effects Determinations

The result of an assessment to determine potential effects of a pesticide's registration on a listed species will result in one of two determinations:

- The pesticide's registered use will have "no effect" on the species or designated critical habitat,
- The pesticide's registered use "may affect" the species or designated critical habitat.

If EPA determines the pesticide "may affect" the species it refines its assessment to determine whether the pesticide's use:

- "may affect, but is not likely to adversely affect" the species or designated critical habitat; or
- "may affect and is likely to adversely affect" the species or designated critical habitat.

A determination that the pesticide "may affect, but is not likely to adversely affect" is made when the effects on the listed species and/or designated critical habitat are expected to be discountable, or insignificant, or completely beneficial. The Services have discussed these terms in guidance as including effects that:

- are extremely unlikely to occur, which cannot be evaluated (discountable);
- would never reach a scale where take would occur (insignificant); or
- are wholly beneficial (completely beneficial).

Consulting with the Services

Informal Consultation

If EPA determines that a pesticide will have "no effect" on a listed species, no further action is required relative to the pesticide's registration, label, or use instructions. If EPA determines that a pesticide's use "may affect, but is not likely to adversely affect" a listed species, EPA will engage the Services in a process called informal consultation. The result of this process is typically a letter in which the Services concur or non-concur with EPA's determination.

REAL-WORLD EXAMPLE 4.2 (cont.)

Formal Consultation

If the Services do not concur with EPA's determination that a pesticide's registered use is "not likely to adversely affect" a species or if EPA determines that the pesticide "may affect and is likely to adversely affect" a listed species, it will engage the Services in a process called formal consultation.

During formal consultation (under 50 CFR Part 402, Subpart B), EPA provides the Services with its detailed assessment of potential risks and its effects determination. The Services review that information and consider it in light of the status and needs of the particular species potentially affected. The Services then provide EPA with a Biological Opinion, a document providing the Services' assessment and recommendations for steps that EPA should take, if any, to reduce or eliminate potential risk to the species.

The Services' Biological Opinion provides the Services' view of whether the pesticide's registered use is likely to jeopardize the continued existence of the species and, if so, describes alternatives to avoid jeopardy. The Services also authorize any "take" (unintended injury or killing of individual listed species) that would otherwise be prohibited, as long as measures to minimize take are implemented.

Implementing Measures to Protect Listed Species

After EPA's risk assessment or formal consultation with the Services, if EPA determines that a pesticide's registration, label, or use instructions should be altered to ensure use of a pesticide will not take or jeopardize the continued existence of a listed species, EPA may require changes to the use conditions specified on the label of the product. When such changes are necessary only in specific geographic areas rather than nationwide to ensure protection of the listed species, EPA implements these changes through geographically specific Endangered Species Protection Bulletins.

4.4.3 The Pollution Prevention Act: PPA 1990

The Pollution Prevention Act was a progressive movement toward a proactive approach – as opposed to a reactive remediation approach – to pollution prevention and source reduction, and moved environmental regulation beyond "end of the pipe" controls (O'Leary et al. 1999). It is different from the other pollution laws in that it is proactive. The PPA works to prevent pollution before it is emitted into the environment through technical assistance, incentives, technology investments, and other programs, and it can be applied to all environmental media (air, water, and waste).

Unfortunately, the Pollution Prevention Act is mostly a reporting regulation without teeth and rarely applied to regulatory polluting situations like the environmental media laws, i.e., it does not have enforceable negative incentives such as fines or other mechanisms included in it as polluters are not necessarily considered violators of this Act. In fact, most of the actions taken by the EPA to implement the PPA include positive incentives, such as rewarding the use of improved technologies, or technical assistance provided to polluters (US EPA 2010). This law marked the EPA's move toward "sustainability" as part of its overarching focus on protecting human health and the environment.

Intent: To protect human health and the environment by stimulating polluters to reduce or prevent pollution at the source through cost-effective changes in production, operation, process and raw materials use.

Major Provisions

- When feasible, pollution should be prevented or reduced at the source.
- When prevention is not feasible, pollution should be recycled in an environmentally safe manner.
- When prevention or recycling is not feasible, pollution should be treated in environmentally safe ways.
- Only when prevention, recycling, or treatment are not feasible should disposal or releases be used.

REAL-WORLD EXAMPLE 4.3 Implementing the Pollution Prevention Act

The Pollution Prevention (P2) Act of 1990 was one of the first pieces of legislation to help reduce the creation of pollution at the source rather than focusing on end-of-pipe pollution control. The P2 Act directs EPA to promote source reduction in several ways, one of them being "identify opportunities to use Federal procurement to encourage source reduction." EPA does this through the work of the Environmentally Preferable Purchasing (EPP) program. The federal government is the single largest purchaser in the world, spending over $650 billion on goods and services each year. The goal of the EPP program is to harness federal purchasing power to catalyze a more sustainable marketplace for all – preventing pollution, reducing climate impacts, improving the health of fenceline communities, and increasing US industry competitiveness. The program helps US federal government purchasers utilize environmental performance standards and ecolabels to identify and procure environmentally preferable products and services, providing a convenient and streamlined way to make sense of the often-complex sustainable products marketplace.

Buying environmentally preferable products has prevented a significant amount of pollution from entering the environment. For example, in 2018 alone, the federal

REAL-WORLD EXAMPLE 4.3 (cont.)

government purchased nearly seven million EPEAT-registered products. EPEAT is an ecolabel that addresses impacts across the entire life cycle of electronic products including design and production, energy use, recycling, and repairability. EPEAT-registered products span from mobile phones, to computers, to printers/scanners, and more. Over the lifetime of these certified products, the procurement will result in the following reductions and savings:

- Reduction of 1.2 million metric tons of greenhouse gasses, equivalent to taking nearly 250,000 average US passenger cars off the road for a year.
- Reduction of 4,400 metric tons of hazardous waste, equivalent to the weight of over 36,000 refrigerators.
- Reduction of 1.8 million megawatt-hours (MwH) of electricity, equivalent to the annual electricity consumption of nearly 148,000 US households.
- All environmental impacts combined result in cost savings of over $182 million.

EPA's EPP program continues to use the P2 Act's authority to encourage source reduction through federal procurement. These environmental results are amplified as the EPP program continues to help federal purchasers identify and procure environmentally preferable products and services across additional purchase categories such as building materials, custodial products, cafeteria services, and more. Additionally, the federal government strives to lead by example and set the precedent for other purchasing institutions – both public and private sector – looking to encourage source reduction through purchasing.

Jenna Larkin MPA-MSES, Environmentally Preferable Purchasing (EPP) Program at the US Environmental Protection Agency

4.5 The "Environmental Media," "Health-Based" Laws Related to "Pollution As Byproducts of the Process of Production"

The second set of "environmental media" laws related to "byproducts of process as pollution" laws pertain to specific "media" – air, waste, and water. These laws employ primary, health-based standards for human health and secondary, welfare-based standards for what humans consider important beyond public health such as the environment. Usually, their major provisions are a result of technology-forced legislation, with health-based standards requiring upgraded design and technology. As well, these laws require some kind of state planning. For instance, air quality requires a State Implementation Plan (SIP) setting out how the state will restore quality in areas exceeding standards. With water quality, states, territories, and tribes are required to submit lists of impaired waters with ways water quality standards can be maintained or restored. Serious attention was paid to providing an economic balance with these laws, as most have some language waiving requirements that are not "economically achievable," i.e., they include consideration of cost–benefit analyses in determining implementation requirements.

Before parsing these laws, there are a few key concepts relevant to both the laws and to the environmental manager that need first to be discussed. For example, there are two generalized types of pollution sources that put emissions into the environment; one is **point-source pollution**. The EPA defines point-source pollution as any contaminant that enters the environment from an easily identified and confined place (US EPA 2015f). Examples include utility energy facility or industrial production smokestacks, discharge pipes, and drainage ditches. Interestingly, air pollution from mobile sources like vehicles is not considered to be point-source pollution even though a discernible source is locatable. As well, drainage from abandoned mining operations is not considered point-source pollution but rather non-point-source pollution.

The alternative to point-source pollution is **non-point-source pollution (NPS)** which generally results from land runoff, precipitation, atmospheric deposition, drainage, seepage, or hydrologic modification. NPS is also called "people pollution" because much of it is the result of activities that people do every day (US EPA 2015d). Common examples of NPS would be:

- Excess agrichemicals (fertilizers, herbicides, and insecticides) from agricultural lands and residential areas, oil, grease and toxic chemicals from urban runoff and energy production
- Sediment from improperly managed construction sites, crop and forest lands, and eroding streambanks
- Salt from irrigation practices and acid drainage from abandoned mines
- Bacteria and nutrients from livestock, pet wastes and faulty septic systems
- Atmospheric deposition (e.g., toxic particulate matter deposited from the airshed into a watershed)

In general, point-source pollution is easier to regulate in terms of the "proof of pollution" and because for NPS it has been considered impolitic to conduct enforcement on drivers, flyers, farmers, and miners. NPS is demonstrably more difficult to track and measure as, in contrast to point-source pollution, it does not enter the environment at a single location but usually from diffuse or widely distributed sources (US EPA 2015f). The lack of standards, permits, and enforcement regarding NPS is one of the biggest criticisms of these laws (Craig and Roberts 2015).

4.5.1 When an Airshed or Watershed Is Not Meeting Standards

Another concept of import is that of **attainment vs. non-attainment or impaired**. A basic concept related to compliance is built into the media laws as to whether or not a geographic area or entity is in compliance regarding meeting specified standards. For example, with regard to air quality if a geographic area is not attaining National Ambient Air Quality Standards (NAAQS) for criteria air pollutants (CAPs) it would be considered to be in "non-attainment" status. (Note that we discuss exactly what NAAQS and CAPs are in the first law listed below.) If a body of water is too polluted to attain water quality standards based on Total Maximum Daily Load (TMDL) standards it would be considered "impaired." Thus, both NAAQS (particularly CAPs) and TMDLs are used as tools to determine if an environmental medium in a specific jurisdictional area is in compliance or not. If that medium is not attaining standards and/or demonstrating "reasonable further progress" toward achieving the standard, it then would be subject to a compliance agreement restricting use (or growth) until the standards are met – such an agreement is often referred to as a consent decree (US EPA 2013g). This is an important political consideration for the environmental manager in that "city managers" or leaders could lose their "tax base" if growth is restricted, in addition to which it can reduce or slow the permitting process of other development plans of a city or municipality.

4.5.2 Air

There is one federal air quality Act that governs ambient air quality in the United States.

4.5.2.1 The Clean Air Act: CAA 1970 and Amendments of 1990

Acts related to ambient air quality had been in effect since the air quality emergencies of the 1950s. These Acts were brought in by health departments prior to the creation of the US EPA. There are a number of different ways of looking at air pollution. Types of air pollution can be differentiated as "ambient" – the air around us, where we always must consider the issue of ecosystem management – or "indoor" – the air created inside of a structure (indoor cooking, the chemical off-gassing of structural components such as paint, mastics, and plastics, pesticides, etc.).

As well, they can be considered natural sources such as from volcanoes, forest fires, plant and animal off-gassing, and dust storms; or from human sources, which are categorized into "major sources" such as transportation, electric plants (coal, oil), industry (steel mills, metal smelters, oil refineries, and pulp/paper mills are some major examples), and "minor sources" such as small engines and wood burners. Air pollution is considered an "emission" and, as mentioned, emissions are either categorized as a point source or a NPS.

Intent: Protect human health and the environment regarding ambient air quality.

Major Provisions

- Defined national goals as ambient air standards from stationary and mobile sources: NAAQS for CAPs for the most sensitive or at-risk Americans (US EPA 2021d).
  - Particulate Matter 2.5 – PM 2.5, that is, particulate matter of >2.5 microns in size
  - SO_2 – sulfur dioxide

  - CO – carbon monoxide
  - NO_X –NO and NO_2 (also referred to as smog)
  - O_3 – trioxygen (also referred to as ozone)
  - Pb – lead
- Directed EPA to set national standards for controlling emissions of toxic hazardous air pollutants (HAPs) for the most sensitive Americans. HAPs "are those pollutants that are known or suspected to cause cancer or other serious health effects, such as reproductive effects or birth defects, or adverse environmental effects" (US EPA 2021g). Examples of the over 187 HAPs include asbestos, mercury, beryllium, benzene, vinyl chloride, arsenic, and radionuclides. Eventually amendments to the CAA set stronger standards for air toxics, with Maximum Available Control Technology (MACT) standards for 187 toxics and required health-based standards for HAPs at a threshold of a 1 in 1 million chance of cancer risk (US EPA 2015b).
- Set limits for emissions from mobile sources such as personal and commercial vehicles, aircraft, trains, marine craft, and small engines and eventually set standards for cleaner fuels.
- Gave EPA authority to set national standards for emissions from new sources of pollution from industrial and energy plants. These New Source Performance Standards (NSPS) would require the Best Available Control Technologies (BACT) in their operating permits. Note that the CAA point-source standards establish technology-based requirements for existing *and* new sources. Existing source permits take economics into consideration. New sources do not. Thus, when modifications to an existing source make it New, a new, expensive pollution-control technology will be required, prompting a conflict with the source operator.
- Directed states to prepare State Implementation Plans (SIPs) describing how they would reduce emissions from existing sources in areas exceeding air quality goals.
- Concentrated on acid rain control (SO_2 and NO_X that produce sulfuric and nitric acid) requiring continuous emission monitoring systems.

The CAA, like most environmental Acts, is always under construction (i.e., CAA – Amendments – CAAA of 1990). It is a very large piece of legislation. Further, the nation's response to global climate change lends frequent and urgent attention to air quality. The environmental manager should always try to keep up with "news" as a professional responsibility.

REAL-WORLD EXAMPLE 4.4 Criteria air pollutants in the Clean Air Act

One key regulatory scheme that the US EPA has established to meet the intent of the Clean Air Act (CAA) is the list of criteria air pollutants (CAPs). CAPs are specifically outlined and regulated by the National Ambient Air Quality Standards (NAAQS) provision of the CAA. The six CAPs are ground-level ozone (O_3), particulate matter of 2.5 microns in size or

REAL-WORLD EXAMPLE 4.4 (cont.)

less (PM 2.5), carbon monoxide (CO), nitrogen dioxide (NO_2), sulfur dioxide (SO_2), and lead (Pb). To regulate these six CAPs, the EPA sets primary and secondary standards (permitting), determines whether areas meet these standards (monitoring and enforcement), and works with areas to attain and maintain the standards (i.e., technical assistance and enforcement). These CAPs have been demonstrated to cause direct harm to human health, and unfortunately, they are produced in myriad human activities from driving cars to electricity production. They are, and likely will always be, a persistent issue for the environmental manager.

Let's look at one of the CAPs up close:

What Is SO_2 and How Does It Get in the Air? (US EPA 2021b)

What Is SO_2?

EPA's national ambient air quality standards for SO_2 are designed to protect against exposure to the entire group of sulfur oxides (SO_x). SO_2 is the component of greatest concern and is used as the indicator for the larger group of gaseous sulfur oxides (SO_x). Other gaseous SO_x (such as SO_3) are found in the atmosphere at concentrations much lower than SO_2.

Control measures that reduce SO_2 can generally be expected to reduce people's exposures to all gaseous SO_x. This may have the important co-benefit of reducing the formation of particulate sulfur pollutants, such as fine sulfate particles. Emissions that lead to high concentrations of SO_2 generally also lead to the formation of other SO_x. The largest source of SO_2 emissions is fossil fuel combustion at power plants and other industrial facilities.

How Does SO_2 Get in the Air?

The largest source of SO_2 in the atmosphere is the burning of fossil fuels by power plants and other industrial facilities. Smaller sources of SO_2 emissions include: industrial processes such as extracting metal from ore; natural sources such as volcanoes; and locomotives, ships and other vehicles and heavy equipment that burn fuel with a high sulfur content.

What Are the Harmful Effects of SO_2?

What Are the Health Effects of SO_2?

Short-term exposures to SO_2 can harm the human respiratory system and make breathing difficult. People with asthma, particularly children, are sensitive to these effects of SO_2.

REAL-WORLD EXAMPLE 4.4 (cont.)

SO_2 emissions that lead to high concentrations of SO_2 in the air generally also lead to the formation of other sulfur oxides (SO_x). SO_x can react with other compounds in the atmosphere to form small particles. These particles contribute to particulate matter (PM) pollution. Small particles may penetrate deeply into the lungs, and in sufficient quantity can contribute to health problems.

What Are the Environmental Effects of SO_2 and Other Sulfur Oxides?

At high concentrations, gaseous SO_x can harm trees and plants by damaging foliage and decreasing growth. SO_2 and other sulfur oxides can contribute to acid rain which can harm sensitive ecosystems.

For the environmental manager, the landscape of CAP monitoring and NAAQS enforcement is rapidly changing. Distributed street-level monitors are rapidly and widely being deployed in cities, and new remote-sensing technologies are changing how environmental management can be and is done in many ways. For example, the recent deployment of the European Space Agency's Copernicus Sentinel-5P satellite has led to the identification of hotspots of various CAPs, and not just in Europe but around the globe. In the US, the National Aeronautic and Space Administration (NASA) and other government entities have developed and deployed pollution-tracking satellite instrumentation as well, as part of the complex US air quality monitoring apparatus (NASA 2021). And even non-government organizations in the US, such as the Environmental Defense Fund, are moving to deploy their own pollution-tracking satellites (Tollefson 2018).

While the technologies available to the environmental manager are seeing rapid change development and our knowledge of what these pollutants do to the human body is evolving, the fundamental situational analysis and planning needed from the environmental manager has not changed. That said, environmental managers need to "keep up with the science" in order to best execute their primary functions: program, resource and political management integrated with permitting, monitoring, enforcing, and providing technical assistance to stakeholders.

4.5.3 Water

There are two Acts that regulate water: the Clean Water Act (CWA) and the Safe Drinking Water Act (SDWA). Though both laws regulate water quality, the CWA regulates surface water quality while the SDWA regulates drinking water which includes both surface and groundwater.

4.5.3.1 The Clean Water Act: CWA 1972

The Clean Water Act was originally called the Federal Water Pollution Control Act (FWPCA) in 1972 and in part was a response to the Cuyahoga River catching fire in 1969 (US EPA 2013f). Its initial goals were to eliminate all discharges into navigable "waters of the United States" by 1985 and to make all waters "fishable and swimmable" by 1983. Navigable "waters of the United States" is an interesting term dating back to colonial commerce where any place a canoe could navigate met the definition. In 1987 the CWA included urban storm sewers and other NPSs. The Waters of the United States (WOTUS) rule currently under debate is crafted to "clearly protect the streams and wetlands that form the foundation of the nation's water resources" (US EPA 2013f).

The "swimmable" goal was eventually met for many but not all surface waters. Unfortunately, many lakes and streams throughout the US still have fish advisories warning susceptible or at-risk human populations (mostly women of reproductive age) not to eat fish caught in those waters.

Water pollution or "effluent" is considered a "discharge" and as mentioned that discharge falls into either a point-source or non-point-source category (similar to air emissions). Common source types include:

- point source
  - industrial production facilities or energy plants
  - publicly owned treatment works (POTWs)
- non-point source
  - agricultural – concentrated animal feeding operations, agrichemical runoff, etc.
  - urban runoff
  - industrial – mining

The major permit for the CWA is the National Pollutant Discharge Elimination System (NPDES) permit. Another major but less common type of permit is Dredge and Fill permits for the waters of the US; these are typically authorized by both the EPA and the Army Corps of Engineers.

Intent: Protect human health and the environment regarding surface water quality.

Major Provisions

- Directed the EPA to set uniform, national limits on effluents for all major sources of discharges into water (O'Leary et al. 1999: 30).
- Set technology-based Effluent Guidelines for industrial and municipal sewage plant dischargers. These guidelines force the use of Best Available Technologies (BAT) that are "economically achievable."
- Set Water Quality Standards Elements. As opposed to the NAAQS in the CAA, these standards are set by the states regarding "use designations" and "mixing zones" and use Water Quality-based Effluent Limitations per state and body of water. These limitations required identification of waters, determination of TMDLs and establish individual limits.

- Established the NPDES permit system.
- Established the Dredge and Fill permit system.
- Began implementing NPS provisions with regard to "runoff" such as with stormwater sewers addressing Combined Sewer Overflow (CSO).
- Developed a financing program for wastewater treatment plants (POTWs).
- "expanded water quality programs for lakes and estuaries including the special ecosystem programs for the Great Lakes" (O'Leary et al. 1999: 32).

REAL-WORLD EXAMPLE 4.5 National Pollutant Discharge Elimination System of the CWA (US EPA 2013c)

Water pollution can degrade our surface waters, making them unsafe for drinking, fishing, swimming, and other activities. The Environmental Protection Agency (EPA)'s National Pollutant Discharge Elimination System (NPDES) permit program, created in 1972 by the Clean Water Act, helps address water pollution by regulating point sources that discharge pollutants into waters of the United States. There are a number of opportunities for the public to participate at various points during the permit issuance process. This fact sheet outlines the NPDES permitting process, highlighting opportunities for public involvement.

The NPDES Permit Issuance Process

Step 1: Permit Application Submission

NPDES permits are issued by the EPA or authorized states. In most cases, the NPDES permit program is carried out by authorized states, but the EPA is the permitting authority in four states (Massachusetts, New Hampshire, New Mexico, and Idaho) and for certain discharges in other states, territories, and Indian Country.

Individual permits: For individual permits, the NPDES permitting process begins when the operator of a facility seeking a permit submits an NPDES permit application to their permitting authority for a new permit, or to renew a permit that is close to expiring. (Permit terms are limited to no more than five years.) After receiving the application, a permit writer reviews it for completeness and accuracy. When the application is complete, the permit writer uses the application data and other information to develop the draft permit and the justification for the permit conditions.

General permits: For general permits, the NPDES permitting process begins when the permitting authority decides to issue the permit to cover a group of dischargers with similar qualities (for example, similar industrial process or materials used) within a given geographical location. In order to obtain coverage under a permit, each facility in that group is generally required to submit a notice of intent (NOI) to be covered under the general permit. In some cases, the individual facilities are automatically covered under the general permit and do not need to submit an NOI.

REAL-WORLD EXAMPLE 4.5 (cont.)

Step 2: Notice and Comment of Proposed Permit Action

Once the permit writer finishes drafting the NPDES permit, the permitting authority initiates a public notice period (of at least 30 days) during which any interested person may submit written comments on the draft permit and accompanying fact sheet and/or request a public hearing on the draft permit. The public notice and comment period is one of the most significant opportunities for the public to become involved in the NPDES permitting process.

Following the notice and comment period and prior to permit issuance, the permitting authority is required to respond to all significant comments and explain any changes made to the draft permit. Responses to comments are publicly available from the offices of the permitting authority.

When EPA is the permitting authority, it must issue a "response to comments" document whenever a final permit decision is issued. A "final permit decision" is "a decision to issue, deny, modify, revoke and reissue, or terminate a permit."

Under federal regulations, when states or tribes are the permitting authority, they are only required to respond to comments when a final permit is issued. Note, however, that state law may require a response to comment document in other situations as well.

Methods for issuing notice: To help publicize the notice and comment period, the NPDES regulations require the permitting authority to:

- Mail a copy of the notice to people who have joined mailing lists by submitting a written request to the permitting authority. The permitting authority notifies the public of the opportunity to be put on the mailing list through periodic publications in the public press and other media.
- Publish a public notice in a newspaper in the area to be affected by the following types of permits: major individual permits, state-issued general permits, and permits that include sewage sludge land application plans; and
- For EPA-issued general permits, publish a notice in the Federal Register.

Process for requesting a public hearing: Any interested person may request a public hearing on a draft permit during the public comment period by submitting a written statement to the permitting authority describing the issues proposed to be raised at the hearing. A public hearing will be held if the permitting authority finds that there is significant public interest. The permitting authority may also decide to hold a public hearing at its discretion.

REAL-WORLD EXAMPLE 4.5 (cont.)

Public notice of a public hearing must be given at least 30 days before the hearing takes place and may be combined with the public notice of the draft permit. Scheduling a hearing automatically extends the comment period until at least the close of the hearing, and the comment period may be extended by the hearing officer by announcing an extension at the hearing.

Step 3: Permit Review, Approval, and Issuance Process

Following the notice and comment period, the permitting authority reviews all comments and, based on an evaluation of the comments, decides whether to issue the permit. States and tribes may provide additional opportunities for public involvement.

For permits issued by EPA instead of by a state, there are some additional steps that may provide additional opportunities for public involvement. Public notice for these processes often occurs concurrently with EPA's public notice of the permit.

Section 401 certification – Under section 401 of the Clean Water Act, every applicant for a federally issued NPDES permit must provide a State Water Quality Certification that the proposed activity will comply with applicable provisions of the Clean Water Act, including EPA-approved state and tribal water quality standards. The federal permit may not be issued unless the state or tribe provides (or waives) section 401 certification. Section 401 requires that each state, tribe, or interstate agency issuing certifications must establish procedures for public notice in the case of all applications for certification and, to the extent it deems appropriate, public hearings.

National Environmental Policy Act (NEPA) – NEPA review applies only to EPA-issued new source permits, and should occur at the same time as an applicant's NPDES permit issuance process. The administrative record for the final NPDES permit must include any NEPA documents.

If a NEPA environmental assessment and finding of no significant impact is prepared, a preliminary finding is made available to the public for comment. If a more detailed environmental impact statement (EIS) is prepared, opportunities for public comment are provided to scope out the issues to be addressed, to comment on the draft EIS, and to comment on the final EIS. Federal agencies are required to make "diligent efforts" to keep the public informed and involved in NEPA procedures, and provide notice to the public of NEPA-related hearings, public meetings, and available documents.

After the permit review and approval process, the permitting authority prepares and issues the final permit. Each person who submitted written comments must be notified of the final permit issuance.

Step 4: Permit Appeals Process

For EPA-issued NPDES permits, individuals who are unsatisfied with the terms of a final permit may petition the Environmental Appeals Board to review the permit. Generally, in order to be eligible to appeal a permit, an individual must have either submitted comments

REAL-WORLD EXAMPLE 4.5 (cont.)

on the draft permit or participated in a public hearing on that permit. Once any appeal to the EAB has been resolved, the agency's final decision may be challenged in court.

Enforcement and compliance of NPDES permits: The public can review publicly available data about whether a permittee is in compliance with its permit. Many NPDES permit holders are required to monitor and report the type and volume of pollutants that they discharge. Monitoring results are typically reported on Discharge Monitoring Reports (DMRs). The information contained in DMRs, including the location of dischargers, what pollutants they are discharging, and the types and amounts of pollutants being discharged, are all publicly available from the permitting authority.

- EPA's website contains a Discharge Monitoring Report (DMR) Pollutant Loading Tool, which can be used to look up information on specific discharges.
- Additional facility information is available from EPA's Enforcement and Compliance History On-line (ECHO) database.

Anyone adversely affected by Clean Water Act violations may bring lawsuits against alleged violators of the Act. In cases where the state or EPA fails to take adequate enforcement actions, citizens can sue for injunctive relief (court orders prohibiting the pollution from continuing), civil penalties, and attorneys' fees.

4.5.3.2 The Safe Drinking Water Act: SDWA 1974

When initially developed, the SDWA had two main objectives, to ensure drinking water is safe and to prevent the contamination of groundwater. As mentioned in Section 1.5 in our Introduction, drinking water quality had been of public health concern from the 1850s in England regarding cholera and was addressed in US public health law in the early 1900s (Barnes et al. 2021). Throughout history, though, humans have been conscious of contaminated water from wells, particularly down-gradient from latrines and even downstream from human habitat. Thus, the SDWA regulates the water being provided through certain size water systems, whether from surface or groundwater or even from a water tower.

It was recognized at the beginning of this century that much of our drinking water from underground sources is contaminated with pesticides and other toxicants (USGS 2006). Some of these contaminants are regulated, and others unregulated but listed under this Act's Unregulated Contaminant Monitoring Rule (UCMR), such as perfluorinated chemicals and endocrine disrupters. Most of the contamination is within acceptable levels; nonetheless, almost none of the water we drink is "pure."

It is interesting to note that "fracking," a fossil fuels extraction process, is not regulated under SDWA and CWA, or other Acts, as a matter of political management These exemptions from the federal environmental laws became known as the "Halliburton loophole," demonstrating the ability for environmental protection to be subordinated to other concerns – in this case, purportedly a combination of energy production and national security – such that certain pollution streams become exempt from regulation (NRDC 2013; NYT 2009).

Intent: Protect human health and the environment regarding drinking water quality, particularly groundwater sources.

Major Provisions

- Developed an underground injection control program to regulate the disposal of liquid wastes underground.
- Establishes minimum standards for state programs to protect underground sources of drinking water from endangerment by underground injection of fluids.
- Required drinking water systems to monitor and treat for contaminants.
- The Act authorizes EPA to establish minimum standards to protect tap water and requires all owners or operators of public water systems to comply with these primary (health-related) standards under the National Primary Drinking Water Regulations (NPDWR) for:
  - microorganisms
  - disinfectants
  - disinfection byproducts
  - inorganic chemicals
  - organic chemicals
  - radionucleotides
- The EPA sets primary drinking water standards through a three-step process: (1) identify contaminants that may adversely affect public health and that occur in drinking water; (2) establish a goal for the level of contaminants below which is considered safe; and (3) establish a maximum permissible level of a contaminant based on the Maximum Contamination Level (MCL) which is deliverable to any user of a public water system (US EPA 2015a).

The 1996 amendments to SDWA require that EPA consider a detailed risk and cost assessment, and best available peer-reviewed science, when developing these standards. State governments, which can be approved to implement these rules for EPA, also encourage attainment of secondary standards (nuisance-related).

REAL-WORLD EXAMPLE 4.6 Highlights of the 1996 Safe Drinking Water Act Amendments

The amendments to the SDWA passed in 1996 led to several substantive changes that clarify what the environmental manager must manage for, and up to what standard. Some

REAL-WORLD EXAMPLE 4.6 (cont.)

changes specify actions to be taken at the federal level, namely by the US EPA, while other actions are required at the state level of environmental management. These Amendments are still in effect today. The major changes included:

- Consumer confidence reports – all community water systems must prepare and distribute annual reports about the water they provide, including information on detected contaminants, possible health effects, and the water's source.
- Cost–benefit analysis – the US EPA must conduct a thorough cost–benefit analysis for every new standard to determine whether the benefits of a drinking water standard justify the costs.
- Drinking Water State Revolving Fund – states can use this fund to help water systems make infrastructure or management improvements or to help systems assess and protect their source water.
- Microbial contaminants and disinfection byproducts – the US EPA is required to strengthen protection for microbial contaminants, including cryptosporidium, while strengthening control over the byproducts of chemical disinfection. The Stage 1 Disinfectants and Disinfection Byproducts Rule and the Interim Enhanced Surface Water Treatment Rule together address these risks. Operator Certification Water system operators must be certified to ensure that systems are operated safely. The US EPA issued guidelines in February 1999 specifying minimum standards for the certification and recertification of the operators of community and non-transient, non-community water systems. These guidelines apply to state:
    - Operator certification programs – all states are currently implementing EPA-approved operator certification programs.
    - Public information and consultation – SDWA emphasizes that consumers have a right to know what is in their drinking water, where it comes from, how it is treated, and how to help protect it. US EPA distributes public information materials (through its Safe Drinking Water Hotline, Safewater website, and Water Resource Center) and holds public meetings, working with states, tribes, water systems, and environmental and civic groups, to encourage public involvement.
    - Small water systems – small water systems are given special consideration and resources under SDWA, to make sure they have the managerial, financial, and technical ability to comply with drinking water standards.
    - Source water assessment programs – every state must conduct an assessment of its sources of drinking water (rivers, lakes, reservoirs, springs, and groundwater wells) to identify significant potential sources of contamination and to determine how susceptible the sources are to these threats.

4.5.4 Solid and Hazardous Waste

There are four waste laws. Two of them, the Resource Conservation and Recovery Act (RCRA) and the Hazardous and Solid Waste Amendments (HSWA), regulate active waste. The other two, the Comprehensive Environmental Response, Compensation, and Liability Act (CERCLA) and the Superfund Amendment and Reauthorization Act (SARA), regulate inactive waste. Active waste means that it is being "generated" from a process, and then further processed for disposal, treatment, and storage of that waste. These two laws are sometimes called "prospective" laws – i.e., forward looking. Inactive waste is waste that has been produced but is no longer being generated and is not being disposed of, or waste that had been generated prior (retrospective) to waste laws or illegally but now must be addressed.

These four laws are particularly contentious regarding the permitting or cleaning-up of facilities that generate, dispose of, treat or store wastes, as citizens perceive their proximity to them as an immediate risk to their health and welfare and/or to future generations (a sustainable development issue) and/or, especially in a disenfranchised community, as an environmental justice issue. We will discuss this more in Chapter 6 on public participation.

4.5.4.1 Active Waste Laws

As we discussed above, active waste laws apply to sites where hazardous and other waste is actively being produced. This is as opposed to sites where waste was once produced and no longer is, but the remaining waste or residue at the site requires that it be managed, i.e., monitored, protected, and/or remediated.

4.5.4.1.1 Resource Conservation and Recovery Act: RCRA 1976 The Resource Conservation and Recovery Act (RCRA) "was an amendment to the Solid Waste Disposal Act of 1965, which was the first statute that specifically focused on improving solid waste disposal methods," and was amended to RCRA to include hazardous waste and introduced pollution prevention as "waste minimization" (US EPA 2021j). The Act establishes the framework for a national system of solid waste control. Subtitle D of the Act is dedicated to non-hazardous solid waste requirements, and Subtitle C focuses on hazardous solid waste. Solid waste includes solids, liquids and gases and must be discarded to be considered waste (US EPA 2021j).

This law addresses active hazardous and non-hazardous waste. It provides a "cradle-to-grave" regulatory framework that covers generation, transportation, treatment, and disposal of hazardous wastes.

Intent: Protect human health and the environment in regard to active hazardous and non-hazardous waste.

Major Provisions

Subtitle D –

- "bans open dumping of waste, and ...
- sets minimum federal criteria for the operation of municipal waste and industrial waste landfills, including design criteria, location restrictions, financial assurance, corrective action (cleanup), and closure requirement." (US EPA 2021j)

Subtitle C –

- Defines hazardous wastes: "A solid waste is a hazardous waste if it is specifically listed as a known hazardous waste or meets the characteristics of a hazardous waste. Listed wastes are wastes from common manufacturing and industrial processes, specific industries and can be generated from discarded commercial products. Characteristic wastes are wastes that exhibit any one or more of the following characteristic properties: ignitability, corrosivity, reactivity or toxicity" (EPA 2021).
  - Ignitability – liquids, non-liquids, compressed gases and oxidizers that are flammable.
  - Toxicity – are hazardous due to the toxicity characteristic, are harmful when ingested or absorbed. Toxic wastes present a concern as they may be able to leach from waste and pollute groundwater.
  - Reactivity – are hazardous due to the reactivity characteristic, may be unstable under normal conditions, may react with water, may give off toxic gases, and may be capable of detonation or explosion under normal conditions or when heated.
  - Corrosivity – include aqueous wastes with a pH of less than or equal to 2, a pH greater than or equal to 12.5 or based on the liquid's ability to corrode steel. (US EPA 2021j)
- Requires tracking of waste from cradle to grave through the use of manifests for the entire production and disposal process
- Sets technical standards and rules for issuing permits for disposal, treatment and storage of wastes.

Note: "Mixed wastes are hazardous wastes which also contain radioactive material. Mixed waste is regulated under the RCRA and the Atomic Energy Act. The hazardous component of the mixed waste is regulated by EPA under RCRA. The radiological component of the mixed waste is regulated by the Department of Energy (DOE) or the Nuclear Regulatory Commission (NRC)" (US EPA 2021j).

REAL-WORLD EXAMPLE 4.7 Hazardous vs. non-hazardous waste: what's the difference?

What Is Hazardous Waste?

Industries produce many types of hazardous and non-hazardous waste, and knowing what to do with waste products can be tricky. Every year, the United States generates about 7.6 billion tons of industrial waste, and it all has to go somewhere.

But what are the different types of waste, and how should you handle them? In this guide, we'll discuss the difference between hazardous and non-hazardous waste, discuss the various categories of hazardous waste, offer examples of both types, explain how to determine if your business produces hazardous waste, and offer tips on what to do about it.

Hazardous waste is waste that poses a severe threat to human health or the environment if improperly disposed of. According to the EPA, a substance is a hazardous waste if it appears on specific lists of hazardous waste or exhibits the established

REAL-WORLD EXAMPLE 4.7 (cont.)

characteristics of hazardous waste. Hazardous waste is regulated under the Resource Conservation and Recovery Act (RCRA).

Listed Waste

Listed wastes appear on one of four official lists: the F, K, P, and U lists.

The F list identifies non-source-specific hazardous wastes from common industrial and manufacturing applications. The K list identifies source-specific wastes from specific areas within industry and manufacturing.

The P and U lists define wastes that consist of pure, commercial-grade formulations of specific unused chemicals. To be classified as a P or U waste, a substance must meet the following three criteria:

1. The waste must contain one of the chemicals on the P or U lists.
2. The chemical in the waste must be unused.
3. The chemical in the waste must be a commercial chemical product. The EPA considers a commercial chemical product to be one that is either a 100 percent pure commercial-grade product or one that is the sole active ingredient in a given chemical formation.

The P list differs from the U list in that it designates acute hazardous wastes – those that are toxic even at low levels.

Characteristic Waste

The EPA considers characteristic waste to be waste that exhibits any of four properties:

Ignitability. Hazardous wastes that demonstrate the characteristic of ignitability include the following: wastes with flashpoints of less than 60 degrees Celsius, non-liquid materials that cause fires, ignitable compressed gases and oxidizers.

Corrosivity. Corrosive hazardous wastes include acidic liquid wastes with a pH of 2 or less and basic liquid wastes that have a pH of 12.5 or more and can corrode steel.

Reactivity. Reactive hazardous wastes include wastes that are unstable under standard conditions, react with water, give off toxic fumes or have the capability to explode or detonate when they are heated.

Toxicity. Toxic hazardous wastes include wastes that are harmful to health when swallowed or absorbed. Toxic wastes are of particular concern because they can leach through soil and contaminate groundwater.

What Is Non-hazardous Waste?

Non-hazardous waste, by contrast, does not pose a direct threat to human health or the environment, but it still cannot be dumped into a trash receptacle or a sewer line because

REAL-WORLD EXAMPLE 4.7 (cont.)

of the risks it could pose. Most of the waste produced in the United States – paper, plastics, glass, metals, etc. – is non-hazardous waste because it is not toxic by nature.

The RCRA considers the category of solid non-hazardous waste to include garbage and other solid materials, but under this definition other substances such as slurries, semisolids, liquids and gas containers are considered solid waste as well.

Because non-hazardous waste is more loosely monitored than hazardous waste, it's difficult to develop precise estimates of how much non-hazardous waste the United States generates every year. However, industry experts believe that, by a large margin, industrial non-hazardous waste is the largest category of waste produced annually, on the order of seven billion tons or more. The mining, chemical, metal, and pulp and paper industries have historically generated large amounts of non-hazardous waste, often in the form of wastewater.

Regulation of Non-hazardous Waste

The law regulates non-hazardous waste. Even though non-hazardous waste is not inherently harmful to humans or to the local ecosystem or wildlife, it could still pose risks and must be disposed of in a controlled, careful way.

The regulation of non-hazardous waste is largely left to state and local governments, though the federal government will supply funding toward these efforts. The EPA can also review and approve state methods. But generally, states are responsible for granting permits, monitoring landfill use, and making sure facilities meet the minimum federal criteria for the disposal of non-hazardous waste.

4.5.4.1.2 Hazardous and Solid Waste Amendments: HSWA 1984 The Hazardous and Solid Waste Amendments "focused on waste minimization and phasing out land disposal of hazardous waste as well as corrective action for releases." Some of the other mandates of this law led to increased enforcement authority for the EPA, more stringent hazardous waste management standards to prevent disposal of untreated wastes into and onto the land, the closing of substandard landfills and incinerators, and the development of a comprehensive Underground Storage Tank (UST) program. In some ways it is this law which forced and funded the states to create their own environmental regulatory agencies in 1985.

Intent: Protect human health and the environment in regard to the contamination of groundwater by the disposal of active solid and hazardous waste.

Major Provisions

- Creation of the Land Disposal Restrictions (LDR) program, otherwise known as the "land ban," because "Uncontrolled land disposal of hazardous waste is threatening to human health and the environment. The LDR program ensures that wastes are properly treated prior to disposal. This makes hazardous waste less harmful to ground water by reducing the potential for leaching of hazardous constituents and by reducing waste toxicity by destroying or removing harmful constituents" (US EPA 2021h).

- Establishment of RCRA Corrective Action Program to require facilities that treat, store, or dispose of hazardous wastes to investigate and clean up contaminated soil, groundwater, and surface water (US EPA 2021h).
- Establishment of permitting deadlines for hazardous waste facilities.
- Regulation of small-quantity generators of hazardous waste and USTs. Note: a leaking underground storage tank is a LUST.
- Requirement for a nationwide survey of the conditions at solid waste landfills.
- Dealt with facilities that close and leave waste behind, through liability, insurance, and corrective-action provisions, thereby preventing a Superfund-like situation.

4.5.4.2 Inactive Waste Laws

Inactive waste laws apply to inactive waste production sites that require management – i.e., monitoring, protection, and/or remediation. There are more than 450,000 inactive waste sites spread throughout the United States (US EPA 2021b), all requiring some form of environmental management.

4.5.4.2.1 Comprehensive Environmental Response, Compensation, and Liability Act: CERCLA 1980 The Comprehensive Environmental Response, Compensation, and Liability Act, "commonly known as Superfund, was enacted by Congress on December 11, 1980. This law created a tax on the chemical and petroleum industries and provided broad Federal authority to respond directly to releases or threatened releases of hazardous substances that may endanger public health or the environment" (US EPA 2015e). It "established prohibitions and requirements concerning closed and abandoned hazardous waste sites; provided for liability of persons responsible for releases of hazardous waste at these sites; and established a trust fund to provide for cleanup when no responsible party could be identified" (US EPA 2015e).

First, to be clear, hazardous waste can generally be considered a hazardous material which is not properly contained and presents a danger to human health and the environment; much like gasoline in your gas tank is a hazardous material but when it leaks out of the tank it is a hazardous waste. Hazardous wastes have been created before regulations, from colonial times to the present day, in the form of tanneries, mines, munitions plants, metal plating facilities, print shops, dry cleaners and many more ancient or more modern endeavors which used, created, and disposed of hazardous materials without regulatory protections.

Our nation is dotted with these sites, with many of the oldest being next to surface waters or over groundwaters where the first American colonies were located; in fact, today there are currently over 2,000 sites nationwide characterized as Superfund sites, and over 450,000 sites characterized as brownfield sites (US EPA 2021b) (these are regulatory characterizations and programs we discuss below). Many of these sites were created before CERCLA and the active hazardous waste laws were created, while others have more recently emerged from more nefarious activities like dumping. This "dumping" is most always the result of knowingly avoiding proper transportation, disposal, or storage of hazardous wastes. Both pre-regulatory and illegal hazardous wastes present issues of accountability by regulators and also of sustainable development for future generations needing to be protected. One might believe that unfunded or underfunded mandates are an issue regarding cleaning up a Superfund site. While there are always resource stresses regarding running a program, funding for the cleaning-up Superfund is provided either by the fund itself or from the "responsible party" (RP).

The fate of a Superfund site simply put is that it gets cleaned up to where the site is "fully protective of human health and the environment" and is "de-listed" or it remains a danger to human health and the environment with requisite remediation, or it becomes a "brownfield site."

CERCLA defines a brownfield site as

> real property, the expansion, redevelopment, or reuse of which may be complicated by the presence or potential presence of a hazardous substance, pollutant, or contaminant. Brownfield sites include residential, commercial and industrial properties. Brownfield sites are basically sites that have been remediated to the extent they can be reused, but normally have use-restrictions such as a residential area or areas where children might live or play.
>
> The safe and sustainable reuse of previously contaminated properties is supported through EPA's Land Revitalization (LR) Program. EPA's LR Program helps remove barriers to contaminated property redevelopment. When a property once again becomes an asset to the community, the redevelopment often brings about new opportunities to protect public health, improve the environment and grow the local economy. (US EPA 2021b)

Intent: Protect human health and the environment in regard to inactive hazardous waste.

Major Provisions

- Directs EPA to identify sites, rank them according to hazards they present, and maintain National Priorities List (NPL).
- Gives EPA authority to identify potential responsible parties (PRPs) and force them to clean up.

The process or program management for enacting CERCLA is to:

- identify the sites
- rank the sites using the Hazardous Ranking System (HRS)
- create and maintain the NPL
- clean the sites up
- recover cleanup costs

- de-list the site based on "once cleanup goals have been achieved and sites are fully protective of human health and the environment, EPA deletes them from the NPL" (US EPA 2021b), and finally
- "site re-use/redevelopment."

REAL-WORLD EXAMPLE 4.8 The Hazard Ranking System for the National Priorities List

The Hazard Ranking System (HRS) is the principal mechanism that the EPA uses to place uncontrolled waste sites on the National Priorities List (NPL). It is a numerically based scoring system that uses information from initial, limited investigations – the preliminary assessment (PA), the site inspection (SI), and the expanded site inspection (ESI) if necessary – to assess the relative potential of sites to pose a threat to human health or the environment. Any person or organization can report spills and environmental violations and petition the EPA to conduct a preliminary assessment.

HRS scores do not determine the priority in funding EPA remedial response actions, because the information collected to develop HRS scores is not sufficient to determine either the extent of contamination or the appropriate response for a particular site. The sites with the highest scores do not necessarily come to the EPA's attention first – this would require stopping work at sites where response actions were already underway. The EPA relies on more detailed studies in the remedial investigation/feasibility study which typically follows listing.

The HRS uses a structured analysis approach to scoring sites. This approach assigns numerical values to factors that relate to risk based on conditions at the site. The factors are grouped into three categories:

- likelihood that a site has released or has the potential to release hazardous substances into the environment;
- characteristics of the waste (e.g., toxicity and waste quantity); and
- people or sensitive environments (targets) affected by the release.

Four pathways can be scored under the HRS:

- groundwater migration (drinking water);
- surface water migration (drinking water, human food chain, sensitive environments);
- soil exposure and subsurface intrusion (population, sensitive environments); and
- air migration (population, sensitive environments).

After scores are calculated for one or more pathways, they are combined using a root-mean-square equation to determine the overall site score.

4.5.4.2.2 Superfund Amendments and Reauthorization Act: SARA 1986 As a result of the "Love Canal" episode, when a residential community near Niagara Falls, New York, was constructed on a chemical waste dump and later found to cause issues for the uninformed residents, it was determined that citizens have "the right to know" if and what they are being exposed to. In fact, Rachel Carson voiced this same concern more than two decades earlier in *Silent Spring.* This law sought to remedy that deficiency in CERCLA by increasing public participation. Further, it required companies to "maintain and make available data about harmful chemicals that are used or stored on a site and report the annual emission of such chemicals" (O'Leary et al.: 39).

Intent: Protect human health and the environment in regard to the community's right to know their exposure to hazardous waste.

Major Provisions

- Added community right-to-know provisions to inform the public of health and environmental risks and to allow them to participate in how government addresses those risks.
- Mandated that emitters create a Toxic Release Inventory (TRI) such that the public and regulators have data available to them regarding risk exposure. Today, there are over 20,000 polluters listed on the TRI (US EPA 2020a).

Thus, the environmental manager as a matter of sustainable development and ecosystem management (if the waste migrated to or from another jurisdiction) would be accountable for having a public participation program as part of their technical assistance/education programming and to be able to monitor the TRI of firms required in the operating permit of their jurisdiction.

4.5.5 Materials Laws That Are "Risk-Based" Where Pollution Is Not a "Byproduct of Process"

The "materials laws" are for pollution that is not a byproduct but rather materials that are intentionally applied to the environment. These are laws that attempt to limit damage to human health and the environment by addressing products our society uses before they go to market. As mentioned at the beginning of this chapter, the "materials laws for pollution that is not a byproduct" and "economically balanced laws" pertain to toxic materials (toxics and pesticides) which impact human health and the environment with or without becoming waste, but rather than being health-based, they are "risk"-based and require a risk–benefit or cost–benefit test. It is implied that our society needs these products and is willing to assume a reasonable risk with the permission of the product to enter the marketplace. Therefore, any product denied registration must be shown to cause "unreasonable adverse effects on the environment." At this point, environmental managers should be reminded of the concept that "science changes," as it appears that many substances and pesticides, once considered to be of little or no risk, with newer scientific discovery turn out to be not only risky but of "unreasonable risk." The professional should know that a permit for TSCA and FIFRA (see below) is called product registration.

4.5.5.1 Toxic Substances Control Act: TSCA 1976

The Toxic Substances Control Act is considered a "catch-all" Act as it includes so many products. The TSCA provides EPA with authority to require reporting, record-keeping and testing requirements, and restrictions relating to chemical substances and/or mixtures. Certain substances are generally excluded from TSCA, including, among others, food, drugs, cosmetics, and pesticides. The Act addresses the production, importation, use, and disposal of specific chemicals including per- and polyfluoroalkyl substances (PFAS), polychlorinated biphenyls (PCBs), asbestos, radon, and lead-based paint (US EPA 2015c).

Intent: Protect human health and the environment in regard to toxic substances.

Major Provisions

- Gives EPA authority to collect information on chemical substances.
- Authorized the EPA to regulate the manufacture, use, distribution in commerce, and disposal of chemical substances.
- Requires industry to test chemicals for harmful effects.
- Regulates the production or use of any chemicals that pose "an unreasonable risk of injury of health or to the environment" before hitting the marketplace.
- Provides a risk-based chemical assessment requiring an economic "balance" of the benefits to society vs. the risks to society.

The TSCA was amended with the Lautenberg Chemical Safety Act in 2016 to require EPA to evaluate existing chemicals with clear and enforceable deadlines; make risk-based chemical assessments; increase public transparency for chemical information; and have a consistent source of funding for EPA to carry out the responsibilities under the new law (US EPA 2015c).

REAL-WORLD EXAMPLE 4.9 The TSCA Inventory (US EPA 2015c)

What Is the TSCA Chemical Substances Control Inventory?

Section 8 (b) of the Toxic Substances Control Act (TSCA) requires EPA to compile, keep current and publish a list of each chemical substance that is manufactured or processed, including imports, in the United States for uses under TSCA. Also called the "TSCA Inventory" or simply "the Inventory," it plays a central role in the regulation of most industrial chemicals in the United States.

The initial reporting period by manufacturers, processors, and importers was January to May of 1978 for chemical substances that had been in commerce since January of 1975. The Inventory was initially published in 1979, and a second version, containing about 62,000 chemical substances, was published in 1982. The TSCA Inventory has continued to grow since then, and now lists more than 86,000 chemicals.

REAL-WORLD EXAMPLE 4.9 (cont.)

As part of EPA's commitment to strengthen the management of chemicals and increase information on chemicals, the agency provides free access to the inventory online.

Inventory Basics: How TSCA Defines "chemical substance"

TSCA defines a "chemical substance" as any organic or inorganic substance of a particular molecular identity, including any combination of these substances occurring in whole or in part as a result of a chemical reaction or occurring in nature, and any element or uncombined radical.

Chemical substances on the Inventory include:

- organics;
- inorganics;
- polymers; and
- chemical substances of unknown or variable composition, complex reaction products, and biological materials (UVCBs).

Chemical substances not on the Inventory are those with uses not regulated under TSCA. The use of these chemical substances is governed by other US statutes on, for example:

- pesticides
- foods and food additives
- drugs
- cosmetics
- tobacco and tobacco products
- nuclear materials, or
- munitions.

What Does It Mean for a Chemical to Be on the Inventory?

For purposes of regulation under TSCA, if a chemical is on the Inventory, the substance is considered an "existing" chemical substance in US commerce. Any chemical that is not on the Inventory is considered a "new chemical substance."

In addition to defining whether a specific substance is "new" or "existing," the Inventory also contains "flags" for those existing chemical substances that are subject to manufacturing or use restrictions.

Determining if a chemical is on the Inventory is a critical step before beginning to manufacture (which includes importing) a chemical substance. Section 5 of TSCA requires anyone who plans to manufacture a new chemical substance for a non-exempt

REAL-WORLD EXAMPLE 4.9 (cont.)

commercial purpose to provide EPA with a Premanufacture Notice (PMN) at least 90 days before initiating the activity.

How Are Chemicals Added to the Inventory?

After PMN review has been completed, the company that submitted the PMN must provide a Notice of Commencement of Manufacture or Import (NOC, EPA Form 7710-56) to EPA within 30 calendar days of the date the substance is first manufactured or imported for nonexempt commercial purposes.

Once a complete NOC is received by EPA, the reported substance is considered to be on the Inventory and becomes an "existing chemical." The agency receives approximately 400 NOCs each year, thus the TSCA Inventory changes almost daily.

Substances reported through exemption submissions and exempt uses that are not subject to reporting do not require an NOC and are not added to the Inventory. Examples include:

- Low Volume Exemptions (LVEs)
- Low Release/Low Exposure Exemptions (LoREXs)
- Test Market Exemptions (TMEs)
- substances used for research and development
- polymers that meet the 1995 Polymer Exemption Rule Amendments.

4.5.5.2 Federal Insecticide, Fungicide, and Rodenticide Act: FIFRA 1947

Environmental managers should always consider that agrichemicals such as pesticides are unique pollutants in that they are *intentionally* applied to the ecosystem. The Federal Insecticide, Fungicide, and Rodenticide Act passed in 1947. FIFRA was originally a labeling act for consumer protection against fraud (see Figure 4.1). Until 1970 it was enacted by the USDA but it was folded into the new EPA which changed the intent of the Act, influenced by *Silent Spring*, to require protection of human health and the environment in 1972. At this point the EPA began setting tolerances or "maximum allowable concentrations" for pesticide residues in food according to the Federal Food, Drug, and Cosmetics Act (FFDCA) (O'Leary et al. 1999).

Many of the USDA employees became "charter members" of the EPA when it assumed authority for FIFRA. It took decades to normalize the culture of this part of EPA from one whose previous mission was "to promote and protect American agri-business" to one with a mission to "protect human health and the environment." Indeed, even today USDA acts as a consulting agency to the EPA with regard to pesticides and this presents a constant conflict

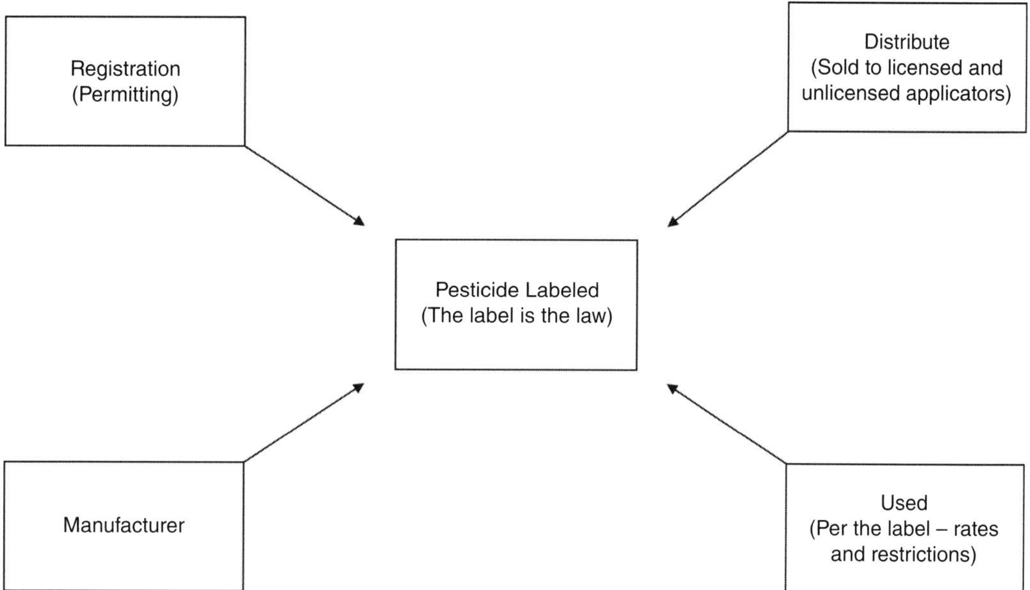

Figure 4.1 The label is the law in FIFRA.

of missions regarding the registration, labeling and continuance of pesticides regarding their effect on pollinator protection, endangered species, wetland protection, farmworker protections, and more. This conflict, where a powerful, cabinet-level agency provides protections and guidance to a powerful industry (American agribusiness), impacts the environmental manager's program, resource and political management.

Because the reader might remember the genesis of environmental management with *Silent Spring* and the concern as to the effect of pesticides on our ecosystem, we will specifically discuss the problems with the environmental management of pesticides in the Case Study at the end of the book, where programs to provide environmental protection from the unnecessary use of pesticides will be illustrated through a pollution prevention strategy known as Integrated Pest Management (IPM).

Intent: Protect human health and the environment in regard to pesticides.

Major Provisions

- Registration of pesticides by Office of Pesticide Programs (OPP) prior to entering commerce.
- During registration, the agency is responsible for:
  - Ensuring pesticides do not pose unreasonable adverse effects on human health or the environment.
  - Prescribing labeling and other requirements to ensure safe use.
  - EPA may also set tolerances (maximum legally permissible levels) for pesticide residues in food under the Federal Food, Drug, and Cosmetics Act (FFDCA).

REAL-WORLD EXAMPLE 4.10 FIFRA in action: Stop Sale, Use, or Removal Orders issued to Amazon.com Services LLC (US EPA 2020c)

On June 10, 2020, the US Environmental Protection Agency (EPA or "the agency") issued an order to Amazon.com Services LLC (Amazon), directing the company to immediately stop the sale or distribution of over 30 different pesticide products and devices on its website that are unregistered and/or misbranded.

Overview of the Case

Amazon, headquartered in Seattle, Washington, is the largest online retailer and distributor of consumer goods in the United States, with nearly 300 million active customers and over 350 million unique products. Many of these products are pesticides and devices regulated by FIFRA. The products subject to the order include personal defense sprays, antimicrobials, disinfectants, sanitizers, dehumidifiers, insect traps, and pest repellers.

Under the Federal Insecticide, Fungicide, and Rodenticide Act (FIFRA), 7 U.S.C. §§ 136–136y, all pesticides must be registered with the US Environmental Protection Agency before being distributed or sold in the United States. "To distribute or sell" is a term defined by FIFRA to include distributing, selling, offering for sale, offering to deliver, shipping, etc. In addition to unregistered pesticides, FIFRA also prohibits the distribution or sale of pesticides and devices that are misbranded. The term "misbranded" applies to pesticides with false or misleading claims and pesticides with labels missing certain required information, for example ingredients, precautionary statements, and adequate directions for use. These provisions are meant to minimize risks to consumers and the environment associated with using the products. Pesticides not registered with the EPA have not undergone the agency's evaluation for safety and efficacy. Pesticides claiming to control viruses, bacteria, or any other human pathogens, are subject to heightened scrutiny and data requirements by the EPA in order to ensure that the product is safe and effective.

Basis for the Order

Beginning in December 2019, Region 10 reviewed information from tips and complaints and conducted its own investigation of pesticide products being distributed or sold on Amazon.com in violation of FIFRA. Based on the results of these efforts, the EPA has reason to believe Amazon had been distributing or selling, and intended to continue distributing and selling, unregistered pesticides and misbranded pesticides and devices in violation of sections 12(a)(1)(A), 12(a)(1)(E) and 12(a)(1)(F) of FIFRA, 7 U.S.C. §§ 136j(a)(1)(A), (a)(1)(E) and (a)(1)(F). Specifically, the EPA documented Amazon was distributing or selling FIFRA regulated pesticides that were not registered with the agency (including some

REAL-WORLD EXAMPLE 4.10 (cont.)

with claims for use against SARS-CoV-2, the virus that causes COVID-19, and pesticides and devices with false or misleading claims on the product labeling.

Summary of the Order

Section 13(a) of FIFRA, 7 U.S.C. § 136k(a), authorizes the EPA to issue a Stop Sale, Use, or Removal Order (SSURO) to any person with a pesticide or device in their ownership, custody, or control whenever there is reason to believe that such pesticide or device is in violation of any provision of FIFRA, or that such pesticide or device has been, or is intended to be, distributed or sold in violation of any provision of the Act.

Pursuant to the authority in section 13 of FIFRA, the EPA ordered Amazon to immediately cease the sale, use, or removal of the products listed in Attachment A to the SSURO under its ownership, control or custody, wherever they are located. The SSURO also prohibits Amazon from using, selling, offering for sale, holding for sale, shipping, delivering for shipment, moving or removing for disposal from any facility or establishment, the subject products for any reason, unless approved by EPA in writing. The SSURO shall remain in effect unless and until revoked, terminated, suspended or modified in writing by EPA.

4.5.5.3 Oil Pollution Act: OPP 1990

The OPP is about protecting the waters of the US, particularly coastal water, and involves a number of federal agencies.

The Oil Pollution Act of 1990 (33U.S.C. 2701–2761) amended the Clean Water Act and addressed the wide range of problems associated with preventing, responding to, and paying for oil pollution incidents in navigable waters of the United States. It created a comprehensive prevention, response, liability, and compensation regime to deal with vessel- and facility-caused oil pollution to US navigable waters.

On March 24, 1989, the *Exxon Valdez* spilled over 11 million gallons of Alaskan crude into the water of Prince William Sound. There were many lessons learned in the aftermath of the *Valdez* oil spill. Two of the most obvious were: the United States lacked adequate resources, particularly federal funds, to respond to spills; and the scope of damages compensable under federal law to those impacted by a spill was fairly narrow (USCG 2021). The OPA was the legislative response to the identified shortfalls.

Intent: To protect human health and the environment with regard to oil pollution in the waters of the US. Note: as the OPA is somewhat limited, contemporary oil spill violations and penalties have been addressed via the Marine Mammal Protection and Migratory Bird Treaty Acts.

The goal of the Oil Pollution Act of 1990 (OPA), 33 U.S.C. 2701 et seq., is to recompense the environment and public for injuries to natural resources and services resulting from an incident involving a discharge or substantial threat of a discharge of oil (incident). This goal is achieved through the return of the injured natural resources and services to baseline and compensation for interim losses of such natural resources and services from the date of the incident until recovery (NOAA 2020a).

Major Provisions

- Set new requirements for vessel construction and crew licensing and manning
- Mandated contingency planning
- Enhanced federal response capability
- Broadened enforcement authority
- Increased penalties
- Created new research and development programs
- Increased potential liabilities
- Significantly broadened financial responsibility requirements
- Spill response plans and pre-positioned response resources
- Imposed strict liability for damages from an oil spill (with financial caps)
- Authorized feds to direct and manage oil spill cleanup operations
- Set requirements for vessels to prevent spills (double hull phase-in) (USCG 2021)

REAL-WORLD EXAMPLE 4.11 The Oil Pollution Act in action (NOAA 2021)

How Does the Oil Pollution Act Work?

Responding, and Improving Our Ability to Respond, to Oil Spills

The National Oceanic and Atmospheric Administration (NOAA) responds to as many as 150 oil spills each year in the United States, ranging from just a few gallons to millions of barrels. The Coast Guard oversees the government's immediate response when spills and chemical accidents happen in US waters. NOAA assists with the science-based expertise and support they need to make informed decisions during these emergencies.

NOAA works with state and federal agency responders to reduce immediate risks to human health and the environment. During a spill, scientists work on the ground, on the water, in the air, and remotely to gather data. In addition, response teams rely on technology like satellites, airplanes, computer models, and other scientific tools. NOAA has designed and built these tools over time, and uses them to predict:

- Where spilled oil will likely go
- How fast oil will travel in bodies of water
- How much oil is spilled
- How it's impacting the environment

REAL-WORLD EXAMPLE 4.11 (cont.)

Meanwhile NOAA contains the oil and collaborates with response teams to protect coastal and marine resources and limit further impacts. Eventually the oil is cleaned up to the best extent possible.

Over 30 years of responding to oil spills, science has improved in all aspects of this work, and it continues to get better.

Holding Polluters Accountable

The Oil Pollution Act states that parties that release hazardous materials and oil into the environment are responsible for the cost of cleaning up the release. They are also responsible for restoring any "injuries" (harm) to natural resources that result. So, the polluter most often pays the bill after oil spills.

Since 1990, NOAA and other agencies, called natural resource "trustees," have worked cooperatively with polluters to arrive at settlements to pay for natural resource impacts. Along with the US Coast Guard's National Pollution Fund Center, NOAA has secured more than $9.8 billion for restoration from 72 oil spill settlements.

Directing Science-Based Assessments of Their Impacts

Oil injures fish and wildlife, and may smother or destroy sensitive habitats, altering migration patterns and negatively affecting life cycles. Oil spills can also lead to beach closures and restrictions on local fisheries, severely impacting local economies.

If it appears that significant impacts may have occurred to natural resources, a Natural Resource Damage Assessment may be initiated. The process is driven by law, science, economics, and public input, and is led by designated federal, state, and Tribal trustee agencies. NOAA works together with the responsible parties to identify both negative impacts to natural resources and lost recreational uses resulting from the incident. NOAA's experts determine the extent of injuries, and – with public input – the best methods and locations for restoration activities.

The rigorous scientific studies necessary to identify the magnitude of injuries may take years. However, this process ensures an objective assessment – and that the public's resources are fully restored.

4.6 Conclusion

The effective environmental manager must be familiar with at least the basics of these 10 laws. It is likely that each environmental manager will become much more familiar with only a subset or potentially even only one of the laws above as they can constitute the bounds of a career quite easily. But it is important to still know the basics of each to understand how they complement each other, their overlaps, and also how they evolve over time. Again, we have provided an essential base foundation of knowledge for these laws, but it is up to the environmental manager to determine how much more detail they need to effectively and efficiently execute their role, whatever it may be.

4.7 End of Chapter Questions

1. Why is the skill of matching legal authority to environmental goals important?
2. What does the term "provision" mean with regard to a law?
3. Name the intent and three major provisions of the 13 environmental Acts in this chapter.
4. How does TMDL relate to criteria air pollutants?
5. What is a SIP and why is it important to tailor it to its location?
6. List the six NAAQS criteria air pollutants.
7. What is a non-attainment area (according to the CAA) and why would a city manager not want her city to become one?
8. How does the EIS relate to the intent of NEPA?
9. Describe the meaning of each letter in HIPPO.
10. What is the major federal permit required in the SPCC stormwater plan?
11. What four categories are defined as hazardous wastes under RCRA?
12. What is the major provisional difference between the SDWA and CWA?
13. Describe the three Phases of Environmental Assessment based on possible inactive hazardous waste contamination of property.
14. What is a "brownfield"?
15. How is TSCA similar to FIFRA in terms of major objectives?
16. Why might you consider agrichemicals unique pollutants?
17. Why was CERCLA amended to SARA?
18. How does the Pollution Prevention Act differ from the other environmental Acts you have studied thus far?
19. With your understanding of the sequence (steps) in implementing pollution prevention, what is source reduction for pesticides?
20. What is the major difference between RCRA and CERCLA?
21. Wastewater discharge permits allow POTWs to comply with what provision of the CWA?
22. According to Porter, what is the ultimate effect of environmental regulations?
23. According to Porter, by what process can well-designed environmental regulations create a competitive advantage?
24. What is the difference between the primary and secondary health-based standards?

5 Navigating the Environmental Regulatory Infrastructure

Skill: *Navigating the environmental regulatory infrastructure* to analyze and utilize governmental relationships. The environmental manager must learn the language of environmental management and how to communicate with their stakeholders, particularly when navigating the complex and often muddled infrastructure of the national to local environmental regulatory infrastructure. Communicating across stakeholders and integrating voices and perspectives from diverse participants, while being open and transparent with all processes, decisions, and action, is a key skill requisite of the effective environmental manager.

5.1 Introduction

Knowing, understanding, and efficiently navigating the regulatory infrastructure of the US is critical to environmental management. Some people have no intention of becoming regulators or public environmental law enforcement officials. However, just like drivers who are not police must understand their responsibilities and the rules of the road, environmental managers and professionals that are not regulators (who constitute a substantial portion of the environmental management community) must understand their responsibilities and the regulatory infrastructure that determines and monitors their actions.

Whether your goal, as a public servant, private sector manager, or not-for-profit influencer, is to protect human health and the environment, keep your company compliant with environmental laws, rules and regulations, improve your industry's competitiveness, or influence environmental policy, you need to understand how the regulatory system works. You need to understand how agencies interface, how they are interconnected, and, more basically, what all agencies and institutions exist for, environmental management. This chapter begins this process by looking at the system or infrastructure in which environmental management occurs. We outline the major components of the environmental management infrastructure in the US, discuss the relationships between institutions at different scales and how they interact, and provide insights on how the complex inner workings of the local-to-national environmental regulatory apparatus can be most efficiently navigated, and even leveraged to the advantage of the manager and the people and ecosystem they are charged with protecting.

5.2 Federalism and Intergovernmental Relations

Our regulatory system in the United States is based on federalism. "Federalism is the sharing of power between national and state governments" (Monk 2013). Our environmental management depends on how and with whom this sharing occurs. What different jurisdictions, and the agencies regulating them and operating within them, are sharing is the *authority* to create and enact environmental policy – the laws, rules, and regulations needed to effectively protect human health and the environment in a jurisdiction.

From the point of view of the environmental manager, the standards to which they are accountable to protect human health and the environment are set by the federal government. And these standards must be met or exceeded by the state or by the regional or local authority as they create their individual standards. This standard setting cannot go in the other direction; it is hierarchical and a system of systems. In other words, as a jurisdiction becomes more specific it can have stricter regulations than the national regulations but not less strict, i.e., a jurisdiction can choose to hold a higher standard than the national standard but it cannot choose a lower standard; the national standard is the baseline requirement. An example of this can be seen with the air quality standards in California's South Coast Air Quality Management District, which may have more strict regulations on emissions than the state of California, which may have more strict regulations on emissions than the federal NAAQS. If the regulations were to go in the other direction it could create the opportunity for a "downward spiral" of decreased regulation as states compete with other states in what has been called a "race to the bottom" (Ambec et al. 2013; Porter 2011). We will discuss this more in Section 5.4.3 later in this chapter but first we must try to understand how the relationships to share authority or power between jurisdictions work.

According to Robert Agranoff, to better understand this relationship scholars began to study intergovernmental relations (IGR) because there exists "an important body of activities or interactions occurring between governmental units of all types and levels" (Agranoff 1986). In other words, what matters is not just that an agency or institution exists, but also how it interacts, coordinates, and co-produces, formally and informally, with other entities. Intergovernmental relations consists of:

1. transcendence of constitutionally recognized patterns of governmental involvement to include varieties of relationships, including national–local, regional–local, national–regional, interlocal, as well as quasi-governmental organizations and private organizations,
2. a human element or the activities and attitudes of persons occupying official positions in the units of government under consideration,

3. relationships between officials involved in their continuous contacts and exchanges of information and views,
4. involvement of all types of public officials – legislators, judges, administrators – at different levels of government as potential or actual participants in decisionmaking processes, and
5. a policy dimension, involving interactions of actors across boundaries surrounding the formulation, implementation, and evaluation of policy. (Agranoff 1986)

Intergovernmental relations are thus a hierarchical, complex set of both nodes – the people, organizations, and institutions that interact – and the interactions between nodes – the relationships and flows of information, policy, and directives between the nodes – that constitute intergovernmental relations. The effective environmental manager has to determine their place in this web and how to best navigate, manage and leverage their connections, that is, their relationships, in this web while concurrently assuring they are in compliance with the statutes they are accountable to given their location in the hierarchy of the system.

5.3 Environmental Management at the Federal Level

The US Environmental Protection Agency (EPA) was a bipartisan policy initiative borne from increasing national concern regarding human degradation of natural resources and air and water quality, inspired by Rachel Carson's (1962) *Silent Spring* (US EPA 2013d). However, this was not a light switch that was suddenly turned on by the US citizenry and government regulators in the 1960s (Montrie 2011). Earlier in the twentieth century the American public understood that without regulation their health was in jeopardy and responded to it, first with formal, national water quality regulations developed in the early 1900s and then with air quality regulation in the 1950s (Barnes et al. 2021). Further, Americans recognized the health and economic consequences of unregulated agricultural practices intersecting with variable climate conditions – specifically drought in this case – as early as the 1930s as a result of one of the largest coupled human–natural-resource disasters in US history, the Dust Bowl in the plains of the American Midwest between 1935 and 1940 (Romm 2011).

President Nixon created the EPA, first by executive action in 1970 and then with congressional approval in 1971 (US EPA 2013d). This new environmental regulatory body inherited the existing federal authorities (laws) regarding public health and environmental management and helped in the development of further authorities to "protect human health and the environment" (US EPA 2013b). The development and implementation of these laws, rules, and regulations were, and still are, influenced and often mutually negotiated by public, private, and not-for-profit activists and advocates at the federal, state, and local level so as to benefit the "tri-sectoral world." In fact, every agency and their field offices are required to have a federal advisory board or committee composed of regulated entities (public and private), not-for-profit advocates, academic experts, and intergovernmental partners, mandated by the Federal Advisory Committee Act of 1972 (FACA).

REAL-WORLD EXAMPLE 5.1 Serving on a Federal Advisory Board

I (M.L.) was on the EPA's Office of Pesticide Programs Federal Advisory Board Act (FACA) chartered committee – the Pesticide Program Dialogue Committee. This "broadly representative" group met with OPP to discuss pesticide regulatory, policy and program implementation issues on a regular basis. In doing so, we had workgroups that separately addressed specific issues such as pollinator protection, farmworker protections, labels, registration review, school IPM, and endangered species ecological standards – to name just some of what I was exposed to over my six-year tenure.

I considered my nomination to and membership on this FACA as an honor, and a privilege to serve. The ability of experts and stakeholders to advise and assist the EPA is one of the longstanding beneficial tools the agency has employed since its inception. The approximately 40 members represented affected agricultural producers (farmers), animal welfare groups, environmental advocates, farmworkers, public health, agrichemical manufacturers, state and Tribal government and federal partners (USDA, USFWS, and the uniformed services). Most members were from the trade associations or NGOs, while individual academics also had membership. One would find similar FACAs throughout the federal government, with the idea to provide expertise and balance regarding regulations, policy and program implementation.

Our semiannual meetings at EPA Headquarters normally opened with a welcome and thank-you from senior EPA officials, then directives on what was needed by the agency, and rules for how we could assist. The meetings normally lasted for two and a half days and it was a tremendous opportunity to understand what OPP was accomplishing, provide advice regarding priorities, render assistance for improved implementation, and network with experts in the diverse stakeholder community the agency had to work with.

The realities of this FACA and, I suspect, all FACAs are:

(1) Even though the agency was grateful for our advice and assistance (in light of our expertise but also the cover it gave them regarding public participation), agency staff who had to spend time managing our workgroups and subsequent presentations were not always appreciative, as participants in policy formulation, of what they viewed as unofficial oversight and suggestions. No one likes to be told how to do their job, particularly by folks from outside.

(2) While the agency attempts to balance membership between regulated entities (industry trade groups), environmental advocates, academics, and governmental partners, the membership can be skewed in favor of the direction desired and pressed for by the administration and their political appointees.

(3) Members of the regulated community are often represented by professional lobbyists (hired guns) who are very good at agenda setting whereas many environmental or public health advocates are untrained, and academics often just don't have a clue.

REAL-WORLD EXAMPLE 5.1 (cont.)

(4) In my experience, agency personnel make every effort to give equal voice to advocates and academics; however, most of these entities do not reside in Washington, DC, and do not have the resources to access agency decisionmakers to the extent trade group professionals do. Agenda setting requires access.

As one of my colleagues observed, "I was incredibly impressed by the EPA scientists involved, unimpressed by most administrators, and utterly shell-shocked at the awesome political power of industry lobbyists. There is a clear need for reform and transference of influence to **academics** with **NO** conflicts of interest."

Finally, my tips to environmental managers who are attempting to manage FACAs or state-level advisory boards are to understand their administration's priorities regarding this service, try to develop a trusting partnership with these individuals who, regardless of affiliation, are attempting to "serve," and do not waste their time. This last aspect was perhaps the most offensive to me as several environmental managers assigned our workgroups "busy work" either out of laziness or unwillingness to be subjected to suggestions requiring change. These strategies are apparent and, like I said, offensive, as our time and service is valuable.

At its founding in 1970 the EPA had 4,000 full-time employees, more than doubling that number to just over 10,000 by 1975 (US EPA 2021e). In 2020, the EPA employed approximately 14,000 people. Over the last 10 years the number of full-time employees has decreased almost 20 percent from 17,278 in 2010 to 14,172 in 2020, with the largest drop between 2012 and 2013. The fiscal year 2020 budget was just over US$9 billion, or less than 0.01 percent of the federal budget for the year.

5.3.1 Navigation Means Knowing Your Mission and Position

The relationships between different environmental and natural resource management institutions are heavily influenced by the mission of each entity and they are often in conflict. Thus, a dynamic is always in place, requiring the environmental manager to have a willingness to work with potentially adversarial others in order to understand the issue at hand and to mutually negotiate an effective solution that satisfies the needs of both while complying with the laws they are accountable to. For instance, the EPA mission of protecting human health and the environment is not "to protect and promote American Agribusiness," to paraphrase the mission of the US Department of Agriculture (USDA 2021). Promoting and protecting the agrichemical industry or agricultural production industry (i.e., farming) can directly harm human health and the environment, requiring environmental managers to work with regulators or representatives from the USDA to accomplish their mission.

As we will discuss below, state environmental agencies, which have a very similar mission to the EPA, often have a higher allegiance to the political leadership of their state rather than the federal agency that has oversight of their work and that regulates them. For example, the commissioner of the Indiana Department of Environmental Management (IDEM) may have a stronger allegiance to the governor of Indiana than the EPA Region 5 administrator they are accountable to as well. The EPA does have authority over the state agency, but it is the political leadership of the state – the governor and the state legislature – that determines the state's trajectory, regulatory preferences, and, maybe most importantly, the budget of state environmental management agencies. This is one of the reasons why political management, including "managing up," is so critical for the environmental manager. And of course, the objectives of the regulated, namely private sector groups, and environmental advocates in the not-for-profit sector can and do conflict with EPA and state-level policy and management. Conflict is a constant companion of, and challenge for, the environmental manager.

Frequently the EPA is the "lead agency" and has principal authority and responsibility for carrying out or approving projects related to protecting human health and the environment. As an executive agency the president, with confirmation from the Senate, appoints the leadership of the EPA and dictates its direction and focus areas. This direction influences the mission, goals, and funding allocation of the EPA, and all federal agencies' environmental practice, and can have positive or negative results for environmental policy and management.

Shifts in focus and funding are most acute, but expected, when there is a transfer of power between presidential administrations, especially when authority is transferred from one political party to another (Kraft 2017). Environmental managers, who are usually civil servants rather than political appointees, are regularly caught in this tide of power transition that can substantially alter their day-to-day operations, planning, and, ultimately, their ability to do their job. While the president and their appointed EPA leadership influence how environmental laws are enacted across the country, they cannot make laws. They can tailor budgets, roll back or create executive orders, and issue other guidance affecting how the environmental manager executes their mission, but they cannot change the laws the environmental manager is accountable to.

Environmental protection is practiced by all federal agencies at some level, a fact reinforced by the passage of the National Environmental Policy Act (NEPA). For instance, the US Department of Health and Human Services manages indoor air quality, drinking water quality, solid waste and to some extent hazardous waste remediation in, on, and near their housing properties (HHS 2015); every branch of the military in the US Department of Defense (DoD) does likewise, as well as manage their installations for surface- and groundwater contamination from fuel bladders or fire protectants, and "ammo dumps" that degrade to hazardous waste sites (DoD 2021); and the US Department of Energy (DoE) does the same regarding air and water quality issues from energy exploration and development. These are examples of federal agencies that conduct environmental management such that they can comply with environmental regulations.

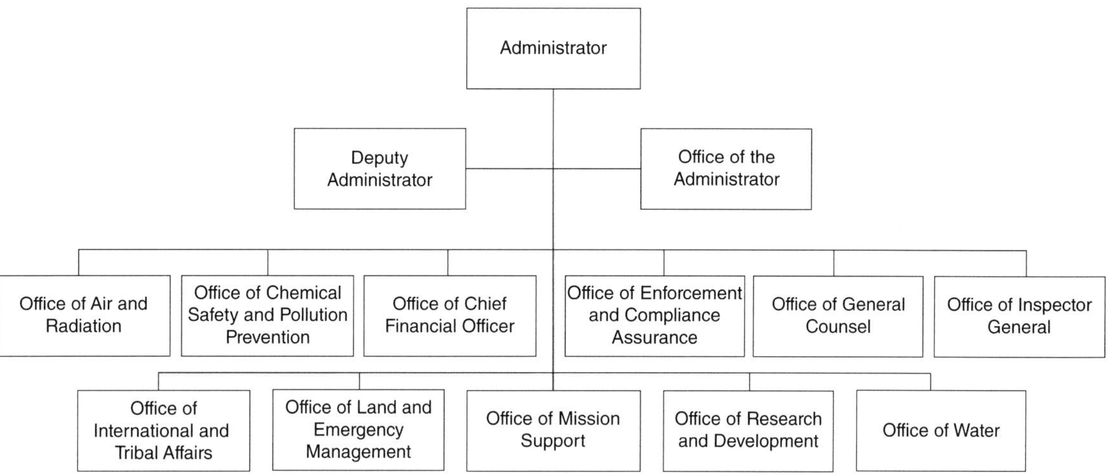

Figure 5.1 EPA organizational chart.

The EPA is the lead agency that implements national environmental policy and provides national oversight of environmental law. The policy created is based on scientific knowledge balanced with social and economic concerns, and when necessary is crafted in consultation with other federal agencies as described above in IGR. Examples of agencies that the EPA often works with would include: the DoE; the Department of Transportation (DoT); the Federal Highway Administration (FHWA); the Federal Aviation Administration (FAA); the Department of the Interior (DoI) and its services such the Fish and Wildlife Service (FWS), the National Park Service (NPS), the Bureau of Land Management (BLM), the Bureau of Ocean Energy Management (BOEM), and the United States Geological Survey (USGS). The USDA, as they did at the founding of the EPA regarding agrichemicals, have a strong relationship with the EPA, as do their subordinate services: the Forest Service (USFS), Natural Resource Conservation Service (NRCS), National Institute of Food and Agriculture (NIFA – formerly the Agricultural Extension Service), Agricultural Research Service (ARS), and the Animal and Plant Health Inspection Service (APHIS).

The EPA Headquarters Office comprises 13 subordinate offices (see Figure 5.1). Each of the subordinate offices has a specific focus or mission. The offices are the Office of the Administrator, Office of Policy, Office of Air and Radiation, Office of Chemical Safety and Pollution Prevention, Office of the Chief Financial Officer, Office of Enforcement and Compliance Assurance, Office of General Counsel, Office of International and Tribal Affairs, American Indian Environmental Office, Office of Land and Emergency Management, Office of Mission Support, Office of Research and Development, and Office of Water (US EPA 2021a). As the titles make clear, some offices are institutional management or internally oriented (e.g., Office of the Chief Financial Officer) and others are environmental management or mission oriented (e.g., Office of Air and Radiation).

Collectively these offices oversee all of the functions and administration of the EPA, and coordinate all of their efforts, programming, and planning with the 10 Regional Offices (discussed in the next section).

As has likely become obvious at this point, environmental management at the federal level is complicated and complex, with many different agencies and their corresponding missions intersecting and, as we have noted, often conflicting. In the end, while conflicts do arise, the ultimate motivation for the civil servants that comprise these agencies is the service they provide to the American citizenry – hence the inclusion of "service" in the title of many of the agencies. The environmental manager will find that keeping this unifying point and mutual goal in mind will help guide them in coordinating and building effective programs with the various agencies they will have to interact with. Knowing what their counterpart managers and regulators are beholden to will only further enable the environmental manager to do their job. We are not promising it will ever be easy, but it will always make determining and achieving their objective easier.

REAL-WORLD EXAMPLE 5.2 Working with Tribal governments

As an environmental manager, you cannot paint every Tribal nation or community with a broad brush of generalizations (this would be not only unethical but also just wrong and lead to poor outcomes management). Similar to other communities we work with as environmental managers, it is important to recognize specific history and relationships with the government (at any level), natural resources, and the emissions source you are regulating, or the program that you are implementing.

However, one thing that is true of all federally recognized tribes in the United States is that the federal government has a trust responsibility to them and treaty rights to uphold for them. This is something that environmental managers at every level of government, in non-profits, or in the private sector are bound by. This relationship is codified by hundreds of years of treaties, many of which the Supreme Court determined have not been upheld over time. So, understandably, there is a long history of distrust from the Tribal perspective. With that history in mind, all our work with tribes must be entered into with the understanding that a tribe's top priority is preserving or regaining its sovereignty. As environmental managers, it is our duty to ensure that we do everything we can to uphold our trust responsibility, while also implementing regulations and programs that stand to benefit Tribal citizens and environmental outcomes (air quality, water quality, etc.). This means that it is our responsibility as environmental managers to ensure that Tribal governments are meaningfully involved early in the decision processes that impact tribes and their citizens.

Another important nuance to consider when working with tribes is that this work is different from the work we do on environmental justice (EJ). While we may work with

REAL-WORLD EXAMPLE 5.2 (cont.)

tribes in an EJ capacity because they face disproportionate harm from pollution or climate change, Tribal governments are different from EJ communities in many important ways. At the federal level, we interact with tribes on a government-to-government basis, codified by treaties, and further emphasized by Executive Order 13175, Consultation and Coordination with Indian Tribal Governments. This means that the level of engagement with a tribe and any level of government should be at the decisionmaker level. However, that is not true of EJ communities, where it is important that we consult and take into consideration unique impacts to EJ communities, where the consultation is not with a sovereign government leader, and EJ communities are directly influenced by state and local regulators. Furthermore, states and local governments are not necessarily key constituencies in Tribal environmental decisions. It's important to learn the structure of the Tribal government you're interacting with, address leaders by their official titles, and provide due respect. We also must respect the traditions they want to bring into the decisionmaking process, such as a prayer or ceremony they'd like you to participate in prior to engaging in discussion.

Tribal governments are excellent partners in carrying out environmental programs because of their unique relationship with this land. They can be considered some of the first environmental managers, have a deep understanding of how humans impact the land (called Traditional Ecological Knowledge or TEK), and are called to preserve it. Where tribes and federal environmental managers can sometimes run into issues is when we don't take into consideration that their relationship with the land goes far beyond human health or pollution prevention but rather is considered sacred. It is then our job, as environmental managers, to find a way to ensure we are respecting that sacred relationship while also building the broader community's knowledge of the current understanding of environmental science and how we are working to ensure tribes have a healthy environment. In many cases where tribes cannot be the environmental manager due to resources or other constraints, that role falls upon the federal government and not the neighboring state, which generally has limited or no authority in Indian country for implementation of environmental statutes.

Jessie Mroz, MPA-MSES, is the current Tribal Coordinator for the Environmental Protection Agency's Office of Transportation and Air Quality, where she also oversees the process for all mobile source regulatory actions.

This work is not a product of the United States Government or the United States Environmental Protection Agency. The author is not doing this work in any governmental capacity. The views expressed are her own and do not necessarily represent those of the United States or the US EPA.

5.4 Environmental Management at the Regional and State Level

The EPA was set up such that policy and planning with regard to enacting US environmental laws would be conducted at their headquarters in Washington, DC, and then implemented by the 10 EPA Regions and their respective Regional Headquarters Offices (see Map 5.1). In addition to the 10 Regional Headquarters Offices, each Region has a varying number of field offices and other facilities that they maintain. For example, Region 6 – which regulates Arkansas, Louisiana, New Mexico, Oklahoma, and Texas – maintains its headquarters in Dallas, a laboratory in Houston, Texas, and field offices in El Paso, Texas, and Tulsa, Oklahoma. Each Region is structured differently based on the geography, ecology, and programs and initiatives of its area of responsibility (US EPA 2020d).

Headquarter managers develop clear policies and strategic plans (program, resource and political management strategies) with and for the Regional Office environmental managers and their partners. Some Regional program personnel conduct programs in the states in their regional area of operations (e.g., Superfund site management). However, most Regional personnel provide oversight in the form of monitoring, enforcement, and technical assistance to state environmental regulators such that they comply with federal environmental laws through rules, regulations and programs developed by the state-level agency.

The Regions use carrots (funding) or sticks (withholding funding or program authorization) as part of their oversight responsibilities, i.e., they can enable or disable the resources

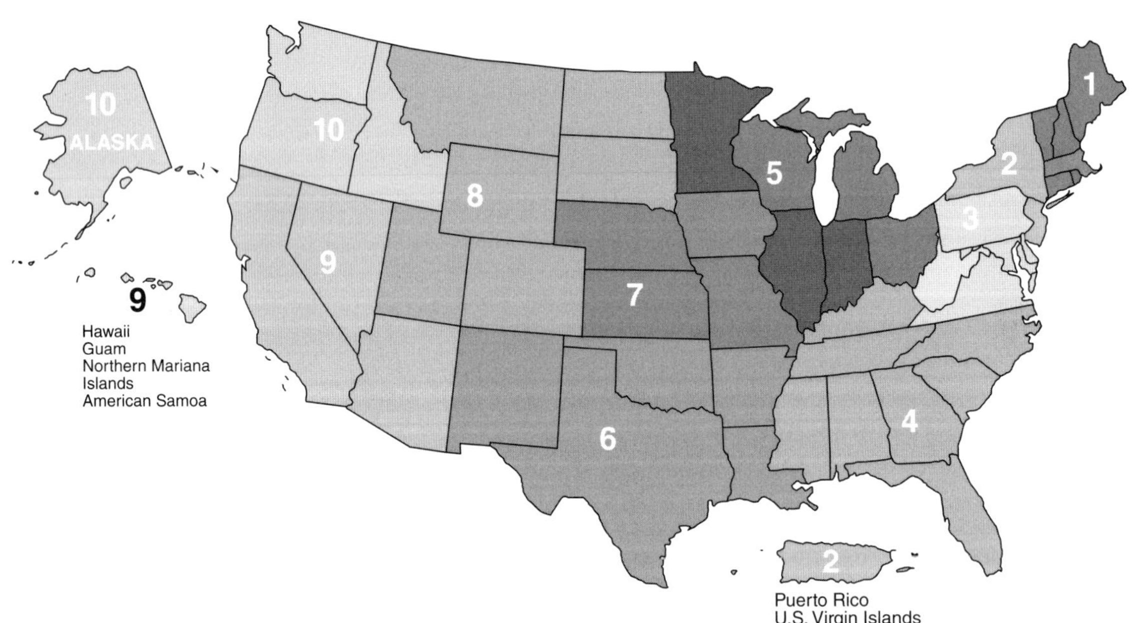

Map 5.1 Each of the 10 EPA regions has a regional headquarters, as shown here (US EPA 2021a).

available to and needed by the state environmental managers for their program and resource management functions. In fact, oversight is regulation. The primary function of the Region is to oversee that the states, tribes, and territories are providing environmental protection that achieves, at minimum, the national statutory standards. (Note that hereafter, when we say states, we mean this to include all 50 US states, seven territories, and the 278 federally recognized sovereign tribes located within the boundaries of the US.)

The program management for the Regions is to authorize (through permitting) the states to meet or exceed federal environmental standards, to assure that these standards are being met (monitoring for compliance), to withhold or withdraw authorization if the states are not compliant (enforcement), and to assist when needed (technical assistance/education) to improve the states' abilities and capacities to achieve the required standards. The resource management comes in the form of monetary, technological, material, or various other forms of support that can better enable the effectiveness and grow the abilities of states (i.e., the carrots). And political management comes in the form of building and maintaining a functional working relationship between the Region and all of the stakeholders within the Region – ranging from state-level environmental managers, to other federal agencies or services, to the citizenry that live there.

5.4.1 Problems with the Regions

While the 10 Regions were similarly set up as part of Nixon's "New Federalism," over time they each have developed their own personality. Some are more subject to urban influence, and others more rural. Some are more industrial, others more agricultural. Some are more environmentally conscious and others less so. In instances, states have even "bullied" Regions into not upholding their responsibilities. This was the case with Michigan not allowing in-person inspections by Region 5 personnel at the onset of the Flint drinking water crisis (OIG 2018). In this case, the Region 5 administration was liable for poor oversight because of the lack of political management. For example, in a post-event investigation the EPA reported that "While Flint residents were being exposed to lead in drinking water, the federal response was delayed, in part, because the EPA did not establish clear roles and responsibilities, risk assessment procedures, effective communication and proactive oversight tools" (OIG 2018: 3). As a result of the fallout for the EPA and their partial responsibility for the failure of effective environmental management that occurred, they have since awarded over $100 million in water infrastructure upgrade grants to the Michigan Department of Environmental Quality (US EPA 2021f).

Another challenge is regional autonomy. Each Regional Office has some discretion regarding prioritizing initiatives from headquarters and thus can suffer from inadequate or inefficient resource management. For instance, a full-time employee (FTE) in Region 5 might be assigned a number of different programs or initiatives, and their superiors, often the Regional Administrator (RA), will use them differently than the RA in Region 3. This lack of uniform goals, coordination, and even cooperation can inhibit the effective and efficient implementation of initiatives (OIG 2007).

A more serious problem for the Regions – and this is not a problem of the Regions per se, but one that the Regions have to deal with – is that an anti-environmental Congress or president can inhibit implementation, by limiting resources for regional operators, as they strive for a deregulatory agenda. That is, they can restrict funding and other resources allocated to specific Regions, and programs or initiatives within those Regions, to effectively deregulate through defunding mandates. Most politicians will not outright oppose a law that protects human health and the environment like the CWA or CAA, but they can make those laws hard if not impossible to enact. The regional overseers sometimes must conduct site visits to ensure environmental standards are being met and enforced by the states. Further, Regional personnel often must work side-by-side with their state partners to implement and provide technical assistance. If human resources and/or travel resources are not available, the program will not be implemented efficiently or effectively (OIG 2007, 2018; US EPA 2021f).

5.4.2 State Environmental Regulatory Agencies

Most states formed their environmental agencies in the early 1980s as the US public became more concerned over environmental problems and as the EPA was able to develop programs and resources for states to assume the responsibilities of enacting relatively new national environmental laws (Sutter 2013). More specifically, after the passage of HSWA in 1984, states were both required and funded to address the issue of hazardous waste, prompting them to develop their own environmental management institutional infrastructure. Since the common mission of these new institutions was to protect human health and the environment, it made sense that many state environmental management agencies grew out of or were modeled on existing state public health agencies and regulatory programs (e.g., IDEM 2021; see Figure 5.2). Building upon existing infrastructure, the states developed their own environmental management processes and practices, but all were accountable to the EPA and their respective state leadership.

Agency names vary. Departments of Environmental Management, Departments of Environmental Quality, and Departments of Environmental Protection are most common. Each is headed by a director, administrator, or commissioner, depending on how the state administration is structured. State environmental statutes normally require the regulatory agency to have "boards" or a commission to provide advice and oversight to each of the media offices. These boards consist of agency officials, representatives from the governor's office, regulated industry representatives, and academic or research scientist experts. It is interesting to note these boards are often "stacked" in favor of whatever the presiding governor and their environmental protection philosophy is oriented toward. For example, in general the boards of industry or economic development–oriented governors tend to appoint more representatives from the industries of the state, whereas more environmental protection–oriented governors tend to include more agency officials and research experts (Koontz et al. 2010). As mentioned, state environmental statutes corresponded with the federal laws and are required to attain at a minimum the standards set out by those laws.

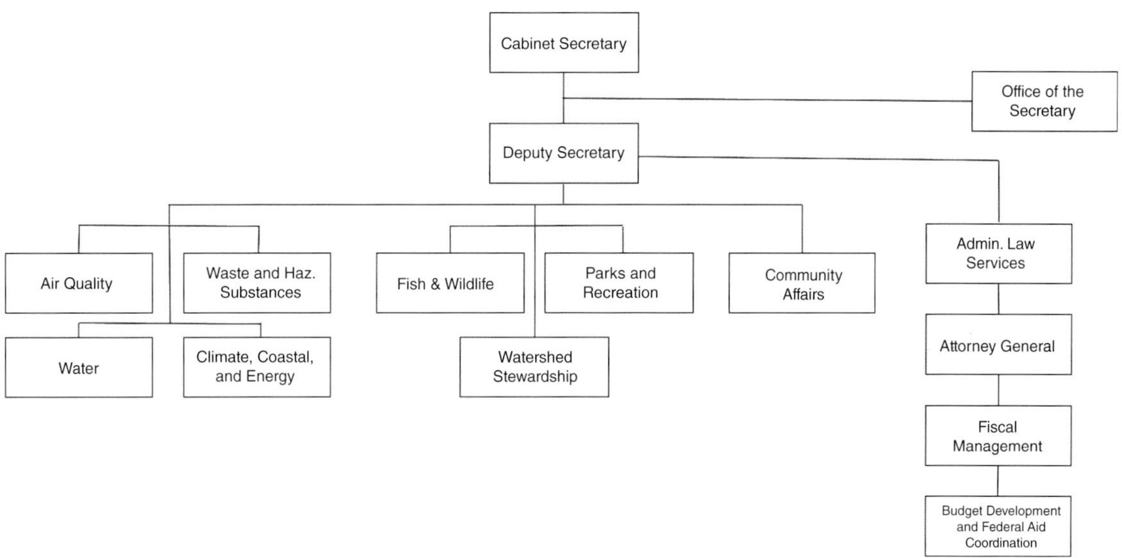

Figure 5.2 Example of a state environmental department organizational chart.

The following, which is the setup of Arizona's state environmental structure, is the typical environmental management structure of most states:

> ADEQ stands for the Arizona Department of Environmental Quality. We administer the state's environmental laws and delegated federal programs to prevent air, water and land pollution and ensure cleanup.
>
> Under the Environmental Quality Act of 1986, the Arizona State Legislature created ADEQ in 1987 as the state's cabinet-level environmental agency. ADEQ is composed of three environmental programs: Air Quality, Water Quality and Waste, with functional units responsible for technical, operational and policy support.
>
> ADEQ carries out several core functions: planning, permitting, compliance management, monitoring, assessment, cleanups and outreach. (ADEQ 2021)

What is also typical is that the state missions and/or visions often include wording regarding "balance" and "costumers" and avoid the term "enforcement" when referring to the agency's relationship with those they regulate, thus setting up a dynamic regarding the environmental management at the federal and state levels.

5.4.3 Problems with the States

The relationship between the EPA and its Regions and the states is an ongoing, understandable, and somewhat predictable problem. These entities strive for sovereignty, often resent the authority of the federal government, and do not believe the EPA has "the local knowledge, political sensitivity, and administrative resources that a state agency would command" (Barnes, Graham, and Konisky 2021: 13). Further, in many ways this autonomy can be good, as states can act as "laboratories of democracy" experimenting with different policy approaches to find ways to effectively reduce pollution without causing negative side effects (Barnes, Graham, and Konisky 2021), though it has been shown that such experimentation still often requires federal prodding for it to transpire (Livermore 2017a). In general, the problems with the states revolve around four issues:

- The downward spiral – which is when states compete with each other to attract jobs (businesses, etc.) by cutting regulatory corners. Examples of how this can be achieved could be interpreting which regulations apply to a permit application, how long it will take to obtain an operating permit, how monitoring for compliance might occur, non-enforcement of violations, and/or providing inordinate technical assistance.
- Funding – where states do not provide funding for functions other than permitting, thereby relegating state agencies to become "permit shops" accountable to applicants rather than the public.
- Rotation or continuity – when state administrations change, often so does the leadership of the state environment regulatory agency, making it difficult to develop or maintain working relationships between the state and region. The discontinuity and shifting of focus that happens in this process can reduce the effectiveness of strategic planning and reduce credibility and confidence with partner organizations.
- Accountability – when the states are not properly implementing the laws they are authorized to enact (Barnes, Graham, and Konisky 2021). This has sometimes been characterized as a "race to the bottom" in certain instances (e.g., Porter 1991), and in others is a result of the adoption of "no more stringent" laws (discussed below).

In recent years several states have passed "no more stringent" laws where state legislators require that state environmental organizations regulate to no more stringent a standard than that required by federal statute (Woods 2020). Not only do these laws apply to state environmental managers and regulations, but they also extend to municipal or city environmental managers and the laws of their respective governing unit such that they cannot be any stricter than federal statutes dictate. Thus, municipalities can be subject to limitations from the state regarding enacting more stringent environmental regulations unless they are in a true "home rule" state (versus partial autonomy under what is called Dillon's rule).

Such legal action is referred to as preemption, that is, where a higher legal authority limits the regulatory power or standards-setting the subordinate authority can determine, an action increasingly used to limit the development of environmental regulatory practices at various scales (Livermore 2017b; Weiland 1999; Woods 2020). Examples of preemption are commonly found in environmentally progressive cities located in anti–environmental regulatory states. "No more stringent" laws and other forms of preemption substantially impact what the environmental manager can and cannot do, and thus they must be aware of such laws, and their development, in their area of responsibility.

5.4.4 Solutions

Several excellent memos from the EPA have addressed specific problems related to its relationship with the states. In 2004 the EPA, along with the Environmental Council of the States (ECOS), established a State Review Framework to highlight problems in their relationship. Then, in 2013, they developed a National Strategy for Improving Oversight of State Enforcement Performance (US EPA 2013a) stating:

Environmental enforcement is an important tool in securing compliance with those laws and regulations. To accomplish this mission, EPA authorizes state, tribal and territorial agencies to directly implement environmental laws. Federal and state regulators work cooperatively together as co-regulators to achieve compliance, with delegated or authorized states performing the vast majority of enforcement across the country. EPA relies heavily on authorized states to implement the day-to-day business of compliance and enforcement programs, with states contributing a majority of the staff and resources necessary to ensure protection of public health and the environment. EPA in turn develops national policies and guidance, many of which set goals for performance to achieve consistency across state programs and establish a level playing field for businesses, states and the public. Strong state performance is therefore fundamental to maximizing compliance and public health and environmental protections . . .

The following three elements of this strategy are aimed at improving state enforcement performance over time:

1. The Escalation Approach to Problem-Solving: A series of escalating steps intended to provide consistent guidance to the regions in their review of and response to state enforcement performance issues.
2. Plans for Addressing Significant Issues: EPA Regions and states should work together to develop plans to address identified significant individual state performance issues.
3. Transparency Efforts: Efforts intended to provide the public with timely, high quality information on state and federal enforcement performance can motivate government to improve . . .

Performance issues are also sometimes caused or compounded by legal or administrative issues that can hinder timely and appropriate enforcement. Examples include:

- A state environmental agency's lack of administrative penalty authority or limits on its statutory or regulatory penalty authorities;
- Issues related to legal resources or approvals needed from independent boards or commissions;
- Other statutory or regulatory impediments;
- Technical limitations (including data);
- Use of enforcement orders to circumvent standards or to extend permits without appropriate notice and comment; or
- Failure to inspect and enforce in some regulated sectors . . .

These unresolved and recurring issues indicate the need for a focused national effort to address them, and include:

- Widespread and persistent data inaccuracy and incompleteness in national data systems, which make it hard to identify when serious problems exist or to track state actions;
- Routine failure of states to identity and report significant noncompliance;

- Routine failure of states to take timely or appropriate enforcement actions to return violating facilities to compliance, potentially allowing pollution to continue unabated; and
- Failure of states to take appropriate penalty actions, which results in ineffective deterrence for noncompliance and an unlevel playing field for companies that do comply.

5.4.4.1 The Barnes Memo

This national strategy was a significant follow-up on the precedent-setting 1984 memo from EPA Deputy Administrator Jim Barnes with his Revised Policy Framework for State/EPA Enforcement Agreements (Barnes 1986). In this memo, Barnes established the framework for the working, oversight relationship between the EPA and the states by issuing the following guidance:

> This section sets forth seven key measures EPA will use, at a minimum, to manage and oversee performance by Regions and States. It summarizes State and regional reporting requirements for: (1) compliance rates: (2) progress in reducing significant non-compliance: (3) inspection activities: (4) formal administrative enforcement actions: and (5) judicial actions, at least on a quarterly basis. It also discusses required commitments for inspections and for addressing significant non-compliance.
>
> In addition, it sets forth State and regional requirements for recordkeeping and evaluation of key milestones to assess the timeliness of their enforcement response and penalties imposed through those actions. (Barnes 1986)

5.5 Local Environmental Management

The EPA works with cities, counties and their associations primarily as a function of providing funding and technical assistance for infrastructure regarding wastewater, drinking water, and hazardous and solid waste. The regulation of air, waste and water of local government is normally conducted by the state except for Superfund functions. Councils of Governments (COGs) are "voluntary associations that represent member local governments, mainly cities and counties, that seek to provide cooperative planning, coordination, and technical assistance on issues of mutual concern that cross jurisdictional lines" (WRCOG 2021).

Normally, municipalities manage their air, waste and water quality through health departments and public works. For example, the city of Phoenix has two entities (see Figure 5.3) which assure their compliance with environmental laws:

> **Environmental Services Division**
>
> The Environmental Services Division (ESD) is part of the Water Service Department and is dedicated to protecting public health and the environment by maintaining compliance with environmental regulations for drinking water, stormwater, and wastewater quality.
>
> Through the effective management of the Water Monitoring, Industrial Pretreatment, Commercial Inspection, Stormwater Management, and Wastewater Monitoring programs, we provide a safe and reliable domestic water supply to all residents and maintain a clean, healthy environment. (CoP ESD 2021)

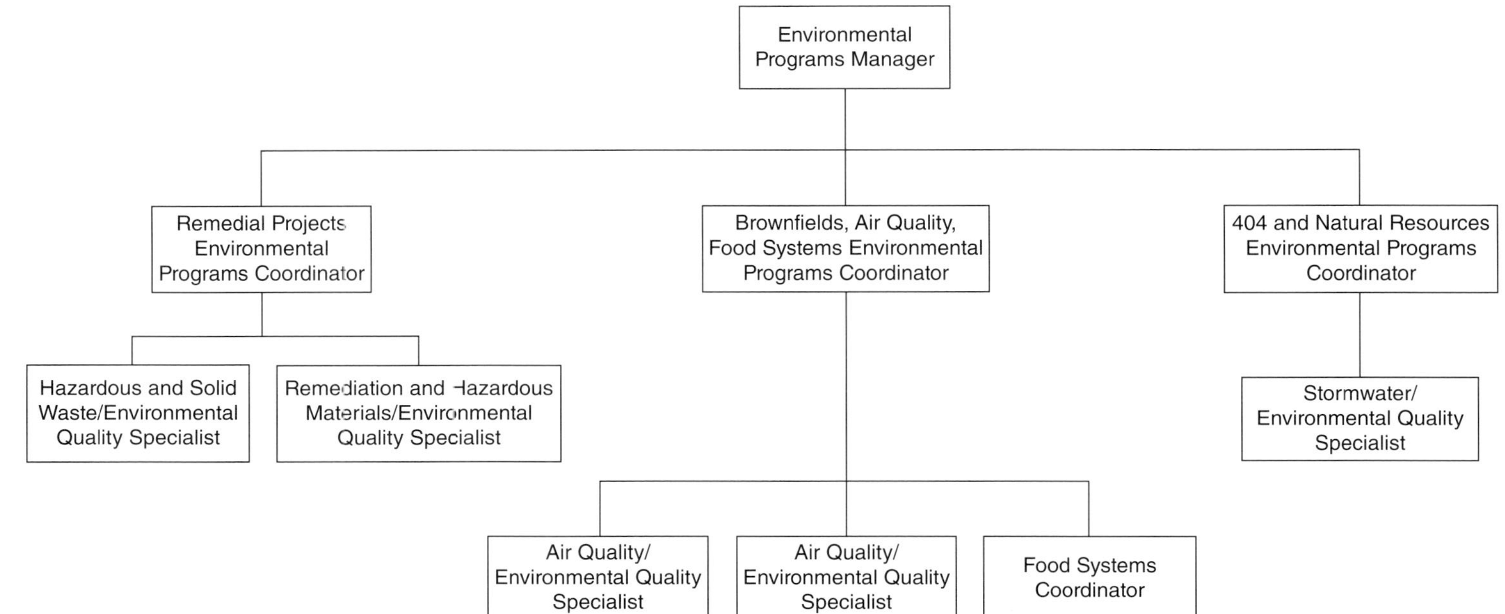

Figure 5.3 Example of a city environmental department organizational chart. (*Note*: 404 = Dredge and Fill Permits.)

Office of Environmental Programs

The city of Phoenix Office of Environmental Programs (OEP) provides independent technical expertise and environmental leadership to City management, serves as a technical and/or management resource for operating departments, and leads in the creation and implementation of city-wide environmental policies and programs to reduce environmental risk and liability in City operations.

OEP manages a number of programs, including: Air Quality, Brownfields Land Recycling, Climate, Food Systems, Hazardous Materials, Waste and Remediation, Sustainable Purchasing, Surface Water, and Wildlife programs. OEP staffs the Environmental Quality and Sustainability Commission. OEP completed a Community Greenhouse Gas Emissions Inventory, implemented the STAR Communities Rating System, and assisted the Office of Sustainability in achieving the LEED for Cities Platinum certification. (CoP OEP 2021)

REAL-WORLD EXAMPLE 5.3 Integrating local, state, and federal environmental agencies: the case of Tres Rios

At the city of Phoenix, environmental managers (EMs) manage the complex interplay of environmental regulations for city projects and operations. This is key to achieving city goals while promoting environmental stewardship and regulatory compliance. The Tres Rios project along the Salt and Gila rivers in Phoenix, for example, engaged multiple local, state, and federal regulatory agencies and required compliance with many environmental laws including NEPA, ESA, CWA, and NHPA. Tres Rios included construction of a levee, man-made wetlands for habitat and effluent treatment, river channel improvements, invasive species removal, native habitat expansion, and trailheads, and currently provides flood protection, improved flow conveyance, 128 acres of public wetlands, and is a Globally Important Bird Area with over 200 species of birds and other wildlife.

Tres Rios was funded through US Army Corps of Engineers (USACE) Civil Works via congressional appropriation, making NEPA an essential planning component of the project. The city developed agreements with multiple agencies including the USACE and local landowner Arizona Game and Fish Department for design and construction. Consultations with USFWS and SHPO as well as coordination with the local flood control district were key. Because the city is the long-term operations and maintenance (O&M) operator, the requirements under which the completed project would need to operate were also important, including: (1) under section 14 of the Rivers and Harbors Act (33 U.S.C. 408), any action beyond O&M within the USACE Civil Works project area could require a section 408 permit; (2) anticipated occasional disturbance to endangered species habitat as a result of O&M led to the city entering a 50-year Safe Harbor Agreement (with associated incidental take permit) with the USFWS under section 10 of the ESA; (3) long-term O&M activities in the river led the city to work with the USACE regulatory branch to define the limits of Waters of the US and obtain a CWA individual section 404 permit; and (4) the

REAL-WORLD EXAMPLE 5.3 (cont.)

discharge of treated effluent to Waters of the US from the constructed wetlands required a CWA section 402 National Pollutant Discharge Elimination System permit from the EPA.

Tres Rios illustrates the importance of EMs being aware of both short- and long-term requirements in order to effectively guide a project through regulatory complexities. The city's relationships of trust with regulatory partners was, and continues to be, of vital importance for this project. These relationships should not wait until a complex project is imminent but must be built over time with habits of proactive communication, acknowledgment, and quick correction of compliance concerns, and partnering on common goals. These actions provide benefits to the regulatory agencies, to the city, and to the community as we then more effectively work together to create projects like Tres Rios.

Tricia Balluff, MS, City of Phoenix Office of Environmental Programs

5.6 Conclusion

The value of integration of environmental management between different agencies, organizations, and authorities at the local, regional, and national levels cannot be overstated. The environmental manager should always view their fellow environmental entities – be they public, private, or non-profit – as partners and should always look for potential partner entities in all of their program planning and execution. There generally is no environmental issue that does not cross a jurisdictional or functional area boundary. Spread loading the work while also ensuring cross-coordination and support, while it can take some extra effort and planning to make it work upfront, is worth the effort in the end to ensure your program, resource and political management plans and implementation are sustained and effective.

AUTHOR'S NOTE 5.1 Intergovernmental management issues today

The problems we have noted with the intergovernmental relationships that are critical to environmental management have been recognized by all parties involved, and several attempts to alleviate them through policy and law have been made mostly with regard to enforcement. Unfortunately, when local, state, or federal leadership is anti-environment and/or is pushing a policy of deregulation, human health and the environment suffer at many levels. Environmental Regulatory Agencies must "play nice" with their partners, though it is natural for agency professionals to believe they better know the issue goals, timeline, and strategies to achieve environmental management goals. Thus, sometimes

AUTHOR'S NOTE 5.1 (cont.)

they do not want to share power, but these relationships work best when there is a partnership, that is sharing and with the realization that all successful partnerships require leadership.

In the past decade many state-appointed environmental officials orchestrated the degradation of environmental protection. Tens of thousands of American lives every year are lost early and unnecessarily to environmental health hazards (Dedoussi et al. 2020; Fuller, Sandilya, and Hanrahan 2019; Pope et al. 2020). Isn't it hypocritical to profess support for personnel in the military and law enforcement pursuing their mission to protect American citizens from harm while at the same time denying support for public servants who provide protection from environmental harm? Americans require assurances regarding our physical security, yet there is a disjunction regarding environmental health hazards. We do not draw this comparison to diminish the vital protections that our armed services provide, but rather to illuminate the salience of environmental health issues incurred by citizens every day, with environmental managers serving as the frontline protectors.

Many public environmental managers are not being allowed or supported to achieve their mission. No matter their politics, all Americans can agree that we want assurance that the water we drink, the air we breathe, and objects we come in contact with (food, soil, toys?) are safe. That assurance can only be given if our environmental protectors can name their "clients." Are those clients the water systems and industries they regulate or the "the public"?

Whereas our environmental protection historically consisted of permitting, monitoring for compliance, enforcement, and technical assistance, many state environmental agencies have been unreasonably downsized (by over 50 percent in some states) and these law enforcers have been relegated to act as clerks in state-run "permit shops" (Barnes, Graham, and Konisky 2021; EIP 2019). By hollowing out our environmental regulatory enforcement capacity we are inviting vulnerability.

States that focus performance evaluations on how many environmental permits are quickly issued, without equally providing resources for monitoring for compliance and enforcement, are competing in a "race to the bottom" (Esty and Porter 2002; Porter 2011; Woods 2020). The "prizes" of this race are the lowest national rankings for water and air quality and increased rates of infant mortality. Increasing jobs by decreasing environmental protection does not work and is not logical; short-term benefits are significantly outweighed by the long-term environmental and social welfare costs. Most economists recognize that well-crafted and -implemented environmental regulations force countries as well as industries to innovate, increasing efficiency and competitiveness. This "race to

AUTHOR'S NOTE 5.1 (cont.)

the top" approach results in economic success in states and nations serious about environmental protection.

This lack of mission-oriented management is not only a result of strategic ineptitude but of malice. Leaders who do not support – let alone respect – their troops cannot expect them to sacrifice. Governors opposed to environmental regulation appoint like-minded environmental directors who not only ignore their mission and legal obligation to pursue it but openly display distaste and disrespect for our public servants attempting to "protect human health and the environment."

As with Flint, Michigan, the EPA is supposed to intervene when states fail to protect citizens from environmental harm. Unfortunately, the EPA (under certain administrations) has lacked the resources in terms of political will and funding (Barnes, Graham, and Konisky 2021; Konisky 2015). Congress knows repealing the environmental protection laws for air, water, and toxic materials would be wildly unpopular. Instead, Congress warned the EPA not to intervene and, in many instances, has made sure it does not have the funding to do so. Further, the morale of our federal environmental managers has predictably been trending to a "petty bureaucrat" culture rather than that of the professional public protector many have been trained for. Is this civil–environmental-manager disjunction not similar to civil–military disjunctions that we have witnessed before? It is certainly as dangerous. Thus, our citizens' health is threatened by state and federal politicians who subscribe to the idea that the non-enforcement of environmental law is good for jobs.

Citizens must become aware that "institutional anorexia" (the pathological downsizing of both the EPA and state environmental agencies, paralyzing regulatory function) is a bureaucratic disease that is dangerous and, like the actual human disease, requires lengthy and expensive convalescence. As we fear with our military and law enforcement professionals, we must recognize that rigorously educated environmental management professionals will leave public service or decide not to serve for the protection of future generations. Finally, strong and caring leadership is needed to allow EPA to assure that states do not sacrifice public health and that it is the citizens who are the most important "clients" of government services (Lame and Marcantonio 2016).

In this chapter we wanted to provide the reader with a better understanding of how regulators work and work together through their intergovernmental relationships to "co-produce" solutions to environmental problems. In our next chapter we will expand on the concept of "co-production" and how communication and public participation is critical to working with our private sector, non-profit-sector environmental advocates and public stakeholders.

5.7 End of Chapter Questions

1. Why is it important to be able to analyze your professional relationships regarding environmental protection?
2. What agency is responsible for enacting most of the US environmental laws in your state?
3. What does preemption mean in terms of the state and municipality intergovernmental relationship?
4. Name three of the six key areas that make up the Policy Framework for State/EPA Enforcement Agreements.
5. What does IGR mean regarding the environmental regulatory infrastructure?
6. What are five major reasons why the US EPA does not feel comfortable with states having more autonomy regarding regulating the environment? And what effect do these "problems" have regarding the concept of "federalism"?

INTERVIEW FROM THE FIELD 5.1 Leadership in environmental management

Ed Fox, JD, Former Administrator of the Arizona Department of Environmental Quality

What allowed you to feel like you were capable of leading a state environmental agency?

First of all, I'm not sure I was competent when I took the job. I spent a decade practicing environmental law and I knew the law, but managing environmental programs is about management ... about people. I used to say all the time ADEQ is just letters, and the building we occupied is just brick and mortar. The agency isn't either of those things. It is in reality the people implementing programs and making decisions every day. It's about the people inside the agency. As a leader if you don't manage the people, if they're not given the tools they need, if you don't hire the right people [with the right] salary and benefits you'll never have an agency that functions well. It won't matter how good your policies are. Because in the final result, it's not about the policies, which are not self-actualizing, but about the people who implement and make the decisions every day.

We were in office during an era where it wasn't like everyone loved us and we certainly weren't given the resources needed to fully do the job required, but I believe that my relationship with the governor shielded the agency from a bunch of political BS, allowing us to do our job without too much interference in our day-to-day functions. This allowed us to maximize the resources we were given. One of the more debilitating aspects of these jobs, as a public employee, is being vilified for having jobs paid for by tax dollars, and "being at the public trough" somehow makes it OK to denigrate the work being performed. The reality, realized by anyone who's been a leader in a public agency, is that

INTERVIEW FROM THE FIELD 5.1 (cont.)

employees take their jobs seriously and work hard. Like any organization, public or private, there are some employees who shouldn't be there, and they should be fired, but most employees take pride in their work. They deserve managers who know how to support them and fight for the resources and an environment that allows them and the organization to succeed.

So, you asked me what were the skills that I brought to the agency that made me think that I was capable of doing it. I was arrogant enough to believe that I knew the law and knowing the law would make it possible for me to implement programs and develop policies that would advance the agency. It didn't take me long to figure out programs and policies are nothing without good people and that I needed a great executive team capable of managing the people who made the decisions every day.

So, what was it about your team? What were you looking for in that team? And of course, you know, we understand that you inherit the team, but that doesn't mean that you keep those people, you know. So, what were the characteristics that brought those people and allowed you to continue to rely on them?

From a management point of view, I think this may have been the best thing that I did. The management team was made up of the assistant directors of Waste, Water, Air and Administration. I remember sitting down one-on-one with each of them when I first came. I had done my homework and gathered information on each including input from environmental groups, the business community and others who had interactions with the agency.

One assistant director I wanted to keep was a challenge. Few people outside the agency liked her at the time and many thought she needed to leave the agency, but she was highly regarded inside the agency. Everyone agreed she was smart and talented. I needed people on my team who were respected inside and outside of the agency, and I decided that if I could help this individual to see where change was needed, we might both succeed. At our one-on-one I presented the individual with the issue, telling her, "There is no question that you're one of the smartest people on environmental policy in the state of Arizona. The problem is, nobody likes you and so nobody wants to work with you. There are a bunch of people outside this agency who have been telling me for a month now that you need to leave. But I need really smart people like you." I gave her a list of 10 things that I believed were needed for the agency to be successful. I told her to look at the list and tell me tomorrow whether you can help me achieve the 10 items. I opened the door to input by saying I was open to discussing the list if she disagreed with any of it. Interestingly, she came back the next day telling me that no one had ever given her feedback on her work style and that she could certainly help with the list. In the end, she was maybe the best policy and management person I ever worked with in terms of analyzing and

INTERVIEW FROM THE FIELD 5.1 (cont.)

understanding the scope of considerations needed to make good policy and management decisions.

One of the assistant director slots was vacant and there wasn't an obvious candidate. Based upon the input of a colleague, I reached out to an individual who didn't have a strong background in the area but who had superior interpersonal and management skills. It may seem like a workaround to some people, but I decided that I needed strong leadership and management skills more than substantive knowledge and that the subject expertise could be addressed by allowing her to hire a deputy. While a bit risky, the strategy succeeded both in and outside of the agency.

This was my first public leadership position and I learned more about leadership from our ombudsman and PIO [public information officer] than I learned from anybody else. As I've said before, the agency is an abstract concept only activated by people, and I remember the PIO always throwing that idea at me and saying, the agency needs a face and as the director, it has to be you! I don't deny having a big ego but taking the job was, for me, about establishing a better-functioning agency that could thereby implement the policies and laws designed to protect public health and the environment, it was never about being "the face." But the need to publicly engage was further reinforced by our ombudsman who had a unique ability to engage with people who I never would have wanted to talk to. His ability to engage a wide range of people was a real eye-opener to me.

What about your executive assistant?
She really kept my feet to the ground in a big way, making me leave the office, walk the halls, and go talk to people, go meet the frontline employees. If she was in the military, she would have been a sergeant. She kept me straight, she kept me organized, she kept me orderly. She kept me grounded. She understood protocol and people and was able to help me steer clear of controversy and problems when the job required meetings with other officials and ceremonies.

What about your receptionist, your gatekeeper?
My receptionist had been there many years before me, and I was lucky she was there. She took the time to understand my priorities and managed access accordingly including my priority to have open office hours for ADEQ employees. She often made time on my calendar to meet with employees. I have no doubt that those people (in and outside of the agency) who took the time to develop a relationship with her had better access than those who just called to talk with me. And as we know, developing relationships with assistants and support staff is often the most effective way to achieve desired results.

To paraphrase Napoleon, leadership is being able to tell people what the situation is and to give hope. Leadership is about defining a desired future state of being and then the

INTERVIEW FROM THE FIELD 5.1 (cont.)

strategy to achieve that state. Getting from here to there requires understanding the current situation and, assuming the desired future state is better, then the strategy should provide hope. I would say however that the strategy and endstate need to be articulated with passion so those following know that you truly care. You can articulate a future state that you want to achieve, but if people don't believe you, if they don't believe you will do what is needed to implement the strategy, they won't follow.

You know, one of the most important things that you did, and I don't know if you ever realized it, was when you would walk into our offices and say, "How am I doing today?"
I learned that from Mayor Koch. I'll tell you what, it's important to get candid off-the-cuff input from trusted advisors. Formal reports and meetings with prepared agendas are important, but innovation and creativity often happens during impromptu chats in the breakroom. The informal "How am I doing?" thing was intended to open communication that might not otherwise happen in a more formal setting.

Encouraging communication at all levels was important to me. It's why I had an open-door policy. One day a young woman came up from the bowels of the agency from some program that I didn't even know existed. It was an exceptionally busy afternoon, and my receptionist was going to schedule her time for another day. I saw her standing there and I had a break, so I said, "it's OK." So, she came in and we sat down and chatted. She was passionate about her work, and she just wanted me to know about it. It was motivating to talk with this young woman and hear her passion. While the work she was doing wasn't going to change the agency, I later learned that her work would feed into a database that could eventually change how the agency addressed certain water quality issues. But on reflection it wasn't the potential outcome of the work that mattered to me, it was the woman's passion. The outcome would take care of itself, but I knew then that we had to take care of her.

What is your opinion about environmental management systems like Total Quality Environmental Management which seemed to evolve to Six Sigma and Lean, then Lean Plus?
Some very techie managers see these programs as problems to be solved, as equations that if implemented will yield a right answer. But the outcome of these programs as applied to environmental management systems can't simply be to identify errors and streamline processes. How the results can advance the mission and goals of the agency and the programs must be the focus. For example, if a permitting program has a backlog and efficiency measures are identified that can make permitting faster and frees up resources but those "freed-up" resources are x-appropriated in the next budget process, then the measures will do little to reduce the backlog. In my experience public environmental agencies

INTERVIEW FROM THE FIELD 5.1 (cont.)

rarely have the needed resources to properly implement programs. So for me, when I think about TQM, I think about it as a strategy and a methodology for getting the job done and freeing up resources so they can be reinvested to achieve greater environmental improvement. At the end of the day, those process improvement programs can and should lead to a better environmental outcome because you can do more.

What is the biggest impediment to TQM, Six Sigma or Lean programs?
A few things.

(1) – I think program managers are reluctant because they are concerned that if the process identifies ways to free up resources, those resources will be taken from the program in the next budget process. An example might be that if you can do the current level of effort with three people instead of five people, then do it with three even if the current level of effort isn't fully implementing the program. Saving money becomes the primary goal rather than more fully implementing the program with existing resources.

(2) – The fear of change and doing something differently. This is always an obstacle to any kind of program that requires a change in procedure or process.

(3) – The failure of leadership to fairly articulate how process and program efficiency can advance the agency mission. A big mistake is not linking them.

How was the intergovernmental relationship with the EPA when you were a state director?
For most of my tenure I was lucky that the Region 9 Administrator was open to developing a strong working relationship. We certainly didn't always agree, but we were each committed to respecting the other and working to find amenable solutions to issues and conflicts that occurred. This type of relationship doesn't always develop in our federalist system where state directors and EPA can have very different politics and ideas about environmental programs, but it is always true that relationships matter even when the politics and policies don't align.

I didn't really have much of a relationship with the folks in DC. I didn't spend a lot of time in DC [EPA Headquarters]. I left that to the political folks. My EPA focus was on Region 9 because during my tenure I believed that the vast majority of EPA matters that would have meaning to Arizona would happen as a result of EPA action out of Region 9.

I want to ask you one last question. What do you see on the horizon that the environmental managers of today will have to deal with?
I believe that climate change is going to change everything that we've been doing. But when you think about all the potential outcomes that could happen, it seems to me that

INTERVIEW FROM THE FIELD 5.1 (cont.)

the situation will require all hands (agencies) on deck and that the right programs may not even exist today. Managers of existing programs will need to be open to change and possibly act within existing programs as the environment evolves while at the same time not overstepping their existing authorities. This will put a lot of pressure on agencies and their employees.

6 Ethical Environmental Management and Communication

In this age, in this country, public sentiment is everything. With it, nothing can fail; against it, nothing can succeed. Whoever molds public sentiment goes deeper than he who enacts statutes, or pronounces judicial decisions.

Abraham Lincoln

Skill: *Communication.* Learn the language of environmental management and how to communicate with your stakeholders ethically to co-produce solutions to environmental issues. Environmental issues are issues because they affect people; thus, it is critical to the environmental manager's program, resource and political management to engage and include stakeholders in their planning and solution design process. How stakeholders should be included will vary based on the scale, effect, and history of the issue at hand, but it will always be in the environmental manager's interest to, in some form or fashion, co-produce solutions.

6.1 Introduction

Whether you are a public servant, private sector manager, or not-for-profit influencer it can seem that working with citizen stakeholders is costly in terms of your resource management and political management. Such engagement and interactions require time, and they mean sharing power at some level. Many managers, particularly technocrats, believe they know what the problems are and how best to fix those problems based on "the science." As well, it is a natural tendency just to provide and implement solutions, even with the best of intentions, to move on to the next problem. Or, as some say, "check off that box."

However, it is a mistake to believe that you can manage a program, particularly a public program, without ethically and authentically involving your stakeholders. All people have different perspectives and values such that any single solution to an environmental issue will likely not wholly address the concerns and preferences of all stakeholders. But by genuinely engaging all stakeholders and facilitating their participation and input in developing a solution, i.e., in developing their program, resource and political management plans, the environmental manager can increase the likelihood that their solution will be well received by the people whose health they are charged with protecting.

In Chapter 5 we discussed the obvious intergovernmental relationships and skills environmental managers must develop and maintain to manage with in governmental infrastructure. Now, we must further understand the concept of "co-production." This chapter begins this process by looking at what co-production is, why stakeholders require it, and how to ethically co-produce and implement environmental solutions.

6.2 Co-production

Co-production means that more than one entity – be it a person, group, or institution – produces a solution or accomplishes a task. It does not mean giving up leadership, but it does mean the sharing of power. The pilot of a plane has the responsibility to fly the plane but does so with a co-pilot. A co-produced environmental solution involves knowing the impacts or implications of accountability, environmental justice, ecosystem management, sustainable development, and unfunded/underfunded mandates *and* it includes a co-produced, ethical response that addresses these environmental issues.

As we have stated, "environmental management is people management." Thus, without environmental managers and pollution producers alike "owning" pollution and without citizen trust that the system is addressing it, environmental management becomes difficult and very expensive. Going back to our tendency not to take the time to engage with and share our power with stakeholders, we must now understand that we can do this right or do it wrong. To "do this right," the environmental manager must incorporate concepts of ethics such as transparency, inclusion, leadership vs. ownership, equity, resource commitment, reliability, among others, into their program, resource and political management. To "do this wrong" often involves choosing expediency and perceived efficiency, especially in the short term, in planning and executing an environmental management plan. The environmental manager can get by with this from time to time to an extent. But if they make a mistake by misjudging a situation, not considering relevant contexts, or simply getting the science wrong, they will have made themselves vulnerable and invited challenges that can at best cost them unnecessary time, money, and trust, and at worst lead to the failure of their program. In the end it is always safest and most ethical, and often most efficient, to take the time to co-produce.

Today's environmental managers should attempt to conduct their programs in an ethical manner, particularly considering the legal trends in civil environmental litigation to use the courts to hold all sectors and responsible managers to account (Anchondo, Gardner, and Hyde 2020). Spending time in court, which as we have said costs program effectiveness, and money on attorneys and/or expensive settlements, can sometimes be avoided if those who are affected by the actions of environmental managers believe they have some "stake" in the solutions. That is, if stakeholders participate and co-produce a solution, they are less likely to levy a legal challenge if the solution proves ineffective for its intended purpose.

6.3 The Public Faces of Environmental Management

Public agencies, at least at the federal and state level, have some type of office of "external affairs" which formalizes their attempts to co-produce with entities outside the agency. The positions within external affairs include, but are not limited to, a legislative liaison, a public ombudsman, and a public information officer. Often the members of external affairs work as a team as their respective expertise overlaps in many situations. These professionals usually report directly to the head of the agency and are considered that individual's "eyes, ears and mouth" with the responsibility to assist and advise the administration regarding what is going on outside the agency and to assure quality control within the agency. Their day-to-day programs are to establish external relationships with stakeholders and other agencies, to monitor external developments pertinent to the agency, and to communicate to give or get information, to affect attitudes and to influence behavior.

The legislative liaison: A legislative liaison provides political management for the agency and is typically an expert on policymaking and policy interpretation. Note that public agencies are not allowed to have "lobbyists" but do have "legislative liaisons"; that is, someone who solicits people, usually political officials or other civil servants, to push for a particular change or action to take place. They develop working relationships with elected officials and lawmakers to provide pertinent information regarding existing or new legislation. They track or monitor legislation that influences the agency they represent and guide and develop the legislative agenda for their agency. Finally, they advise agency personnel regarding their political interactions such as public statements or engagements with political officials and other public figures.

The public ombudsman: The public ombudsman is someone within an agency who deals with people outside the agency who have a problem with that agency. They typically have four roles. First, an ombudsman acts as an anchor such that external entities (individuals, businesses, and communities) do not get lost in the bureaucracy and "left on hold." Second, an ombudsman also acts as a facilitator to ensure the interactions of their agency with external entities achieve the desired goal. Most of these interactions have to do with a "managed," i.e., pollution producer, or other entity seeking to obtain a permit and technical assistance from their agency, but sometimes the ombudsman can facilitate discussions or negotiations about monitoring for compliance and enforcement conducted by their agency. Third, the ombudsman's role as an educator is to teach external entities, including those mentioned above but also tribes and municipalities, how to navigate the regulatory system to the effect they desire. Fourth, the ombudsman acts as a referral service to make sure the external entity contacting their agency is interacting with the correct person or office. Ombudsmen often have agency experience as environmental managers but must have some expertise in organizational management and communication; that is, experience with intra-agency administration and communication. Ideally every member of an environmental management organization that interacts with external audiences should try to perform as an ombudsman, or at least have some training and understanding of the role of the ombudsman to better enable their interactions with stakeholders.

The public information officer (PIO) is often considered the public relations professional inside a public agency. Part of their job is political management and part is technical assistance or education. They typically have roles which begin by being the spokesperson (the "talking head") for the agency, particularly with regard to media appearances and the delivery of public statements. Unfortunately, the PIO sometimes can become the agency "spin doctor" who "spins" negative stories in a positive direction for the agency. This professional has the responsibility to be an ethical communicator and information resource to the public and the press. Thus, this role requires expertise in working with the press. Like other members of the external affairs team, they must develop working relationships with external entities with whom they communicate and share information. In this instance, those individuals are normally reporters or journalists, and, like the legislative liaison, the PIO must understand and abide by the culture they are working in. As in all relationships, reporters require and rate ethical communication. That communication must be accurate (never lie to the press!), useful, on time regarding meeting their deadlines, and fair (e.g., as in when it comes to exclusive stories, they should be rotated through the press corps without favoritism). If one breaks these rules of ethics they can expect to be punished in the press or "burned."

In addition to these more political management responsibilities, the PIO also orchestrates and distributes publications to educate or technically assist the external entities that the agency they work for regulates. These publications are often developed on digital platforms and located on the agency's webpage, but can and should also be in the form of hard copy and in the language required by the public audiences, i.e., translated into all the languages of the stakeholders regulated and/or protected by the agency. Finally, the PIO develops, updates, and provides internal information for the personnel of the agency to keep agency personnel up to date on relevant policies, practices, or happenings within their area of responsibility.

6.4 Stakeholders

Agency plans and strategies should be co-produced with goal-oriented partners who have a shared interest in developing a productive plan to solve an environmental management problem and in producing an effective solution. Not all stakeholders are interested in participating in producing a solution, but most expect and want to feel that they have been heard and included. The effective environmental manager will have to learn to navigate these differing types of stakeholders to forge useful and productive partnerships for co-production with some stakeholders while still being inclusive of all stakeholders. The partnership for co-production is not for changing or influencing the stakeholders but for producing an ethical, effective, and mutually understood solution.

6.5 Sources of Environmental Conflict

In all environmental management programs, some stakeholders will want more environmental protection while others will want less as the list of stakeholders includes both pollution producers and citizens alike. The result of these positionalities is often conflicting values, preferences, concerns, and desired future use of the environment among the relevant stakeholders the environmental managers must work with. The three primary sources of environmental conflict are:

- differences in values and worldviews – such as the ethics of Pinchot, Muir, and Leopold (discussed in Chapter 1) where there is conflict regarding how humans should address natural resources, or Carson challenging the dominance of humans polluting the ecosystem, or even the more fundamental differences as to whether money can solve all problems or nature does not care about money.
- conflicting interests – this refers to the "frame of reference" of the stakeholder. Often, the regulated polluter wants less regulation because they want to generate more money or political will, whereas the stakeholders assuming the risks of pollution want more regulation/enforcement in order to lessen the risks to their health and environment including to the value of their property.
- the technical uncertainty that surrounds various courses of action – perhaps the best example for this conflict is global climate change. Is it real and what causes it? While most trust the obvious answers to these questions based on current science, they involve the understanding of whether a risk exists, the effects of the risk and the best available solution to the risk. Our profession dictates we understand that "science changes," which is why laws like TSCA and FIFRA require periodic review of products which are considered safe. Indeed, what affects our health and environment and how we address those threats has changed as the science improves.

It is the responsibility of the environmental manager to attempt to understand what the source of conflict is as they attempt to produce environmental solutions with stakeholders. Always, the manager charged with protecting human health and the environment should try not to take the passions of their stakeholders personally and to be as equitable and open as possible with the programming and attention throughout the solution development process. In today's climate, many if not all environmental issues can become political problems or challenges. The environmental manager is charged with protecting human health and the environment without regard to the political and social affiliations or affinities of their stakeholders and their own political and social positioning, always keeping their mission – again, protecting human health and the environment – and the mandates they are accountable to, at the center of their work.

6.6 When Co-production Is Not Working

When the generally informal process of co-production breaks down, sometimes it is helpful to employ a more formal method of solution development such as Environmental Dispute Resolution (EDR). EDR is "a set of technologies, processes, and roles that enable parties to a dispute to reach agreement with the help of a neutral third party known as a mediator" (O'Leary et al. 1999: 193). The employment of EDR can be more time- and resource-consuming than informal consultation and co-production with stakeholders. Its benefit is that it can reset the solution development process and often has a clear framework for proceeding that can be pointed to and shared with stakeholders.

It is quite possible that you might end up in court, at which point you will have to rely on whatever resources are necessary to achieve your mission. But first let us introduce perhaps the most important set of skills any environmental manager must learn and use – communication.

6.7 Communication

Communication is the "process whereby participants create and share information in order to reach a mutual understanding" (Rogers and Kincaid 1981: 63). Notice the word "agree" does not appear in this definition. The goal is not necessarily to get everyone to agree but rather to obtain – or more correctly, to create – a shared, mutual understanding. The three fundamental reasons we communicate are to: (1) give or get information; (2) affect attitudes; (3) influence behavior. Probably the last thing a manager should do with stakeholders is to attempt to get them all to agree or be in perfect alignment, as this is rarely achievable, if not impossible, and often not necessary to achieve a sufficient and effective solution to the environmental issue at hand.

The goal of reaching a mutual understanding through sharing information might result in a change of attitude and influence behavior, but it is a process, not a single action. This communication process has some basic, fundamental parts, as modeled long ago by Harold Lasswell (1948), that "answer the following questions: Who, Says What, In Which Channel, To Whom, With What Effect?" (Lasswell 1948: 117). Figure 6.1 is Lasswell's original depiction of this process.

While useful to understand the key components of communication, Lasswell's linear and delimited – that is, simple and non-dynamic – model is less than representative of how communication actually looks. A more real-world model would be Kincaid's "convergence model" (Figure 6.2), where participants converge on a mutual understanding through bidirectional feedback (Rogers and Kincaid 1981). Through multiple iterations of giving and getting information and sharing of perspectives, preferences, and concerns, and often the correcting or refining of interpretations of these between groups, a mutual understanding coalesces or is homed in on. Note that Kincaid's model only depicts two participants, but more often than not the environmental manager will be co-producing and communicating

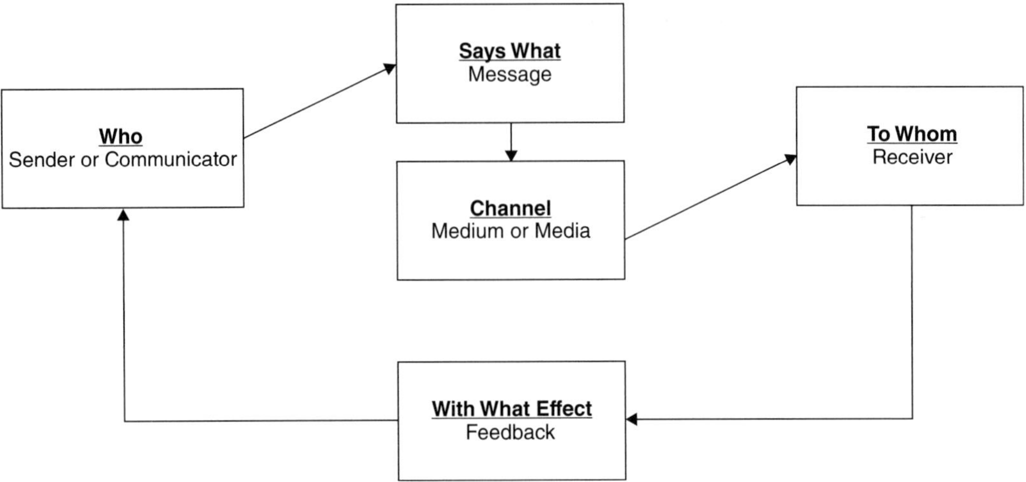

Figure 6.1 Lasswell's communication model.

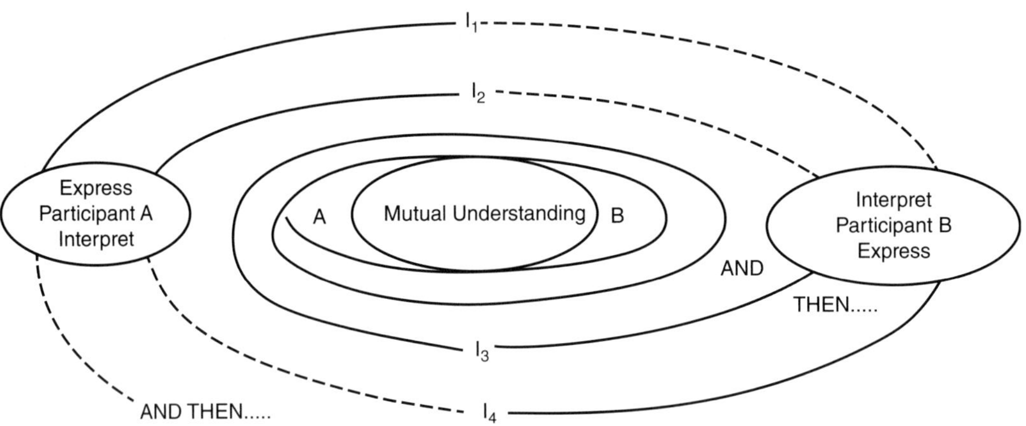

Figure 6.2 Kincaid's convergence model of communication (Rogers and Kincaid 1981).

with multiple participants, making the model and flows of information and perspective-sharing much more complicated and complex.

Communication is rarely easy. Today there are so many ways of transmitting messages, from social media to digital meeting platforms such as Zoom, that effective sharing and interpreting of information between managers and stakeholders has become more easily actioned yet substantially more complex. What is more, with new methods and means of communication available, many stakeholders expect timely (read: immediate) responses to their inquiries or statements about an event or incident that affects them or the area they live in. While the tools of communication have thus grown, the ease of execution has not necessarily done so, with many barriers to communication still as present as ever.

6.7.1 Barriers to Communication

The number one job of the manager with regard to communicating is to not create barriers to communication as participants converge on a mutual understanding. However, barriers always occur, whether you create them or not, and the environmental manager must be aware of the differences between their audiences (industries, municipalities, and NGOs) regarding size, diversity, and legal constraints. Thus, the environmental manager is much like a skilled physician in that they must be able to diagnose and treat (or overcome) barriers in order to effectively communicate a "cure" (Garnett 1992). Ideally, barriers should be considered prior to communication, but when it appears conversations are just not working, your first "go to" is to identify, understand, and bridge the barriers and then see the results for your communication process. Though not by any means encompassing all of the possible obstacles to effective communication, there are eight basic barriers to communication that are common to our profession. We list them here, and provide some strategies and helpful contexts for overcoming these barriers.

Language – Clearly, if the message is in a language that the receiver is not fluent in it represents a barrier. It is easy to diagnose this barrier if it is a foreign language – that is, the material or communication method is not presented in the primary language of the recipient – as is the remedy of having an interpreter. However, this barrier is mostly one of inherent or implicit profession or agency culture – we have our own language! We use acronyms like NEPA, NPDES, BAT, CAP, FACA and we speak in technical terms regarding chemistry (e.g., ozone), hydrology (e.g., infiltration), engineering (e.g., scrubbers), and even legal terms (e.g., primacy). Environmental managers have trained diligently to learn the lingo of their trade and, like any profession, like to use it and even think in terms of it. Like any culture or profession, we are not only comfortable using this language but enjoy using it as it defines us. This new language which we are quite comfortable using, or more so allows our culture to feel superior, can become a "curse of knowledge," making our messages for permitting, enforcement and technical assistance hard or impossible for stakeholders to understand.

To overcome this barrier, one must assume our audience does not understand our language and be prepared to not use unfamiliar terms, or to explain what those terms mean and even provide a glossary of terms and acronyms. Communicating with stakeholders is not like communicating with adjacent agencies or other environmental managers. The environmental manager is best served by assuming no prior environmental science or management knowledge among their target audience, especially in public forums and statements. This does not mean "dumbing down the message"; it means meeting your stakeholders where they are at and providing them the best opportunity to produce a shared understanding with you.

Prejudice – When the sender or producer of communication feels superior to or does not like the receiver, or vice versa, the message suffers. The most effective communication happens when both sender and receiver have an open and non-predetermined mind such that messages can be created and interpreted objectively. One of the more common prejudices environmental regulators have is that they are communicating with an entity that wants to pollute and that they perceive does not care about the environment. Conversely, the regulated entity might view the regulator as a petty bureaucrat or as someone who is unconcerned about their livelihood. Worse, citizens might view regulators or industries as corrupt or uncaring, of which there have indeed been instances, which is why people sometimes make such assumptions even in situations where there are no indicators of that being the case.

Overcoming prejudice always begins with finding common ground and empathy. Without care – that is, the care to learn, to create understanding, and to build a relationship – prejudice or otherwise biased perspectives will persist. This can be a laborious and uncomfortable process, but without addressing it the trust and confidence that an environmental manager needs, from either those they manage or those they protect, to do their job effectively cannot be gained.

Hierarchy – Similar to prejudice, hierarchy can impede objectivity, thereby interfering with the message being communicated and understanding of the target audience. It is common for mid-level permit writers sitting across a table from company vice presidents to feel/react to a power differential. Or for high-level agency officials to address community members and not display the desired deference to their position, that is, they do not connect with the community where they are because of a feeling of superiority or authority. Overcoming hierarchical barriers most often has to do with understanding your position. Regulators are to serve and protect the public, the public is their charge and is what grants them authority. Remembering that the public is the prime concern can help the environmental manager navigate and reduce the emergence of issues of hierarchy when it works against their mission to protect human health and the environment.

Distance/proximity – Human communication is approximately 90 percent non-verbal and 10 percent verbal (Mehrabian 1971). While our professional communications are more often verbal or written, non-verbal communication, e.g., body language or demeanor when delivering communications, is still critical to the interpretation and transmission of messages. Thus, though we have several different media available to us today, from social media options to agency webpages, it is sometimes best to have in-person communication so participants can more accurately converge on the message because they can read and respond to each other's non-verbal messaging. To overcome this barrier, arrange for in-person meetings as much as possible or use digital interface products like Zoom or Google Hangouts. In general, and as appropriate, the more intimate our communication the less likely it will be misunderstood, and the better chance the environmental manager has of creating a productive and enduring relationship with their stakeholders (Rogers and Steinfatt 1999).

Distraction – Distractions can come in different forms, such as emotional, physical, and psychological. We all know how difficult it is to have a productive conversation when we are worried, there is too much noise, technology is not cooperating, we are stressed, or we are simply just tired. The professional should plan to minimize distractions by considering communication logistics such as timing or venue and establishing "rules of engagement" such as prohibiting cell phones or side discussions. To help guide plans to reduce distractions, it is helpful to keep in mind the requirements for active listening:

- Hearing – exposure to the message
- Understanding – when we connect the message to what we already know
- Remembering – so that we do not lose the message content
- Evaluating – thinking about the message and deciding whether it is valid
- Responding – when we encode a return message based on what we have heard and what we think of it. (Rogers 1995)

By facilitating these elements, the environmental manager can develop a setting that maximizes the likelihood that their message will be heard, understood, and internalized such that stakeholders are informed and are able to engage effectively in further discourse. Here are some additional suggestions for what *not* to do when you are communicating:

- Tune out
- Be critical (or, if the situation requires you to be, do so tactfully)
- Overreact
- Listen only for the message
- Take too many notes
- Distract others
- Become distracted
- Avoid difficult material

Frame of reference – is "where are you coming from" or what is your mission and what is their mission. We discussed this in Section 6.5. An example of this creating barriers might be when a professor's frame of reference is to have students learn the course material and the students' frame of reference is to get a passing grade in the course. If the students want to learn the course material, barriers are lowered. To this end, the professor might try using a more stimulating and entertaining communication. In general, the professional needs to know the participants' frame of reference, understand that frame of reference, and respect them even if you do not agree with them.

One specific concept related to frames of reference that environmental managers should become aware of is collective cultural consciousness This concept refers to the feelings of a culture or mutually identifying group based on what that culture or group remembers (Rogers and Steinfatt 1999). A related concept is that of historicity; that is, the history and past relationship of a group, a site, and/or a management program and all stakeholders involved. General examples of grossly negative past contexts that influence collective cultural consciousness and historicity today in the US are slavery or the US government's treatment of Native Americans in the past, which have substantially shaped the relationships between these groups and the US government today. The past (and ongoing, though to a lesser extent) negative interactions between these cultural groups and the US government foment(ed) distrust and uncertainty that still persists in different forms and intensities.

Some cultures and groups have been communicated with poorly, and others quite sufficiently and clearly, regarding environmental situations, and the living and socially transmitted memory of what actually happened versus what they were told would happen can shift over time, but is not soon forgotten. Clearly, past injustices relating to environmental protection create a charged and complex barrier to communications today with environmental managers. Thus, environmental managers must understand the frames of reference of their stakeholders to effectively address these barriers and overcome them to further their management program.

Information overload – Information overload can be both a distraction and a hindrance to effective communication. Messages that are too dense either with technical language or detail can prevent the target audience from hearing and internalizing the primary message. The environmental manager should thus focus on clear and concise communication, placing particular emphasis on relevance to the stakeholder and any points of uncertainty. Stating points of uncertainty clearly and upfront can prevent all sorts of follow-on problems because you make clear where things might change or where an unexpected outcome might emerge, preparing people for that possibility. Being clear and concise holds true for public announcements or statements just as it does for the crafting of written materials and digital presentations.

These messages, while sometimes requiring voluminous and technical information to both achieve transparency and disclose data for stakeholders that need it for their actions, must be crafted such that the audience does not get overloaded when reading it. Using headings, bullets, and bold type to break up and or give direction to the message are helpful techniques. Another technique is to think about what you would say to a child to explain the situation. Importantly, this does not mean you are "dumbing down the message," but rather it forces you to not assume any prior knowledge on the part of the stakeholder and to be clear, concise, and effective in delivering your message.

Faulty communication skills – Faulty communications skills often are related to an environmental manager's ability, capacity, and practiced proficiency to encode and decode messages, particularly when it comes to face-to-face or public engagements. Some of these barriers can be physical in nature, such as a speech impediment, a habit of staring down at notes, or distractingly moving while presenting a message. Other barriers are more psychosocial or intrapersonal, such as what is commonly called "stage fright" or being "camera shy." And some people are just not as good as others at connecting with other humans. This does not make them a bad person just a limited communicator.

To overcome this barrier, the environmental manager must know their skills, strengths, and weaknesses, and those of their subordinates and superiors. Then the manager must either work to build and cultivate improved proficiency in weak areas or organize their taskings (e.g., who will deliver a message versus who will write it) based on their knowledge of these competences. The professional should anticipate these barriers and have the proper personnel or technology available to maximize the effectiveness of their communication in all circumstances to provide the most accurate, useful, and open messaging to their stakeholders. It is better to know and acknowledge your capabilities and limitations than to hide from them or ignore them, especially when communicating about issues as potentially charged as public health and protection.

6.8 Trust, Credibility, and Ethical Communication

The effective environmental manager must build credibility with and earn the trust of their stakeholders no matter who they are. To accomplish this first and foremost requires the environmental manager to communicate in an ethical manner. There are four basic principles of ethical communication (presented here in question form).

6.8.1 The Four Basic Principles of Ethical Communication

Is it Open? Communication must be open to all stakeholders. Openness equates to transparency, but also includes availability and understandability. Governing takes place in a glass house, and all activities of governance should be subject to the sunshine. Further, the public has the right to access government information. In fact, there are Sunshine Laws or Open Meeting Laws requiring government officials that are meeting to conduct the business of the people to do so in an open manner (with specific exemptions, e.g., closed door meetings on national security matters). For example, there is the federal Freedom of Information Act of 1967 (FOIA) which requires government to, upon formal request, provide documents to the public (US DoJ 2021). There are also state-level FOIA equivalents in every US state, albeit with varying rules, processes, and exemptions (NFOIC 2021).

As environmental managers it is interesting to note that Carson's *Silent Spring*, and later the Superfund Amendment and Reauthorization Act (see Chapter 4), addressed the importance of a community's "right to know," particularly about what they were being exposed to. These two documents, informal and formal, respectively, fundamentally changed public perceptions, awareness, and requirements for openness regarding environmental issues in the US (Dunn 2012; Konar and Cohen 1997). As we have said, there are several exemptions to the freedom of information laws discussed above, often due to one or more of the following considerations: national security; ongoing criminal investigations; proprietary protections; personnel and other privacy matters; and certain activities of agencies that regulate financial institutions. Barring these instances, environmental managers can and should seek to make open to the public all information that they generate – and bear in mind that if they do not it will likely become public information whether they intend it to or not.

Is It Accurate? Information provided in our communication must be factual; that is, based on the best available evidence, though recognizing that as methods and analytical processes change so too do results. Knowingly providing inaccurate information is lying, and the consequences of that for trust are almost always negative. Dishonesty is the quickest way to lose the trust and confidence needed between an environmental manager and the stakeholders they protect. Even when one unwittingly provides inaccurate information, credibility is damaged. It is always wise to remind stakeholders that "science changes" and that our decisions are based on the best science as we know it, but it can change. But if the mistake is from improper or insufficient monitoring or enforcement, the environmental manager is accountable for such shortfalls and cannot assuage this by blaming "the science."

Credibility is a priceless asset to scientists and officials. It is hard won and easily lost. Public perceptions may not always be correct in their portrayal of information received. These misperceptions can create difficulties and a sense of unfairness for the environmental manager. But regardless of the inequity here, the onus is on the manager to proactively navigate and manage their relationship with the public – that is why most environmental managers are also called public servants.

REAL-WORLD EXAMPLE 6.1 Overcoming barriers to communication

The most rewarding project I worked on as a state employee was soon after I was hired. IDEM and EPA were working together to investigate a messy corridor of sites with a commingled chlorinated solvent groundwater plume, multiple responsible parties, and several blocks of homes potentially affected by vapor intrusion. Although a few years later these sites became "mine," I didn't yet have the experience to be the risk assessor for the project. The neighborhood was predominately Spanish-speaking, and the project manager invited me along to translate. The delay in the project at this point was not being able to communicate with the residents to request access to sample the air in their homes and respond to their concerns.

Being an empathetic listener and good communicator is essential in environmental management. It's difficult to gain sampling access to people's homes when they have a range of unanswered concerns: (What is vapor intrusion? Who is going to pay to sample the air in my home? Will this be a headache for me to deal with if chemicals are detected and more sampling or remediation is needed? Am I going to get cancer?) When you show up in a neighborhood to talk to residents, you may be met with overconcern or lack of concern.

Communicating risk involves convincing some people that they should be concerned and others that they shouldn't overreact. Without putting in the effort to listen to and communicate with every stakeholder in this project, the residents either wouldn't have granted access to sample the air in their homes or would have been afraid of what they didn't understand, site delineation wouldn't have progressed, and it would have been difficult to identify sources of contamination and responsible parties to eventually remediate this corridor of sites. The most rewarding part was the cooperation and appreciation from the residents after we took the time to communicate effectively and involve them.

In working for the state, I have seen a variety of communication approaches – some more effective than others – to moving a project forward. The most efficient project managers proactively call or communicate directly with consultants and the rest of the project team (geologists, chemists, engineers, risk assessors, etc.). Others make slow

REAL-WORLD EXAMPLE 6.1 (cont.)

progress simply because they don't know how to communicate to manage their piece of a project. Reviewing a lengthy site report and only responding with a formal comment letter is usually not the best way to communicate with another human being.

Environmental management is accomplished through personal interactions. The best piece of advice I could give to new hires in environmental management is to consider your communication approach. Interactions on the site management level may seem small, but they all add up to determine how effective an agency is overall.

Stephanie Redick, MPA-MSES

Is It Fair? If information is provided to select audiences and not to all, the playing field is not level. If information is generated with one group of stakeholders in mind and not the needs or protection of all stakeholders, it is not fair. If information is shared in a way that is only accessible to or understandable by a portion of the stakeholders, it is not fair. If one stakeholder group or entity is given information before another, it is not fair. These sources of bias and resulting inequity impact human health and the environment on a broad scale but also affect non-health community matters such as property values, environmental service provision, and city planning decisions, among other factors. The environmental manager must always remember that "no one likes a cheater" and environmental management has a long history of environmental injustice (Barnes, Graham, and Konisky 2021; Cole and Foster 2001; Konisky 2015; Shrader-Frechette 2002, 2007).

REAL-WORLD EXAMPLE 6.2 Ten ways to lose trust and credibility (US EPA 1996)

1. Don't involve people in decisions that directly affect their lives. Then act defensive when your policies are challenged.
2. Hold onto information until people are screaming for it. While they are waiting, don't tell them when they will get it. Just say, "These things take time," or "It's going through quality assurance."
3. Ignore people's feelings. Better yet, say they are irrelevant and irrational. It helps to add that you can't understand why they are overreacting to such a small risk.
4. Don't follow up. Place returning phone calls from citizens at the bottom of your "to do" list. Delay sending out the information you promised people at the public meeting.
5. If you make a mistake, deny it. Never admit you were wrong.
6. If you don't know the answers, fake it. Never say "I don't know."

REAL-WORLD EXAMPLE 6.2 (cont.)

7. Don't speak plain English. When explaining technical information, use professional jargon. Or simplify so completely that you leave out important information. Better yet, throw up your hands and say, "You people could not possibly understand this stuff."
8. Present yourself like a bureaucrat. Wear a three-piece suit to a town meeting at the local grange and sit up on stage with seven of your col leagues who are dressed similarly.
9. Delay talking to other agencies involved or other people involved within your agency so the message the public gets can be as confusing as possible.
10. If one of your scientists has trouble relating to people, hates to do it, and has begged not to, send him or her out anyway. It is good experience.

Is It Useful? If the information is not provided in a timely manner, or is delivered in such a way that the other participants do not have reasonable access to it, it is not useful. If the information provided to the public is poorly developed and delivered, it is not useful. If the information is not relevant to the situation or issue which it is intended to address, it is not useful. While these and other types of issues with environmental management information generation might seem common sense, they unfortunately commonly occur. Producing information is not an end in itself; informing decisionmaking such that people can best navigate an issue and aiding their understanding is the goal. As an example, environmental agencies are required to provide notices of permit applications and sometimes enforcement actions to the public. Often these notices are provided such that the public does not have time to comment or is not aware of their publication, and the information becomes not useful, among other issues. Unless information is readily available and distributed and enables timely, informed, and effective decisionmaking and understanding, it is not useful – or at least is not maximizing its potential usefulness.

These four concepts are considered and built into most rules and processes that environmental agencies must follow. For example, many agencies today have strict notification procedures regarding when and where a notice is posted and how it is to be distributed. That said, it is important that while government must provide information, often it must be requested (and sometimes paid for) by interested parties, such as when making a FOIA request. Not all materials, be it data collection and analysis or internal memos of environmental management agencies, are distributed to the public, as this would lead to the information overload that we discussed earlier – not to mention the efficiency-harming burden it would place on managers.

When human health and the environment are in danger, and therefore the work of environmental managers "becomes" newsworthy, investigative reporters from mass media outlets are often there to let the public know whether we are doing our jobs in a competent

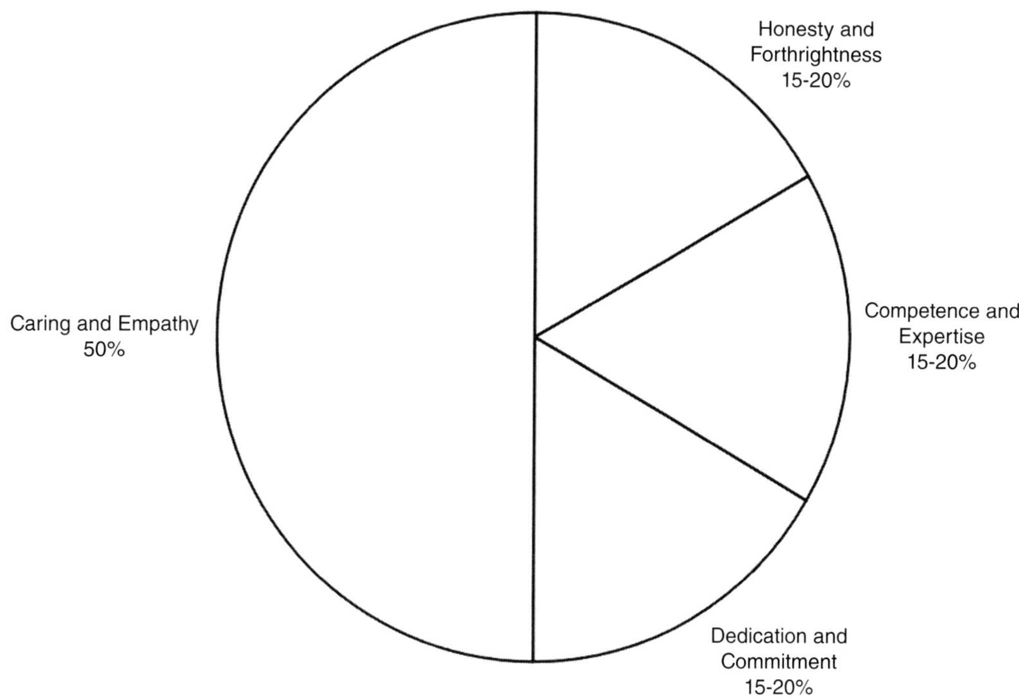

Figure 6.3 Factors that inspire trust and credibility (TPC 2015).

and ethical manner (Figure 6.3). Thus, environmental managers should adopt and adhere to a personal, if not a professional, code of ethics that requires all communication to be open, accurate, fair, and useful. Most agencies explicitly state their code of ethics regarding information generation and distribution, and the environmental manager should be familiar with that of whatever agency or institution they work for.

REAL-WORLD EXAMPLE 6.3 American Society for Public Administration's Code of Ethics

1. **Advance the public interest.** Promote the interests of the public and put service to the public above service to oneself.
2. **Uphold the Constitution and the law.** Respect and support government constitutions and laws, while seeking to improve laws and policies to promote the public good.
3. **Promote democratic participation.** Inform the public and encourage active engagement in governance. Be open, transparent and responsive, and respect and assist all persons in their dealings with public organizations.

REAL-WORLD EXAMPLE 6.3 (cont.)

4. **Strengthen social equity.** Treat all persons with fairness, justice, and equality and respect individual differences, rights, and freedoms. Promote affirmative action and other initiatives to reduce unfairness, injustice, and inequality in society.
5. **Fully inform and advise.** Provide accurate, honest, comprehensive, and timely information and advice to elected and appointed officials and governing board members, and to staff members in your organization.
6. **Demonstrate personal integrity.** Adhere to the highest standards of conduct to inspire public confidence and trust in public service.
7. **Promote ethical organizations**. Strive to attain the highest standards of ethics, stewardship, and public service in organizations that serve the public.
8. **Advance professional excellence**. Strengthen personal capabilities to act competently and ethically and encourage the professional development of others.

6.9 Public Participation

The participation of the public in the development of environmental management programs is critical in today's world. Co-production of solutions through public participation ensures stakeholder integration and buy-in, factors critical for the environmental manager to do their primary job: to manage. Public participation is part of how environmental managers co-produce ethical and effective solutions to environmental situations.

Currently, most public participation with government agencies, particularly at the federal and state level but sometimes at the city level as well, is facilitated by professional contractors – usually working out of the appropriate office or section (air, waste, water, etc.) – and in consultation with the agency's external affairs team. "Public Participation is any process that works to understand people's values and uses their input to make better decisions" (J. Godec, pers. comm. 2021). According to the EPA:

> Public participation can be any process that directly engages the public in decision-making and gives full consideration to public input in making that decision. Public participation is a process, not a single event. It consists of a series of activities and actions by a sponsor agency over the full lifespan of a project to both inform the public and obtain input from them. Public participation affords stakeholders (those that have an interest or stake in an issue, such as individuals, interest groups, communities) the opportunity to influence decisions that affect their lives. (US EPA 2014b: 3)

6.9.1 Fundamental Understanding of Public Participation

The EPA provides further guidance, what it calls its "Fundamental Understanding of Public Participation Principles," by establishing the five principles for building the trust and credibility necessary for public participation. The five principles for creating understanding through public participation are:

- Clear, defined opportunity for the public to influence the decision
- Management commitment to fully consider public input in decisionmaking
- Engagement of the full range of stakeholders from the community, including vulnerable populations and marginalized communities
- Focus on building relationships between and among stakeholders
- Creating and sharing truthful, comprehensive, and clear information." (US EPA 2014b: 89)

Members of the International Association of Public Participation (IAP2), and many of the agencies they work with, use the Public Participation Spectrum as their formal process to achieve the goal of public participation. This spectrum has five components that form a process that these participation facilitation professionals employ to help determine the level of influence that the public might have in a decision. The five components are: inform; consult; involve; collaborate; empower. Each component, generally but not always proceeding in sequential order, has a goal and a promise to the public:

Inform

Goal – To provide the public with balanced and objective information to assist them in understanding the problem, alternatives, opportunities and/or solutions.

Promise – We will keep you informed. [In and of itself simply ***informing*** the public is not public participation because no influence is suggested. But you can't engage the public at any level without adequately and successfully ***informing*** them (J. Godec, pers. comm. 2021).]

Consult

Goal – To obtain public feedback on analysis, alternatives and/or decisions.

Promise – We will keep you informed, listen to and acknowledge concerns and aspirations, and provide feedback on how public input influenced the decision. We will seek your feedback on drafts and proposals.

Involve

Goal – To work directly with the public throughout the process to ensure that public concerns and aspirations are consistently understood and considered.

Promise – We will work with you to ensure that your concerns and aspirations are directly reflected in the alternatives developed and provide feedback on how public input influenced the decision.

Collaborate

Goal – To partner with the public in each aspect of the decision including the development of alternatives and the identification of the preferred solution.

Promise – We will work together with you to formulate solutions and incorporate your advice and recommendations into the decisions to the maximum extent possible.

Empower

Goal – To place final decisionmaking in the hands of the public.

Promise – We will implement what you decide.

The Public Participation Spectrum can be employed by environmental managers just as it is by facilitation professionals. And the spectrum is neither unlike nor divergent from the other concepts and frameworks we have presented, such as open, accurate, fair, and useful information production. The environmental manager will have to choose which process best fits their own needs and those of their stakeholders, and also conforms to the standards of the agency or organization that they work for. In the end, each of these concepts and frameworks is only a guide, not a rigid mandate. Each case, each manager, and each set of stakeholders is different in any given situation and thus these tools should be tailored to the contexts at hand.

What we have been defining and explaining with regard to ethical communication and public participation is meant to provide the knowledge and skills such that the environmental manager, who also understands complex and volatile issues and trends (recall Chapter 2), can co-produce ethical solutions with stakeholders. We conclude this chapter with a discussion of why co-production and accompanying public participation drive our program, resource and political management.

6.10 Risk Communication

Most people do not have a shared understanding of what risk is. The environmental manager, as an expert in both science and policy, generally views risk differently than the public. Environmental managers are trained to deal with risks objectively and dispassionately whereas the public tends to view risks personally and sometimes passionately (J. Godec, pers. comm. 2021). While the focus of this section is on risk communication, as opposed to risk assessment, it is important that we first create a very general understanding of what risk is, fundamentally, for the environmental manager.

Risk is broadly defined as "the possibility of loss or injury" (Merriam-Webster 2021f), or more specifically, the likelihood of harm from a specific hazard (TEF 2020). Estimating risk requires assessing the probability that a hazard will occur or that a person will be exposed to a hazard and the potential or likely effects of that hazard (Kammen and Hassenzahl 2001). The definition risk analysts most often use is $Hazard \times Exposure = Risk$. The level or estimate of risk is then the result of the possibility of occurrence and the likely impacts (i.e., damage or harm). Risk evaluations, often called quantitative risk assessments (QRAs), are used to estimate the potential and likelihood of harm to human health (US EPA 2014a). Quantitative risk assessments model and estimate risk and are often based on analyses of samples from the environment (e.g., soil samples to analyze for heavy metal contaminants), determined health effects of specific pollutants, expected levels of exposure, and population demographics. The determined level of risk to human health is then compared to the appropriate health standard to determine whether the risk exceeds or is within the regulatory requirement.

REAL-WORLD EXAMPLE 6.4 Quantitative risk assessments: formulas, definitions, and application

Risk Assessment Process

The US EPA risk assessment process consists of four steps: (1) hazard identification; (2) exposure assessment; (3) dose-response assessment; and (4) risk characterization (US EPA 2014a). Hazard identification is the process of determining whether exposure to a stressor increases the likelihood of a specific adverse health effect. Exposure assessment is the process of determining the quantity of a pollutant that people are exposed to, and the duration of exposure. Dose-response assessment is the process of determining the health problems associated with different exposure rates. Risk characterization is estimating the extra risk of health problems for the exposed population.

Average Daily Intake

The average daily intake (ADI) – sometimes referred to as the average daily dose – is the average intake of a particular pollutant from soil via one of three intake pathways, which are ingestion, inhalation, and dermal intake. ADIs are calculated using exposure parameters that are differentiated by age, body size, respiration rates, and other factors. As a result, children have different exposure parameters than adults. Standardized exposure parameters are developed by different environmental regulatory agencies for modelling purposes.

The ADI equations include three intake pathways: ingestion, inhalation, and dermal. The ingestion ADI equation is:

$$ADI_{ing} = \frac{C \times IR \times EF \times ED \times CF}{BW \times AT}$$

The inhalation ADI equation is:

$$ADI_{inh} = \frac{C_s \times IR_{air} \times EF \times ED}{BW \times AT \times PEF}$$

And the dermal ADI equation is:

$$ADI_{dsr} = \frac{C_s \times SA \times FE \times AF \times ABS \times EF \times ED \times CF}{BW \times AT}$$

REAL-WORLD EXAMPLE 6.4 (cont.)

The parameters used in these formulas are defined in the table.

Parameter
Body weight (BW)
Conversion Factor (CF)
Dermal Absorption Factor (ABS)
Dermal Exposure Ratio (FE)
Exposure Duration (DE)
Exposure Frequency (EF)
Ingestion Rate (IR)
Inhalation Rate (IRair)
Particulate Emission Factor (PEF)
Skin Surface Area (SA)
Soil Adherence Factor (AF)
Average Time (AT)-carcinogens
Average Time (AT)-non-carcinogens

Hazard Quotient

Non-carcinogenic hazards are characterized by the hazard quotient (HQ) (US EPA 2001, 2014a). HQ is a unitless measure that is expressed as the probability of an individual suffering an adverse health effect. It is defined as the quotient of ADI or dose divided by the toxicity threshold value, which is referred to as the chronic reference dose (RfD). The HQ formula is:

$$HQ = \frac{ADI}{RfD}$$

The HQ is developed for each of the three pathways if there is an RfD for that metal and pathway (not all metals are transmitted via all three pathways).

Hazard Index

The hazard index (HI) is the sum of all HQ_ks for a given sample site (see formula below) (US EPA 2001, 2014a). For n number of pollutants, the non-carcinogenic effect on the population is as a result of the summation of all the HQs – i.e., each pathway – due to individual pollutants. If the HI value is less than 1, the exposed population is unlikely to experience adverse health effects. If the HI value exceeds 1, then there may be concern for potential non-carcinogenic effects.

$$HI = \sum_{k=1}^{n} HQ_k = \sum_{k=1}^{n} \frac{ADI_k}{RfD_k}$$

REAL-WORLD EXAMPLE 6.4 (cont.)

Carcinogenic Risk

For carcinogens, the risks are estimated as the incremental probability of an individual developing cancer over a lifetime as a result of exposure to the potential carcinogen (US EPA 2001, 2014a). Risk is a unitless probability of an individual developing cancer over a lifetime. ADI_k and CSF_k are the average daily intake and the cancer slope factor (CSF), respectively, for the kth pollutant, for n number of pollutants. The CSF converts the estimated daily intake of the pollutant averaged over a lifetime of exposure directly to incremental risk of an individual developing cancer. The total excess lifetime cancer risk for an individual is finally calculated from the average contribution of the individual pollutants for all the pathways. $Risk_{(ing)}$, $Risk_{(inh)}$, and $Risk_{(der)}$ are risks contributions through ingestion, inhalation and dermal pathways. The equation for the carcinogenic risk via each pathway is:

$$Risk_{pathway} = \sum_{k=1}^{n} ADI_k CSF_k$$

The equation for total carcinogenic risk, i.e., the sum of each individual pathway and pollutant for a given sample, is then:

$$Risk_{(total)} = Risk_{(ing)} + Risk_{(inh)} + Risk_{(der)}$$

The total carcinogenic risk is estimated for both children and adults. The interpretation of the value is the probability that an individual will develop cancer in their lifetime due to exposure to the pollutants in the soil sampled. This probability can then be used to develop the ratio of people likely to develop cancer – e.g., "1 in X number of people will develop cancer."

It is important to note that quantitative risk estimates are just that, estimates. And they often are not holistic estimates because they do not measure all the forms of harm that an environmental issue can potentially inflict on a community (Kuehn 1996; Marcantonio, Field et al. 2021; Ranco et al. 2011; Shrader-Frechette 2002; Solomon et al. 2016). For example, QRAs produce an estimate of health risk to an individual, such as what the risk is of a person experiencing cancer in their lifetime due to exposure to lead in their drinking water. But there are other contexts of risk, particularly what is at-risk and pathways of risk, that must be considered when making risk-based decisions or assessments. The person at risk of lead in their water, while an individual, has loved ones who would be harmed if their health was jeopardized by unsafe water. The knowledge that your water could be contaminated with lead, because for example there are still over 6.1 million lead service lines in the US today that provide water to approximately 22 million people, can also produce harm through anxiety, stress, or other effects that are not accounted for in most risk models.

The limitations of QRA and other risk estimates and understanding how the assumptions of those procedures match or diverge from the perceptions, values, and concerns of stakeholders are important to consider when making risk-based decisions, communicating risk, and understanding what is at-risk.

6.10.1 Categories of Environmental Risk

Environmental risk is often separated into two categories for regulatory purposes: risk to human health; and risk to the ecological resources. This categorization results from and corresponds to the primary and secondary health-based standards in environmental laws (Barnes, Graham, and Konisky 2021; O'Leary et al. 1999), that is, protecting human health first (primary standards) and then the environment (secondary standards). The EPA uses a process (see Figure 6.4) that combines environmental and social science, engineering, and economics to determine the impact of pollution or hazardous substances on human health and the environment; to assess the initial and residual environmental risk based on our environmental management programs; and to consider those risks versus benefits to our society. The technology available to mitigate pollution, the monetary cost to abate or reduce pollution, and the costs and benefits to society of the product or service produced with pollution as a byproduct are all considered when establishing the primary and secondary health standards (US EPA 2001, 2014a). Risk assessment and analysis is a critical skill for an environmental manager. But, just as important is the ability of the environmental manager to communicate risk to the public.

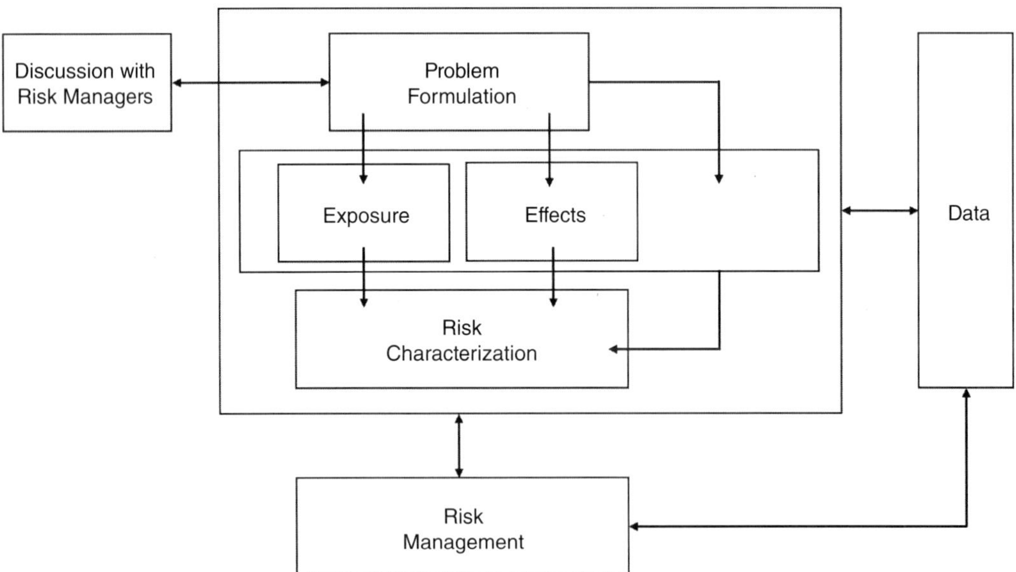

Figure 6.4 EPA's risk paradigm.

6.10.2 The Basic Reasons for Doing Risk Communication

There are three basic reasons why environmental management programs incorporate risk communication: (1) to warn the public about risk; (2) to reassure the public about a specific risk; (3) to facilitate stakeholder decisionmaking regarding or in response to risk prioritization. Warnings are used when environmental managers recognize that there is a risk to a community. Whether that risk is reasonable or not, and may be imminent or not, the community has the right to know what that risk is and what the possibility of it occurring are – and the environmental manager has the mandated responsibility to communicate that risk. But also, the environmental manager wants the community to pay attention! This is sometimes referred to as "Precautionary Advocacy," meaning the environmental manager preempts conflicts by proactively engaging, warning, and providing relevant information to the public about a risk before or as it transpires, as opposed to retroactively addressing it after damage or harm is done. Initial warning messages are best kept short (7–12 words), interesting, and memorable, and the communicator should remain "on message" and repeat the message to maximize public attention, awareness, and engagement (J. Godec, pers. comm. 2021).

Communicating reassurance, as opposed to a warning, is more about addressing outrage and/or crisis communication to assure the community that the risk is recognized, and that the crisis is being managed (TPC 2015). Augmenting the definition of risk that we discussed earlier, crisis communications professionals define public perceptions and understanding of risk as *Risk* = *Hazard* + *Outrage* (Sandman 1993). John Godec, co-founder of The Participation Company and the International Association of Public Participation, suggests that when dealing with outrage or trying to get the outraged entity to "calm down" the risk communicator should: (1) listen actively and empathetically; (2) allow people to vent; (3) apologize for mistakes and past inadequacies; (4) stake out the middle, i.e., do not get sucked into a polarized debate. Outrage normally involves and stems from stakeholders' perceptions of control or voluntary risk vs. lack of control or involuntary risk. When stakeholders feel not in control of their own safety or that they have been involuntarily put at risk, outrage is likely to ensue. The specific perceptions and concerns of stakeholders regarding control and voluntary participation should be addressed in the environmental manager's risk communication and through public participation opportunities. Engaging the public and allowing the community to prioritize their decisions empowers the community to make the decisions most fit to meet their needs and facilitates co-production of programs to implement those decisions.

6.10.3 Basic Problems for Risk Communication

Environmental managers often face two problems when attempting risk communication: the need for more statistical evidence; and the need to decode the technical messages regarding statistical evidence to the layman (O'Leary et al. 1999). Whether due to resource scarcity or simply an unexpected or unanticipated risk, environmental managers often find that they do not have the statistical evidence of a risk that they would prefer to have to make strong statements about an issue. The lack of evidence forces the manager to explain the uncertainties surrounding and issue and why they cannot provide more certain answers. When explaining uncertainties, or even when sufficient evidence is in hand, the manager must decode the technical jargon of quantitative risk analysis to provide a clear and concise message that people can understand but that also cannot easily be misinterpreted by the receiver such that later they may feel they were lied to or misled.

6.10.4 How Communities Perceive Risk

How a community perceives risk will determine how they will act and their willingness to co-produce a solution with an environmental manager. Thus, the environmental manager must first understand and consider some basics regarding risk perception: risk means different things to different people; people tend to ignore or discount discrete, familiar, voluntary, and low-probability risks; feelings of control and opportunities for participation influence attitudes to acceptability of the risk; people evaluate risk as members of a community; and trust is an important influence on risk perceptions (O'Leary et al. 1999). Looking more closely at these perceptions:

- Risk means different things to different people. Clearly, some individuals will jump out of a plane while others would not. Some people will smoke or tolerate being in proximity to smokers while others will not, and some people will choose to live next to a freeway with increased exposure to PM2.5 while others will object to such a freeway being constructed in their community. Each person's understanding, preferences, and willingness to accept or decline risk are different and often based on their individual knowledge, experiences, and worldview.
- People tend to ignore or discount discrete, familiar, voluntary, and low-probability risks. One example students might relate to is drinking alcohol. It is quite normal on university campuses and voluntary. Yet, overdoses of this toxin contribute to thousands of deaths and sexual assaults on students every year (NIAAA 2021). Another example is that of voluntary pesticide exposure by millions of American households even when no pests are present in their homes or schools (HUD 2006).
- Feelings of control and opportunities for participation influence attitudes to acceptability of the risk. Recall the modified definition of risk: *Risk = Hazard + Outrage.* When entities feel they have no control and are not included in decisions that affect their lives they become outraged, and the negative perception of the risk(s) is increased. Conversely, if entities are given the opportunity to co-produce solutions, they are often more accepting of risk, or at least more knowledgeable of the risks they are taking.
- People evaluate risk as members of a community. As with how certain communities perceive vaccines or with how a farming community perceives pesticides, risk perception and evaluation is culturally mediated. Risk perceptions are influenced by the values and norms of a community that developed and evolved over time and transmitted between

individuals and even generations. In this light, current risk perceptions and preferences are cultural artifacts with a historicity tied to them.

- Trust is an important influence on risk perceptions. Environmental managers make decisions and communicate information that literally involves and affects the health and well-being of their stakeholders. How much an individual, group, or community trusts the individual environmental managers responsible for their area, and the agency or environmental institution that the manager works for, directly affects how they will perceive risk and their confidence in the management programs relevant to them. If they are distrusting of the manager or the institutions, they are much more likely to feel at-risk, and vice versa.

Risk communication, as a subfield of communication, has developed many tools to assist the environmental manager's assessment and decisionmaking regarding warning, reassuring, and helping communities prioritize for their decisionmaking. One particularly useful tool is the Risk Perception Matrix (Figure 6.5) to help the environmental manager decide what approaches to take, and when, regarding their risk communication planning. The matrix has four quadrants representing high- or low-risk situations paired with high and low outrage (i.e., concern) of the public about that risk. The environmental manager can use this tool to frame or locate a particular environmental risk with regard to actual risk and perceived risk to determine their best way forward in communicating about that risk to the public. For instance, when the potential risk is high, but public concern, emotion, and/or fear of that risk

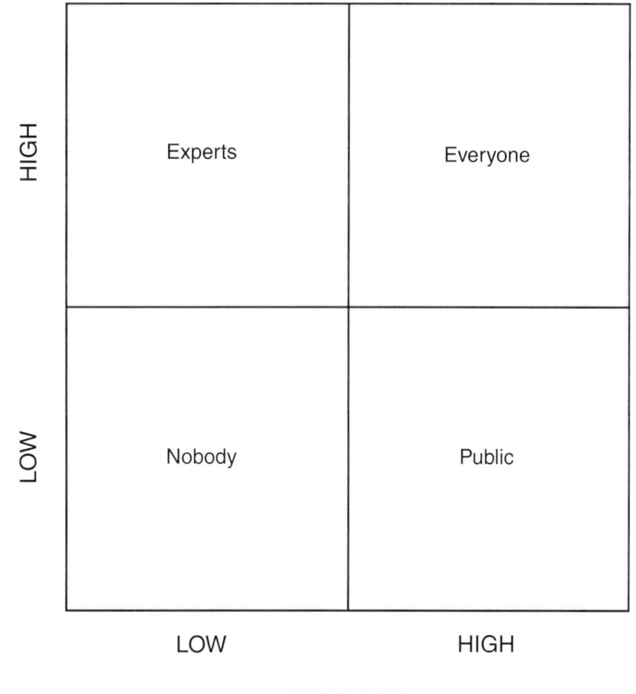

Figure 6.5 Risk perceptions: who cares?

is low, managers will be concerned even though the public is apathetic (J. Godec, pers. comm. 2021). Recognizing that public awareness and concern is low suggests that the risk communicator will have to focus more program resources to communicate with the community and utilize the appropriate message to warn the public and aid their decision-making. Conversely, there are circumstances when risk is low but concern, emotion, and/or fear is high, and the public will be concerned and managers will be apathetic (J. Godec, pers. comm. 2021). Again, in this instance the environmental manager communicating risk would want to concentrate their message on reassuring the community to inform their decisionmaking.

6.10.5 Steps for Successful Risk Communication

Once they have determined the level of risk at hand and the "pulse" of public perceptions toward that risk, the environmental manager will finally have to develop their risk communication plan and program. We have presented considerations that must be taken into account and types of risk communication (that is, to warn, reassure, or inform decisionmaking), but ultimately the environmental manager has to develop their risk communication program for the issue at hand. To do so, there are five principles to guide them in their risk communication program construction and to successfully implement the plan (Garnett 1992). These steps are just a helpful tool to manage with, condensing all that we have discussed in our risk communication discussion section so far into a concise, easy to recall package (think: CAUSE). The five principles are:

- Credibility – establish why you should be paid attention to, rely on the trust you have developed with the community or include ways to build trust if it is in doubt or damaged, and make clear the tools, resources, and evidence that you have available and allocated to your program for the issue at hand.
- Awareness – embrace and directly and widely address the risk at hand in an open, fair, accurate, and useful manner such that the message is clear and only interpretable as you intend it to be.
- Understanding – understand the situation, be understandable, and co-produce a mutual understanding with your stakeholders when communicating risk.
- Solutions – when possible, have ready and offer to stakeholders (as opposed to unilaterally implement) co-produced options for solving the problem.
- Enactment or Implementation – plan for, commit to, and communicate how you will implement the solution to the risk at hand in an accountable manner.

Note that the last step, implementation, or delivery of the co-produced solution, is required. The environmental manager, to be successful, cannot just communicate effectively, they must also execute their function as a manager. Our next module and subsequent chapters are intended to show the budding environmental manager the skills needed to be an effective, efficient, and strategic environmental manager.

Ultimately, the Risk Perception Matrix and the five steps for risk communication messaging, along with the other concepts that we presented earlier and many more that have been developed and taught elsewhere by researchers and practitioners of environmental management, are just tools available to help the environmental manager in their communication programming. In the end, a tool is only effective if employed appropriately. Thus, it is up to the environmental manager to assess situations and employ the right tool for the job, and to do so proactively, to achieve their mission of protecting human health and the environment.

6.11 Conclusion

Ethical environmental management and communication are often the most difficult aspects for the environmental management profession to navigate. More often than not, varying ethical considerations have to be weighed and balanced such that not every stakeholder or impacted party is going to feel that a program is in fact ethically sound or that it has been effectively communicated to them in terms they are comfortable with and fully understanding of. The pitfalls for the environmental manager are many. Only through iterative engagement and relationship development over time can the environmental manager build the trust and credibility with relevant stakeholders such that they are given the benefit of the doubt and create a buffer that allows for discussion to happen before tensions mount when an unexpected or unaccounted-for issue arises. The building of relationships and credibility for the environmental manager is a laborious process but invaluable for effective management.

REAL-WORLD EXAMPLE 6.5 Taking risks and bureaucratic paralysis

Environmental managers can experience a problem of pandemic proportions, best described as "bureaucratic paralysis." When fresh out of college you are ready to conquer the world. But is the world ready for your fresh ideas and unbridled enthusiasm? Once hired you learn some seasoned managers are like briars digging into your legs and slowing you down. Some get comfortable doing things the way they have always done them and resent the passionate upstart, especially if change is involved.

It is human nature to be "risk adverse." Environmental issues are usually complex, and actions come with unintended consequences. Managers do not like being responsible for negative outcomes, and their fear can cause paralysis. Most everything good that has been accomplished in this world came because someone was willing to take a risk. Our society is losing this kind of leadership. When risks are encountered, we tend to punt to upper management or to a committee, resulting in bureaucratic

REAL-WORLD EXAMPLE 6.5 (cont.)

paralysis. Doing nothing seems safe, but we fail the environment when we fear taking action to correct problems. Don't wait until situations become dire, or disaster strikes, before taking corrective action.

Flour mills with water-powered wheels were common across the country more than a century ago. Dams were built to pond the stream and route the water to power the wheel. Many caused problems, such as the idyllic millpond and Zarelli dam in western Washington, which blocked Coho salmon from migrating up Clover Creek to spawn. Decades after the mill ceased to operate, the dam and mill wheel remained in place, blocking salmon. Nobody was willing to solve the migration problem because the area residents wanted to keep the beautiful pond. Biologists did not want to risk upsetting the apple cart, especially since many believed it was too late to save the salmon run. One day tragedy struck, and a young boy lost his life playing at the wheel.

Tragedy was a motivator for taking risks that should have been taken prior to the awful event. The Pierce Conservation District (PCD) decided it was time to remove the dam and all traces of the mill and restore the ability for salmonids to migrate up Clover Creek. At first the project was met with resistance every step of the way. "No" is often the answer to any question when change is involved. Two streamside homeowners' associations fought the dam removal permit because they didn't want to lose the pond. Area scientists came up with many reasons why the project would fail, including that the streambed would leak water and go underground when the dam was removed, allowing years of accumulated decaying matter to be scoured from the bottom.

PCD accepted the challenge and risk. They held community meetings to showcase a design of three successive weirs that would maintain the pond yet allow salmon to easily negotiate the weirs at all water levels. They met individually with those who resisted, and one by one provided solutions to every objection. Bentonite was strategically placed to seal the streambed, and the "ugly" concrete weirs were carefully textured to promote algae growth for a natural appearance. Ingenuity and compassion eventually won community support, and the dam was removed.

Shortly after the project was completed, neighbors lined the banks with their lawn chairs and watched the return of adult Coho salmon sliding over the weirs on their way to spawn.

Don't be risk-adverse. Don't let bureaucratic paralysis allow you to fail the environment. Doing nothing is often a more damaging risk than doing something.

Marc Marcantonio,
Former Fisheries Biologist and Salmon Habitat Recovery Manager, Pierce Conservation District

6.12 End of Chapter Questions

1. Why is it critical that environmental managers hone their communication skills?
2. Define communication per Rogers.
3. Name the four concepts that define ethical communication.
4. What is Public Participation?
5. According to O'Leary, what are the three sources of environmental conflict?
6. What is the technical definition of risk?
7. What is the social definition of risk?
8. What are the five basic steps of risk communication?
9. List the three basic reasons why environmental managers communicate risk.
10. What are the two major difficulties for communicating risk?
11. How would an environmental manager lose trust and credibility?
12. Name exceptions to FOIA.
13. What is the name of the set of laws that requires the notification of meetings where public business is conducted?
14. What is an ombudsman?
15. What is a PIO?
16. You are the head of the Stormwater Runoff Compliance unit of your state's environmental agency. One of your jobs is to implement rules meant to reduce stormwater runoff (silt and toxics) into the lakes and streams of the state by the construction industry. Today you are meeting with 30 of the leading developers in the state. These men are wealthy, powerful, over 40, mostly white, believe they are helping with the economy of the state and are tired of environmentalist, "no growth" people slowing progress. You want to communicate the following message: "Developers can gain a better image as stewards of the land, and actually reduce their costs, if they would begin to use some new technologies designed to reduce stormwater runoff."
    LIST & DESCRIBE Garnett's barriers to your communication with this audience.
17. What kind of program, resource and political management would you need to establish a "co-production environmental ethic" with a community concerned about the imminent siting of a hazardous waste facility near their neighborhood? (Hint: Think about what environmental issues could be present, the stakeholders involved, specifically who in your agency needs to be involved, and what should be the result of this effort.)
18. What kind of program, resource and political management would you need to establish a "co-production environmental ethic" with a community concerned about the management of toxic materials leaking from a copper mine? (Hint: Think about what environmental issues are present, the stakeholders involved, specifically who in your agency needs to be involved and what should be the result of this effort.)

7 It Begins with a Plan: Strategic Planning and Diffusion of Innovations

A goal without a plan is just a wish.

Antoine de Saint-Exupéry, *c.* 1940

Skill: *Strategic planning and diffusion of innovations.* The effective environmental manager must be able to conduct strategic planning and be able to diffuse innovations into the communities and regulated entities that they work with. Strategic planning is a skill not often taught to, yet often expected of, managers. And getting people to try new ways of doing this is not easy and requires effective planning to accomplish. You can develop the most effective solution possible for a problem but if you cannot diffuse it, i.e., get stakeholder buy-in, and do not create a plan to support it – logistically, operationally, or otherwise – then it will not work.

7.1 Introduction

Implementation is delivering on your program goals. When building your program implementation plan to achieve your stated goals, the five issues – Accountability, Ecosystem Management, Environmental Justice, Sustainable Development, and Unfunded or Underfunded Mandates – common in environmental management must be explicitly considered. Your professional performance is based on your ability to effectively design and implement your plan and to efficiently achieve the aims of your plan. By conducting a thorough planning process that considers the five issues, and includes integration of co-production, ethical communication, and the other factors we have discussed in previous chapters, the environmental manager will substantially increase the likelihood of achieving their program goals and ultimately fulfilling their mission.

7.1.1 The Three Questions That Must Be Asked *and* Answered before You Begin

To quote an often-screamed military adage, "Prior Proper Planning Prevents Piss Poor Performance" (the Seven P's). Some people use the term Six P's and leave out the term "proper." However, not all planning is effective and useful planning, whether because it is not detailed and actionable, is of incorrect scope, or has other common issues – so proper planning is needed, not just any planning. By conducting a planning process and building a plan, the environmental manager will be able to sufficiently answer the three questions which must be asked and answered in order to successfully implement any program (Starling 2005):

What action is to be taken? This seems simple, yet often managers can define their goals but not what to do to accomplish them. Or their goals are too broad and undefined to be able to say and evidence, at the end of their program, whether they have actually achieved their goal or not. In fact, many programmatic failures begin with the manager producing the wrong answer to this question and they then waste much of their programmatic resources aiming at the wrong or an unachievable goal. The action to be taken should be detailed enough that a subordinate or a superior could read it and know what the intent and purpose of the goal are, and it should be measurable in some way. Because the environmental manager has considered the five issues when developing their plan, they should also be able to answer, for example, how the action they are taking impacts disenfranchised communities or future generations.

Who will take this action? The answer to this question not only addresses what person or organization will implement the plan but, more importantly, what person or organization has the authority and responsibility to take action. Environmental managers are often asked to solve problems they might be willing and able to address but are not allowed to take action on because they lack the formal authority. This lack of authority is sector-related – i.e., relates to whether they are public, private, or non-profit environmental managers – regarding company policy, advocacy boundaries, and environmental laws, as well as other limitations from jurisdictional or other border boundaries. But most importantly, identifying who or which agency must take the action, and explicitly notifying them and tasking them with that action, makes them unambiguously responsible for carrying out that action. It removes the ability for a person or agency to divert responsibility by formally tasking them with an action (and then, of course, the effective environmental manager must plan to support and empower them to take the action).

Do they have the capacity (resources) to take action? The environmental manager needs to consider whether the entity authorized and tasked to take action has basic resources – funding, time, political will, etc. – such that they can satisfactorily achieve the goal they expected to accomplish. This question is about administrative commitment and empowerment. As a former career military officer, bureaucrat, and financial management professor aptly put it, "Unless Congress allots your program a line item in the budget, they are blowing smoke!" (F. Sackton, pers. comm. 1987). This sentiment addresses the unfortunately common issue of unfunded and underfunded mandates.

When politicians or CEOs at any level state that something is a priority, the environmental manager must ask (out loud or to themselves) and be able to answer, "where's the money?" As well, the environmental manager must ask and answer questions about other resources needed by the authorized entity to accomplish its task. For example, does the person or organization have enough clout to conduct the action? Do they have the programmatic bandwidth and internal resources to accomplish the task, or is the expectation to "do more with less"? And does the higher authority requiring this action really have your back in terms of political will and the will to see the task through?

In other words, the "resources" necessary to accomplish the task can be broad-reaching, hence the need for a thorough examination of them through a systematic planning process. It is not only bad form but also bad management and leadership to assign someone a task that they cannot accomplish given their resources. Their failure is then really your failure to plan. This chapter will discuss the concept of strategic planning by illustrating three types of plans: the strategic plan, the strategic communication plan, and the diffusion plan.

7.2 Strategic Planning

Strategic planning was originally a military concept, which has since been adopted and widely employed by the management community (USMC 2018). Today there is a multibillion-dollar industry around strategic planning facilitation and consulting (CG 2021), a fact that hints at the inadequacy or absence of strategic planning training and capacity within many organizations. Developing a strategic plan is intimidating to environmental managers entering the profession. Often professionals are required to develop and present plans on a regular basis without understanding why they are important because in many instances they are part of, and treated as, a formal process of "filling in the blanks." Further, many employees have not been educated on what strategic planning is, its power, and its ability to help ensure your success as a manager. In this section we will address strategic planning as a simple and familiar concept, then increase the complexity by applying the framework to a familiar topic, and then conclude with more formal analytical and planning tools.

As most people have experience with both successful and unsuccessful journeys and the failures were likely related to planning issues, thinking about traveling is a familiar and appropriate way in to thinking about strategic planning. A strategic plan is a "roadmap" of the journey a traveler must take in order to reach their destination. Like any well-done roadmap, it not only indicates the route from point A to point B but also accounts for the expected mileage, speed limits, rest stops, gas stations, etc. such that the traveler can estimate the costs of reaching the destination, any alternative routes that can be taken, which mode of transportation would be most efficient and safe, and a reasonable time of arrival. The question for the environmental manager becomes "what is the most efficient and effective journey from program initiation to program success?"

7.2.1 Why Is Strategic Planning Important?

It is important to understand why strategic planning is critical to our mission, as this helps motivate us, and the people we manage and lead, to conduct the planning process, a process that can be long and arduous but rewarding if done right. Strategic planning is important to the environmental manager because: it identifies and incorporates objectives (subgoals) and strategies (to achieve goals) that overcome the mistakes of previous programs with similar mission-oriented goals; by answering the "who" question(s) it can aid in designating and properly empowering leadership; it reduces uncertainty by putting a roadmap in place; and it facilitates better intergovernmental coordination and a path to partnership with stakeholders.

7.2.2 Plan Development

More questions! Developing a strategic plan ultimately begins with answering two fundamental questions: "What do you want?" and "How are you going to get it?" More specifically, what is your mission and your program's goals and objectives, and which strategies will your program utilize to achieve them? There are many more components of a strategic plan, but working with these two simple questions is an instructive way to initially conceptualize strategic planning.

For example, if your mission were to become a mountaineer and your starting goal was to reach the summit of a particular mountain, where would you begin your planning? One starting point would be to conduct analysis of the terrain from the trailhead to the summit. Presumably, the analysis would tell you which peak or saddle you would have to hike up first in order to go to a higher peak which would then lead to the summit. Your analysis should also tell you which routes are harder, easier, or impossible for reaching the summit, and make you aware of the skills, energy, time, and equipment needed for each of the options. Each peak or saddle on your proposed route is an objective which leads to another objective until your goal, summiting the mountain, is achieved. How you will get to each subpeak and ultimately the summit depends on your strategies. Based on your analysis, you will have to prepare your equipment, skills, and other coordinating measures and then execute your plan. You will need to train and pack for the journey, hike up the lower routes and objectives, and climb to the summit.

Simple Strategic Plan

1. Goal – Reach the top of the mountain
2. Objectives – Each subpeak or saddle on the way to the summit via your determined route
3. Strategies – Physically train, prepare equipment, hone field skills

This example is intentionally simple to make clear that really we all engage in strategic planning in our everyday lives, with greater or lesser success depending on how we go about it. Obviously for the environmental manager strategic planning is a more complex and substantial matter than simply determining what you want and how are you going to get it.

It involves the questions: Why do I want to achieve the goal, what stimulated that desire, i.e., what is the ultimate purpose? What strategies are best matched to fulfilling that purpose and achieving the desired endstate? And what objectives and goals will best set me on the path toward achieving the desired endstate, my mission? Note the sequencing of planning as well, starting broad with the mission of becoming a mountaineer and then defining goals, objectives, and strategies to achieve the mission. Most strategic planning starts broad and then focuses in so that all of the components are nested together, i.e., they fit together and are mutually supporting. Also, as we will discuss in more detail later, it is important to realize that no action, when executed, will go exactly according to plan, so it is critical to build redundancy (having built-in backups or goals that overlap on difficult areas) and flexibility (the ability to adapt your plan to changing or unforeseen contexts) into your strategic plan.

At this point, instead of mountaineering it might be useful to work with an example you will surely be familiar with, your career. Some call it your "career path" and usually it involves more than your desires but, rather, family, friends, your community, and even your ethics and values. We all have aspirations, whether they are to be rich and famous (admittedly not the usual aim of an environmental manager), to have and support a family, to change the world, to promote justice, or any combination of ambitions. In terms of strategic planning, these aspirations are **vision** statements. A vision statement is a declaration of the broad intent of your plan and direction. For example, your career path strategic plan vision statement might be "To change the world to be more equitable and just." For a small-city environmental protection agency, a vision statement might be "To produce a safer, healthier, and more inspiring environment for the children of our town than the one they were born into." Usually, for environmental managers, the scale of the vision statement correlates with the size and scope of authority of the planner or planning agency, striking a balance between reasonability and boldness.

From the vision statement a mission and mission statement are derived. Our **mission** is usually more focused and practical, with an actionable and achievable task and purpose. Sticking with the career path planning example, your mission might be to secure an enjoyable job that is well paying and environmentally oriented such that you can provide for your family but also care for the environment. The mission of the small-city environmental protection agency might be to "Protect human health and the environment to ensure that environmental quality and safety are consistently increasing in our town." The mission statement is thus nested in and aligned with the vision statement but provides more defined parameters such that more specified and measurable goals and objectives can be developed for the strategic plan.

In large part, what determines the "career" choice in your career path, what preferences and concerns most drive your decisionmaking about profession, is one's **values**. For example, what you value influences your work ethic: How hard do you want to work to be eligible for a position, how much do you want to sacrifice? There are always tradeoffs between time spent working and time spent recreating, with family, or other activities. Values also impact the decisions you make about how you will get your desired career:

Are you willing to obtain what you want in an ethical manner? You and your organization's values play a critical role in strategic planning because they shape where you will place your focus or emphasis, what biases or blind spots may or may not emerge in your planning, and a host of other influences. It is important for the environmental manager to determine and reflect on these values before, during, and after building and executing their strategic plan.

To guide your way down your career path and to keep your career progressing you will develop **goals** and **objectives**. While a mission is a subcomponent of the vision, a goal is a subcomponent of the mission, and an objective a subcomponent of a goal, forming a nested hierarchical system. Goals and their corresponding objectives may be structured and pursued sequentially (one after another) in your plan, or it may be that you can pursue them concurrently (at the same time). For example, your plan might be to first get an entry-level position, move up to a manager position, and then finally fulfill your mission to become an agency administrator. These goals are sequential because one leads to the next. Now if you add another goal of getting a higher academic degree in management along your career path, you could choose to sequence that goal – meaning you plan to do it between let us say your entry-level and manager positions – or you could pursue that goal while also pursuing another goal, e.g., enrolling in a degree program while still working in a full-time entry-level position. And within the goal of completing a higher degree, you may develop objectives such as scoring high grades in environmental economics or building your professional network. Like goals, objectives can be sequential or concurrent. It all depends on what your analysis of the planning situation is and what best facilitates you accomplishing your mission most efficiently and effectively.

To achieve your objectives and goals, and ultimately your mission, you will plan for and carry out different strategies. Strategies can be specific to a single goal or objective, but they can also aid in the accomplishment of multiple objectives and goals at once. In our mountaineering example, the strategies for reaching the summit were to physically train, prepare equipment, hone field skills. In your career path plan you might plan to stay up to date on the state of your field by regularly reading a professional publication (e.g., *Environmental Management*) or networking across agencies and organizations. Strategies are not an end in themselves but facilitate you moving toward and achieving your objectives, goals, and the mission.

Other than the analysis you would conduct to get an idea of what careers are available and what it would take to be successful at, let us say, environmental management, at some point you would want to know "how long is this going to take?," "is this working?" or "how can I do this better?" Thus, a strategic plan includes the development of some metrics for evaluation. In the career path example, it might be indicators such as pay scales for jobs, time between promotions, and level of satisfaction with your career. Whatever indicators you determine will best indicate the progress of your plan, you will want to plan to measure them iteratively (multiple times throughout the execution of the plan) and have a plan for adapting to or integrating what they indicate.

Too often managers "fall in love" with their plans and fail to adapt a plan when it is no longer the best path to achieving their mission. This is especially true when a strategic plan has a long lifespan, when there are big changes in the contexts the plan was built for, or when the mission itself changes. A strategic plan is best thought of as a living document that can and should be adapted as new information is gathered, learning has occurred, or the situation has changed. Expecting this to happen and planning for it upfront is an environmental manager's best bid for success.

Strategic Plan Outline

- Plan Summary with background and analysis of possible careers/paths.
- Vision Statement – "To change the world to be more equitable and just."
- Mission Statement – "To work in a career that protects human health and the environment."
- Values – belief that humans are an integral part of the ecosystem and are responsible for its preservation; that individuals should ethically achieve a standard of living to become a self-sustaining, contributing member of their family and community.
- Goal – become a scientist qualified to conduct and manage environmental assessments
  - Objective 1 – obtain a Bachelor's degree
    - Strategy 1 – participate in class
    - Strategy 2 – write papers
    - Strategy 3 – study for evaluations (quizzes, exams, etc.)
    - Strategy 4 – complete undergraduate level internships
  - Objective 2 – obtain graduate degree(s) that indicate a balanced mastery of science and policy
    - Strategy 1 – participate in class and work groups
    - Strategy 2 – write extensive papers and essays
    - Strategy 3 – study for evaluations (quizzes, exams, etc.)
    - Strategy 4 – complete graduate level internships
  - Objective 3 – obtain professional experience
    - Strategy 1 – complete internships (overlap is OK)
    - Strategy 2 – volunteer in career-related endeavors
    - Strategy 3 – develop a career-related network
    - Strategy 4 – obtain an entry-level position
    - Strategy 5 – develop a reputation for performance excellence/work ethic
- Evaluation and Plan Improvement – formal annual reviews of the status of goals and objectives; annual self-review of personal financial plans and resource allocation, identification of key obstacles or challenges in previous period; discussing with friends, family, peers and mentors about your career progression (note that program evaluation and quality assurance/quality control will be discussed in Chapter 9); revisit and revise plan quarterly, semiannually, or annually to integrate feedback and lessons learned.

Strategic plans (and addenda) for environmental organizations are often hundreds of pages long with many tables and matrices. Formal analyses of what the problems are and their origins, concerns for implementation including barriers to communication, evaluations of past programs, partner relationships and even the current mission and values, among many other factors, should be conducted and discussed prior to building the plan. Conducting a robust information-gathering and analysis process prior to developing the plan is critical because it helps everyone get on the same page and understand the broader contexts within which the plan will exist, and will help root out any (well, most) conflicts between planning participants early in the process. There are structured analytical tools that are commonly used by strategic planners. An example is the PESTLE process which considers external factors related to Political, Economic, Social, Technical, Legal, and Environmental variables (Sridhar et al. 2016).

Also, commonly used is the SWOT analysis which considers the organizational Strengths, Weaknesses, Opportunities, and Threats that the organization enacting the plan might have and how that affects and shapes their mission and strategic plan for that mission (Eisner, McNamara, and Gregory 2018). Whether using PESTLE or SWOT or any of the other frameworks common in planning today, it is important to remember they are simply guides to initiate and develop the early stages of the planning process. In the end, these tools are just like any other tool, they take time, effort, and experience to learn and use well, they are only as good as the user is, and they can aid the planning process if used properly but produce damage if used improperly. It is up to the environmental manager to learn and master their utility.

Ultimately, the environmental manager must adapt, think on their feet, and adjust their plan accordingly because, as Field Marshal Helmuth von Moltke put it, "No plan survives contact with the enemy" or as boxer Mike Tyson paraphrased Joe Louis in 1987, "Everyone has a plan until they get punched in the mouth." Therefore, professionals must learn, remember, and practice the concepts and skills to be prepared for a "change of plans." Finally, no matter how good and how detailed a strategic plan is, if it is not implemented effectively it becomes just another "doorstop." Building a strategic plan is a key part of an action, but you still must take action.

7.3 Developing a Strategic Communication Plan (SCP)

Communication is critical to management, whether as a matter of coordinating a process of co-production or implementing a program. Many managers believe they are good communicators, mostly because they have been communicating all their lives or because they have written so many papers. This belief is sometimes called the "communication paradox" where a person believes they are effective at communicating simply by virtue of having substantial experience communicating (Garnett 1992). In fact, communication skills require logic and practice and, as it turns out, many communications are not logical or practiced. Think about resolving a family or roommate dispute or think about which professor provided adequate direction for assignments. What could be a simple misunderstanding (roommate dispute) or what should be clear and concise instruction (from a professor) ends up a mess because of a lack of effective communication. The goal is not just to communicate but to effectively communicate our environmental management plans. In short, "think before you speak."

One way to develop this professional skill is to use a strategic communication plan. The reason for this strategic communication plan is to have a logical and contextually tailored communication strategy to accomplish your mission. Two fundamental communication benefits to this plan are better messages and better listening – as one can avoid being distracted by having to come up with basic messages while one should be listening. Further, it should act as a guide for someone to follow if you cannot personally conduct the communication. As such, when building your strategic communication plan, you will want to keep in mind, "Could someone else read my plan and still accomplish the mission?" Strategic communication plans, in general, should be simple but also should not allow for the manager to "wing it." Strategic communication plans have five sections where each section feeds into the next in linear or sequential flow: Management Situation; Audience; Message; Media; and Feedback (Garnett 1992):

The Management Situation – Why are you communicating? This should be a simple sentence – use supporting/background information once it is stated. Define your management situation – what are you trying to accomplish? To inform, persuade, to change behavior, to influence attitudes? Or, specifically as with risk communication: To warn; To reassure; or to allow prioritization for decision making?

What is your frame of reference? Identify yourself, position, role and mission.

Audience – Conduct an "audience profile" to identify and know your audience. This can be individuals or a group, they can be regulated entities or their trade groups or they can be the public or environmental advocate groups or they might be the press – but get their names and titles right! There is particularly important, but often mistaken, addressing of people in certain cultures when communicators immediately create barriers of prejudice and hierarchy. People deserve their titles!

Be sure to know why "they" are your audience and how they pertain to your management situation. The effective environmental manager needs to anticipate your audience's arguments and concerns in your message and feedback.

Message – Quote your "message" or "talking points." This should be a message from you, not just a description of a program, etc. Information overload is a real problem here. Be careful not to create barriers either through jargon, prejudicial statements, or other issues common to messaging. (See Barriers to Communication in Section 6.7.1 in Chapter 6.)

A well-crafted message answers a series of questions:

- Wording – are you using the correct (by definition) or most appropriate (by context) words?
- Content – are you including the appropriate information to address the management situation? Is your information accurate, useful, fair, and open?
- Length – is your message long enough to address the situation but not too long so as not to create information overload?
- Style – does the message meet your organizational norms while also being best tailored to meet audience expectations and needs?
- Organization – is the message organized with the proper introduction, body, and conclusion such that it flows logically and can be easily understood?
- Tone – does the message have the appropriate tone to stimulate and be internalized by the receiver?
- Analysis – is the message understandable and logical in terms of what it is saying and how it is saying it?

- Timing – is the message useful in terms of its delivery time within the management situation? (Garnett 1992)

Media – match multiple media (email, social media, in-person meetings, power point briefs, etc.) to targeted audiences to make sure they receive your message in the most effective form for them, i.e., the format that is most likely to ensure they receive and understand your message. It is always a good idea to provide your audience with hard copies of written and/or supplementary materials, or more commonly these days access to a digital archive, as it provides some comfort related to being able to review and understand your message.

Feedback – It is always best to develop strategies for obtaining feedback from those you are communicating with such that you can ensure you arrive at a "mutual understanding."

The feedback mechanisms you include should match your management situation in timeliness (how quick you gather feedback), breadth (how many people you intend to and need to get feedback from), and appropriateness (the feedback you get addresses the most important issues or concerns of the management situation). For example, if your object is to persuade people to take a particular action, how do you know they received the message and are persuaded to take that action? How can you solicit feedback to know whether you need to take further action, and what that action is, to persuade them? Feedback is how the environmental manager can adapt their message to ensure it is received and understood.

A strategic communication plan can be integrated as a component of your overarching strategic plan, it can be used for specific management situations that arise in your day-to-day work, or it can be used to address crisis events (as discussed in Chapter 6). Whatever the context, the general principles remain the same and it always requires proficiency and practice by the environmental manager to effectively employ, i.e., train with it and use it often.

7.4 Diffusion of Environmental Innovations

Diffusion of innovations is a major program area for many environmental managers because managers are often "change agents," defined as "an individual who influences clients' innovation-decision in a direction deemed desirable by a change agency" (Rogers 1995: 28). A change agent's success is determined by their efforts to diffuse innovations and their effectiveness in doing so; their relationship with the community; their compatibility with the community's needs in terms of having a compatible frame of reference; and their empathy (Rogers 1995).

Diffusion as a skill is very much desired by potential environmental management employers, as managers must frequently get people to adopt new permit schemes, new monitoring technologies or requirements, new enforcement policies and technical assistance for pollution prevention programs, etc. Diffusion is to a community what adoption is to an individual (and for the purpose of this book those terms will be used interchangeably), i.e., managers get individuals to adopt innovations and they diffuse them into communities. Diffusion theory is "the process by which new ideas or practices (called innovations) are communicated to, and either adopted or rejected by, members of a social system over time" (Rogers 1995: 5).

7.4.1 Elements of Diffusion

There are four elements of diffusion: Innovation; Communication Channels; Time; and the Social System. In our discussion of risk and strategic communication in previous sections, we addressed communications channels involved in message transmission. In this section we will provide more information on the three other elements of diffusion.

An innovation is "an idea, practice, or object that is perceived as new by an individual or other unit of adoption" (Rogers 1995: 11). It takes little imagination for the environmental manager to consider how many innovations they are constantly trying to get regulated communities or the public to adopt as scientific information dictates the need, and technologies allow, for new ways to protect human health and the environment. From instituting new waste management practices to asking people to simply use less, environmental managers are frequently working to diffuse innovations – some old, some new, and some reinvented. Indeed, environmental management in many ways is about changing how communities think about the environment and what behaviors they can adopt to have a more positive effect.

Time in a diffusion process includes the period from the first introduction or reintroduction of an innovation to a community to the ending of efforts or the diffusion process by the environmental manager. How much an innovation was diffused – that is, how many people or the percentage of the community that adopted it – over that period is the rate of diffusion. Clearly, if a manager can increase the rate of diffusion (Figure 7.1) they become more efficient resource managers by saving time and funding and allowing them to move on to the next program.

7.4.2 Rates of Diffusion

Rates of diffusion are determined by attributes of an innovation; the type of innovation decision people or communities have to make; communication channels utilized; the nature of the social system the innovation is diffused into; and the extent of change agents' promotion efforts (Rogers 1995).

A social system is the complex of formal and informal institutions, practices, social norms, and relationships that form between individuals and groups. These components of a social system, and social systems in general, are dynamic such that they are always changing and evolving. Change can be initiated from internal shifts called endogenous force, or from external influences called exogenous force. Environmental managers are often most like an exogenous force acting on a social system and trying to diffuse an innovation into that system, but they can leverage and even sometimes be an "insider" in the system depending on the management situation. Understanding at a minimum the basic and most influential components of the social system at hand is critical for the environmental manager to be an effective agent of change.

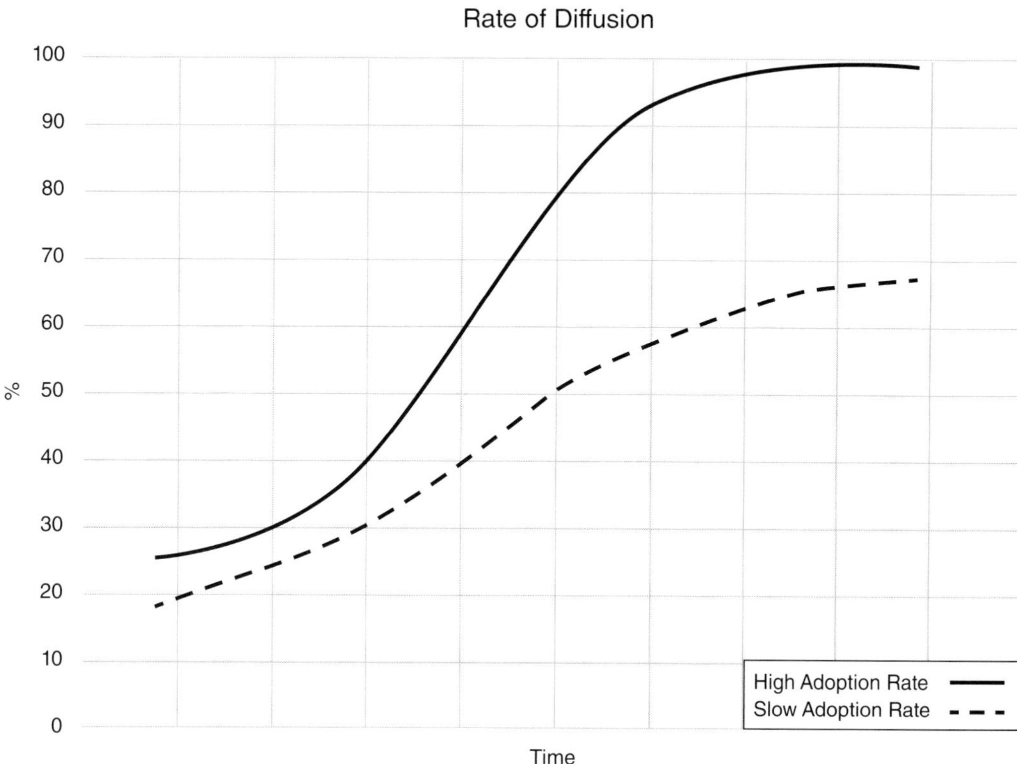

Figure 7.1 Rate of diffusion.

7.4.3 Attributes of Innovations

There are five attributes of innovations: relative advantage, compatibility, trialability, observability, and complexity. Relative advantage, compatibility, trialability, and observability are considered to have a positive effect on diffusion, while complexity tends to have a negative effect on the rate of diffusion of an innovation. It is important to look at the characteristics of the environmental innovation itself to understand how the community might diffuse it or how they might respond to your diffusion program.

- Relative advantage is the advantage of the new technology or policy over the current one used by the community, and the environmental manager must make the case for its advantages.
- Compatibility relates to whether or how well an innovation melds with the current practices, processes, and perceptions of the target community. Some innovations can be more easily integrated by communities, and some have higher or lower rates of preconceived notions (or political "baggage") surrounding them. And some communities might have an existing program that they appreciate that is similar to a new program and thus be resistant to it being replaced.

- Trialability is whether an innovation can be "tried out" or tested by the community before extendedly or permanently adopting it. A tried-and-true method that change agents have used is having a "pilot program" such that the innovation can be demonstrated without risk to the adopter.
- Observability is when the adopting unit can observe the effects of changing their behavior, i.e., the change is easily and often quickly viewable by the adopters.
- Complexity is usually considered a negative attribute of an innovation. A complex innovation may have multiple parts to it that each require action on behalf of the community; it may have secondary effects on other aspects of the community other than just the intended environmental component; or any other of a range of factors that can make an innovation not so simple. The greater the complexity of an innovation, the more thorough and well executed a diffusion plan will need to be to successfully diffuse the innovation.

The environmental manager must plan to accentuate the positive attributes of any innovation they want a community to adopt and mitigate the negative attributes of the innovation to alleviate the perceived risks of adoption.

7.4.4 The Process

Typically, individuals or communities must go through a process to change their behavior, whether that behavior is using a new tool, reducing their use of some material, or any other action. Sometimes the change process might be a coercive, top-down-driven action but often it is more efficient and effective for communities to voluntarily change, especially if the change is demonstrably beneficial for them. In fact, communities can often improve the innovation and increase its rate of adoption when they "own it" and have a vested interest in its success. Change agents sometimes refer to this ownership as "reinvention" or adaptive innovation such that the proposed innovation is refined by the end user.

There are three basic types of innovation decisions: **optional** – when the choice to adopt or reject is made by an individual independently of decisions made by other members of a system; **collective** – when the choice is/must be made by consensus (like a vote); and **authority** – when the choice is made by relatively few individuals in a system who possess power, high social status, or technical expertise. Optional and collective decisions tend to have a more decentralized diffusion process whereas authority decisions are more centralized. Which type of decision must be made in a particular case depends on the context. In some instances, particularly where there are no binding laws or policies, individuals, households, and communities can choose whether or not to adopt the innovation, e.g., buying LED light bulbs for your home instead of incandescent bulbs. But in other instances, there are binding statutes that require entities to adopt the innovation, e.g., the Best Available Technology standards of the Clean Air Act require producers of certain pollution types to adopt new pollution reduction technologies as soon as they are developed and commercially available. The environmental manager will have to determine what decision type is most appropriate for their innovation and build that into their diffusion plan.

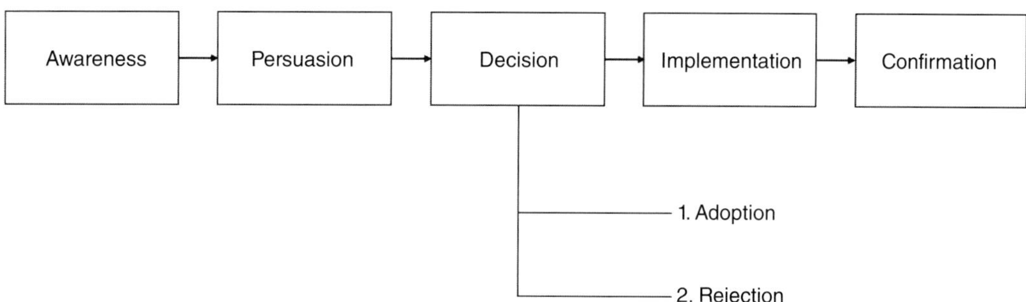

Figure 7.2 The innovation/decision process.

7.4.4.1 The Innovation/Decision Process

The innovation/decision process (see Figure 7.2) has five stages: Awareness (or Knowledge); Persuasion; Decision, Implementation; and Confirmation. In thinking about how the innovation/decision process relates to the planning tools and processes we have presented, each of these stages can, for example, be considered an objective with the goal of diffusion within a strategic plan, with a corresponding strategic communication plan to facilitate diffusion. In this way, the different planning tools and stages that we have presented can be nested to ensure cohesiveness and integration of programs within the environmental manager's overarching plan. The environmental manager will thus have to manage multiple ongoing processes and plans at any given time, and will do best to make them as mutually supporting as possible.

Awareness Stage The Awareness stage begins when individuals in the community become aware of an innovation and how it might work. Typically, awareness occurs through exposure to a mass media technology such as the Internet, television, radio, or the print media. At this point the sender transmits the message, "Here is a new solution to a problem (old or new)," with the intent of communicating that the innovation will meet the community's need – for instance a better permitting policy, monitoring standard, waste minimization program, etc. At this stage the primary audience could be the decisionmakers in the community or local change agents, or it could be the individual community members that might adopt the innovation.

Persuasion Stage The Persuasion stage is where the adopting unit develops an attitude toward the innovation and the environmental manager is actively trying to shape the attitude that is developed. The change agent communicates the evidence that the innovation will match the community's need through interpersonal or mass media channels to convey the benefits of the innovation. The potential adopter sees the innovation not only as a new idea or practice, but also evaluates the potential risks and uncertainties from its application (Downs and Mohr 1977). If the attitude toward the innovation is positive, it could lead to a change in behavior (that is, cursory adoption or trialing of the innovation), whereas, if the attitude is negative, it could lead to rejection of the innovation (Lambur et al. 1985).

At this part of the persuasion stage, interpersonal communication begins to dominate the process. The audience is presented with the characteristics or attributes of the innovation it is being asked to try. The environmental manager should provide a message for accentuation (if positive) or mitigation (if negative) of each attribute. Here are a couple of examples of messages which could be sent to audiences: (1) "This new testing standard will not only identify new pollutants but also new contaminates which will present a liability for your industry" (thus providing a relative advantage); and (2) "By incorporating these BMPs with your current practice of quality control, you can increase your compliance without increasing your costs" (thus showing that the best management practice [BMP] innovation is compatible with the industry).

Decision Stage The Decision stage is where members of the community decide whether they will adopt the innovation for use in their program, practices, or other actions. The change agent needs to provide a platform or venue such that the adopting unit can say, "Yes, you have convinced me/us to try out your new solution," or the alternative, "You have not convinced us yet, either tell us more or we walk away." However, a more important exchange is taking place within the communication network of the adopting unit. This is the communication between early adopters and later adopters. This is the stage where phrases like "It's just the right thing to do," "It's a no brainer" need to be facilitated by the change agent as they will decide the fate of the adoption. Many people base their decision on what other people in their industry or in their community are doing, so, if people are adopting the innovation, the change agents should help spread the word within the target audience's networks to promote other entities adopting the innovation as well.

Implementation Stage The Implementation stage is perhaps the most important in the innovation/decision process. Rogers (1995) contends that actually using the idea or practice represents a concrete test of applicability. Reinvention – or the refinement or tweaking of an innovation to best fit the end user – may occur at any stage, and it is during implementation perhaps more than any other stage that reinvention becomes operational (Rice and Rogers 1980). This "handholding" stage, where the change agent guides the use and tailoring of the innovation, shows the community *how* to use the innovation and to discover their own, individual best practices with the innovation. Implementers utilize messages akin to "Yes, you are using this innovation correctly" or "No, but here is how you can make it work properly" to guide the initial use and trialing of the innovation. At this point, it is critical to avoid information overload but also to ensure best practices are established early to set adopters on the best trajectory for sustaining use of the innovation.

The implementer may recognize, and should accept, that adopters will utilize various coping methods, including reinvention, to make the innovation tailored to them. Nurturing may best describe what the change agent must do at this stage, that is, guide and facilitate the reinvention and practice with the innovation. The change agents need to use interpersonal communication during this stage to guide the use of the innovation. Some actions that they can use to survey the use of the innovation include site visits and inspections, for example as part of a compliance assistance program. If issues are discovered, remedial measures to improve or correct usage should not only be suggested but observed and critiqued as well. Site visits such as these are also conducted to facilitate communication both to the change agents from the community and by the change agents to the community. Often this exchange of information and practices between change agent and community will extend beyond the actual visits as they work in partnership to improve the innovation and its use.

As with relationships between the regulators and regulated community, the environmental manager must make themselves available for questions from members of the community and be open to all community members about the information they have shared. Often the most important phrase used during this handholding stage is "Making this work is a process, not a miracle." The change agent must remind their audience that change is an ongoing process which they have been through before. This is critical communication and makes the audience more comfortable.

Confirmation Stage The Confirmation stage occurs when adopters assure themselves that the decision to adopt was correct and worthwhile. Adopters may decide at any time to discontinue the innovation if it no longer meets their needs or is no longer perceived as the best mode or method available – albeit, if the use of a specific innovation is required by law, then the adopter quitting the innovation may become out of compliance (unfortunately this does happen). While "location, location, location" are touted as the three essential aspects for the successful Realtor, "feedback, feedback, feedback" might be the analogous triad confirming the continued use of an innovation for the environmental manager. Baseline data and program evaluations are critical to this confirmation because they allow the change agent to measure the diffusion rate and impact of the innovation and can help guide changes in the diffusion plan to increase adoption.

"Pats on the back" using mass media or interpersonal communication are necessary and often helpful. The EPA has recognized successful program implementation by companies, communities, and cities with award plaques, special certifications, and other symbols of formal recognition. The plaques contain phrases such as, "... for your successful efforts in protecting the children of ..." These plaques or letters of congratulation are presented to the adopting decisionmaker and help confirm their adoption of the innovation as the correct and "good" course of action. I (M.L.) often remind my students and colleagues that "a $100 plaque can be worth $10,000 of program funds." Further, a good change agent is often successful at obtaining news coverage for the award ceremony to help improve the public recognition of both the innovation and the entity adopting it, creating positive sentiment or "branding" for both parties.

REAL-WORLD EXAMPLE 7.1 Diffusion and Iowa farmers

Diffusion of innovations is critical in voluntary conservation on working agricultural lands. The Corn Belt of the United States is dominated by input-intensive agriculture and is one of the most altered landscapes in the world. Agriculture as an industry is also, for the most

REAL-WORLD EXAMPLE 7.1 (cont.)

part, exempt from or devoid of regulation. As the primary land use across much of the US, effective ecosystem management across a variety of resource concerns – water quality, biodiversity, pesticides, climate change mitigation – requires working with farmers, landowners, and other agricultural stakeholders to willingly implement conservation practices and land use changes.

As early as 1986, the Soil Conservation Service readily recognized that economic factors fail to account for all decisionmaking. More recent research demonstrates that while financial (economic) incentives are important, there are other motivations for and barriers to conservation adoption, most notably technical and social factors. Addressing only one of these barriers to conservation is insufficient; economic, technical, and social barriers must all be addressed for effective diffusion of innovation.

There is much debate over which information-sharing pathways (e.g., government agencies vs. extension vs. crop advisors) are the most effective. However, recent research and anecdotal evidence support the simple yet innovative idea that farmers learn best from other farmers. Internal sources (other farmers, relatives, neighbors, or friends) enjoy greater prestige and spatial proximity than external sources (government agencies), resulting in more effective information transfer. A deliberative and supportive farmer-to-farmer model not only fosters effective transfer of new ideas and practices, but also facilitates true innovation.

Social factors are often overlooked when designing programs intended to promote and accelerate the adoption of innovations, and yet social factors are especially critical in agriculture. Individuals rarely make decisions in a vacuum; rather, decisions are made within a complex network of communities and are "products of shared common interests, collective participation, and concerns based on mutual trust." These building blocks, cumulatively, comprise the concept of social capital, a sort of social carrying capacity that both facilitates and limits changes in behavior within a community.

Social capital is not only beneficial but usually necessary for the uptake and adoption of new management practices and technologies. For a practice, technology, or innovation to be adopted, it has to be positioned, marketed and perceived as socially and economically desirable. Farmers and landowners are more likely to implement a new technology or practice when that idea emerges from a social network or group with shared experiences and mutual trust. Designing programs that are farmer-led can promote and facilitate collective action and collaborative decisionmaking and ultimately enables the widespread adoption of conservation practices.

Jorgen Rose, MPA-MSES, is a program and policy coordinator for Practical Farmers of Iowa, a 501(c)(3) non-profit that works to equip farmers to build resilient farms and communities through farmer-led education, outreach, and on-farm research

7.4.5 The Nature of a Social System

Audience is a critical element in diffusion, and the attitude of the change agents' audience will influence the rate of diffusion. In diffusion theory, a social system is defined as a set of interrelated parts – institutions, practices, people, and places – that are engaged in joint problem-solving to accomplish a common goal (Rogers 1995: 24). One of the ways election campaign experts successfully manage their resources is by concentrating on the "mushy middle" as opposed to voters who have already decided for (early adopters in terms of diffusion) and voters who have decided against (laggards) the candidate innovation. Similarly, the environmental manager as a change agent can assess the social system into which they are working to diffuse an innovation and identify the people in the "mushy middle" who might be most open to adopting the innovation.

7.4.5.1 Who Influences Social Systems?

Social systems, communities, organizations, and individuals normally are influenced by peers and peer groups (whether it is high school cliques or our industry competitors) and opinion leaders. Opinion leaders are determined "by the degree to which an individual is able to informally influence other individuals' attitudes or overt behavior in a desired way with relative frequency" (Rogers and Kincaid 1981: 123). These individuals, and their potential influence on the diffusion of an innovation, are important whether the planned diffusion pathway is *vertical*, a top-down approach using authority and opinion leaders to directly orchestrate change, or *horizontal* where diffusion occurs within and between a community of peers that are influenced by opinion leaders.

Opinion leaders are sometimes called "bell cows" by change agents, in reference to when a farmer identifies a cow the other cows follow to or from the barn and they put a bell on her so as to recognize when the herd is on the move. No doubt some human opinion leaders would take exception to this reference, but it does help to understand the concept. Opinion leaders are typically identified by the change agent with regard to their external communication type and frequency, their accessibility, socioeconomic status, innovativeness or willingness to become early adopters, and the community organizations they belong to or hold leadership positions in. Accounting for and coordinating with opinion leaders is an important strategy for a change agent, especially if they have limited contacts or little rapport with the community they are targeting with the innovation. Of course, aligning with opinion leaders can also get a diffusion program into trouble because if they fall out of favor, or are only opinion leaders for a portion of the community, the diffusion rate will likely slow, plateau, or reverse. As with all things, care and caution is needed in choosing partners.

Community members normally are most comfortable diffusing an innovation based on what others in their community are doing, a tendency not unrelated to the idea of "group-think." This concept is called *homophily* which is when people listen and talk to people like themselves. However, often communities also recognize the value of outside experts, which is related to *heterophily* or the tendency for people to be interested in people unlike themselves. The balance between homophily and heterophily in a community will vary, and the change agent should assess this during their planning process. Paraphrasing the old saying, "a consultant is someone who carries a briefcase and is from outside a 50-mile radius," experts who are more educated, well-read, "cosmopolite" individuals of higher socioeconomic status may be valued and listened to, but it takes work, trust-building, and evidence of benefits.

It is always helpful when the change agent and their championing opinion leaders have ties to the adopting community but they also take advantage of the *strength of weak ties* where "more useful channels for gaining useful information are from one's more distant (weaker) acquaintances – a 'bridge link' from a homophilic clique where most individuals know the same thing to another clique" (Rogers and Kincaid 1981: 248). These individuals, who are recognized or respected within even distant groups or who bridge multiple communities with exceptionally large numbers of social ties, are often referred to as hubs or connectors and prove to be critical leverage points in the diffusion process (Goldenberg et al. 2009). Most of us remember individuals in high school who were able to move in and out of cliques and were respected by all. Similarly, hubs or connectors, even with distant or weak ties in a community, can be effective catalysts for a change agent trying to diffuse an innovation.

7.4.6 Models of Diffusion

The diffusion process can break down at any stage, which would be a program failure. In our experience, whether trying to diffuse an Integrated Pest Management program in a school system (M.L.) or evaluating the effectiveness of water treatment programs in rural Zambia (D.M.), one of the most common failures of diffusion stems from the change agent not properly considering and harnessing the contexts and access points of the social system they are targeting. Often in technical assistance or education programs aimed at diffusion the environmental manager believes and/or finds it more expedient to simply provide information to the adopting community and call the job done. This is what is called the *hypodermic needle model*, where information regarding an innovation is "injected" directly into the social system, ignoring the role of opinion leaders and community contexts (Bineham 1988). Under certain circumstances it can be effective as powerful, mass media messages to vulnerable audiences can work, for example, with propaganda or commercial advertising. However, a more successful and sustainable model for diffusing environmental innovations is the *two-step flow model* (Nisbet and Kotcher 2009; Robinson 1976).

In step 1 there is a transfer of information from media or other sources provisioned by the change agent to opinion leaders, initiating knowledge and exposure of the innovation. In step 2, with the help of a change agent, there is a spread of interpersonal influence from opinion leaders to their followers where the culture of the community's behavior is more strongly affected. The environmental manager, acting as a change agent, must be able to analyze the situation and the target community to determine which model might work best and thereby use their resources effectively. In certain instances, the inject model may be most appropriate – for example, when an innovation is only marginally different from the current practice or tool that it is replacing – but more often than not the *two-step flow model* is the better bid for success.

To review, the sequential roles of the change agent diffusing an innovation are:

1. Develop a need for change (awareness stage)
2. Establish an information exchange relationship (awareness stage)
3. Diagnose problems (awareness stage)
4. Create an intent to change in the client (persuasion and decision stages)
5. Translate an intent into action (decision and implementation stages)
6. Stabilize adoption and prevent discontinuance (confirmation stage)
7. Achieve a terminal relationship. (Rogers 1983)

7.4.7 Sustaining the Innovation

Condensing the process of diffusion concepts into sequential roles for the change agent/environmental manager provides a logical end point for the agent, an indicator of when to move to their next priority program. One measure of successful diffusion is the achievement of a critical mass of adopters. Critical mass "is the point after which further diffusion becomes self-sustaining at the community level and occurs at the point at which enough individuals in a system have adopted an innovation so that the innovation's further rate of adoption becomes self-sustaining" (Rogers 1995: 343). Reaching critical mass is, of course, always dependent on how the change agent targets individuals respected in the community or opinion leaders in a social system's hierarchy for initial adoption. In other words, a critical mass is not only determined by the number of people who adopt an innovation but also who the adopters are and their influence and impact within a social system.

The sustainability of an innovation will be increased if members of the community perceive the innovation as "inevitable, desirable or that critical mass has already occurred … everyone else is doing it" (Rogers 1995: 183). Sustained use of the innovation post-diffusion can be further ensured through the provision of incentives – be it public recognition as we talked about earlier or through subsidies paid to adopters – for the community to continue using the innovation.

In addition to critical mass as an indicator, other measures of the success or failure of a diffusion process are the rate of adoption within the community and if, and how quickly, other communities adopt the innovation. And, maybe most importantly, the change agent must measure the results or impact of the intended innovation's adoption – has the environmental manager's program, monitoring, enforcement, or technical assistance been improved? Has the target pollutant or other environmental hazard been reduced or mitigated? Has human health and the environment improved? And have there been any unintended effects? While diffusion is about getting people to adopt an innovation, the ultimate pursuit is better human health and environmental outcomes, so the focus of measuring diffusion cannot be limited to the use of an innovation but must extend to the effects achieved by the innovation. These measurements begin at program inception, and the environmental manager should be able to measure them throughout the process to track progress, adapt their program, and identify points of improvement and points of success. It is also important that the change agent have a system in place to measure for sustainability after their implementation is complete.

The implementation and confirmation stages of diffusion are critical to avoid discontinuance. "*Discontinuance* is a decision to reject an innovation having previously adopted it" (Rogers 1995: 190). This can happen when an innovation is replaced with another innovation, or even a return to the original practices, because the community becomes disenchanted with the innovation (Rogers 1995). In this case, if uncertainties go unaddressed or if the community is unsure how to maintain the innovation, they may choose to abandon the innovation. And they may even advise others *not* to adopt the innovation, further perpetuating discontinuance.

The change agent should plan for the consequences of both a successful and failed diffusion. Change agents usually want to reach critical mass. While this is a logical desire, they need to be ready to deal with the consequences of a widely adopted innovation. Consequences are changes that occur in an individual, or a social system, due to the adoption or rejection of an innovation. Consequences of diffusion can therefore be desirable or undesirable, directly related to the innovation or not, and anticipated and/or unanticipated. Therefore, a plan should be in place to troubleshoot problems with the innovation or diffusion, problems that can be detected through frequent measuring of indicators (discussed above) and feedback from the community. These "glitches" will be anticipated by some entities but could come as a shock to others, whether arising from known unknowns (things we know we have uncertainties about) or unknown unknowns (things that we do not account for or anticipate at all but often arise and challenge our best-laid plans). The change agent should attempt to develop a mechanism to monitor these consequences, especially those that arise after the main efforts of a diffusion process have been completed, such as establishing or utilizing an existing community oversight group which meets periodically, and which has a vested interest in sustaining the innovation. This group can monitor and report on any issues that arise as the innovation progresses over time and as other alternative innovations become available as well. To maintain communication with this group the change agent or agency should establish a strategic communication plan with the group they establish.

Achieving a terminal relationship with the community is complex as the change agent should always provide a venue to monitor the sustainability of the innovation, and for intervention, should discontinuance occur. All environmental managers should understand that relationships with communities never end, as there probably will be future innovations to diffuse in that community and therefore the networks developed should be considered valuable resources. It is much easier to build upon a solid foundation with a community than to rebuild connections repetitively.

7.4.8 The Diffusion Plan

Using components of the strategic plan and the strategic communication plan, the environmental manager can take a transferable and strategic approach to diffusing environmental innovations. Below is a description of a diffusion plan outline which incorporates the concepts discussed in this section. It can and should be developed as a program to diffuse an innovation into a community to ultimately change the behavior, processes, or tools of that community, be it an industry, human community, or other pollution-producing entity.

Plan Introduction – This defines your role, mission and values as a **change agent**, and the relevant authorities (laws and policies) and agency or organization you represent.

The Role of the Environmental Manager – to plan for and guide the adopting community through the innovation/decision process: Awareness, Persuasion, Decision, Implementation, and Confirmation. For each stage you will take into account your strategic communication: management situation, audience, message, media, and feedback. Additionally, you will need to ensure your innovation reaches critical mass and is sustainable (that is, it will not be discontinued by the community). If your innovation becomes an integrated feature for all current and future pertinent environmental behaviors, it would be considered a successful diffusion program.

Problem – This could pertain to any environmental practice or behavior that is no longer the most effective and efficient or that is causing unnecessary or unacceptable risk to human health, with a corresponding new tool or practice that you are trying to get a community to adopt – e.g., updated standards, monitoring techniques, updated BMPs, compliance agreements, pollution prevention programs, etc.

Innovation – a detailed description in terms of how it is new or perceived as new by the community in relation to the environmental problem and is it a policy (soft innovation) or a technology (hard innovation)? Describe each attribute in terms of how the change agent and the community might perceive it and what it is actually expected to do.

Community (Social System) – This could be a regulated industry (e.g., energy, auto, pesticide, etc.) or municipal group (e.g., city wastewater treatment plants) or public communities at a national, state, or local level which would diffuse an environmental innovation (e.g., plastic recycling, an energy conservation practice, Integrated Pest Management). The community could also be a political community in which an innovation could be diffused, such as a policy innovation that requires support to make into law. This special type of diffusion would probably fall under "policy entrepreneurship" (see Chapter 11).

- What type of innovation decision would be most efficient and effective? Optional for each member of the community, authoritative mandate or collective (voting)?
- How would this innovation be diffused in this community – centralized or decentralized – and which model of diffusion would work best – hypodermic needle or two-step flow?
- Who or what positions are considered opinion leaders in the community?

The Process – each stage of the innovation/decision process (Awareness, Persuasion, Decision, Implementation, and Confirmation) is a separate but related management situation, sometimes with a different audience where specific messages, possibly different media and feedback techniques will be utilized. Thus, a strategic communication plan will be required for each of the five stages.

Metrics – rate of diffusion begins at program inception. Thus, the environmental manager should be able to measure the adoption by community members throughout the process. Further, they should be able to measure the results and effects of the intended innovation's adoption – has programming, monitoring, enforcement, or technical assistance improved? Has the pollution hazard decreased? Have human health and the environment improved?

Stage 1: Awareness – the objective is to inform the community that there is an innovation that can address an environmental problem in that community.

- Management Situation: To communicate there is an innovation that can address an environmental problem in that community.
- Audience: They could be early adopters and/or opinion leaders but more likely social media, the press or a trade organization or NGO. Identify them, define and describe them, and get to know them.
- Message: The message will vary in length and tone, but in general terms would be "this community needs to address a specific environmental problem and this innovation is the solution to this problem." This message or talking point should literally be written with quotes for the change agent to present.
- Media: The message will be transmitted to the community mass-media-like communication channel. While interpersonal communication is known to be more effective, at this stage the more people who know about this innovation, the better.
- Feedback: The use of survey, focus group, or other engagement methods with community members or opinion leaders can aid in understanding how the message was interpreted and if it was effective. Change agents can work with external affairs personnel to get feedback from their different audiences, if their organization or agency has such personnel.

Stage 2: Persuasion – the objective is to provide the community with evidence that can alleviate the perceived risks of adoption.

- Management Situation: To accentuate the positive attributes of the innovation and mitigate the negative attributes of the innovation via providing evidence of each to the community.
- Audience: The primary audience upfront is the identified potential early adopters and/or opinion leaders.
- Message: The messages will vary in length and tone but in general terms would be "this innovation is better than what the community is doing now," "it is compatible with how the community currently behaves," "the community can try it and comment on it before it commits to it," "the community will be able to observe how the innovation is working,"

and "the community will be provided with assistance (technical assistance, incentives, etc.) to diffuse the innovation." Again, these messages or talking points should literally be written with quotes for the change agent to present to maintain consistency and clarity of messaging.

- Media: The message will be transmitted to the audience via interpersonal communications and/or mass-media-like communication channels. As well, supplemental materials (handouts, references, etc.) should be provided at this stage to promote confidence and assurance.
- Feedback: Surveys, personal inquiries, and a question-and-answer session by the change agent are used to provide feedback at this stage. Again, change agents could work with external affairs personnel to get feedback from their different audiences.

Stage 3: Decision – the objective is to provide a venue for the community members to either adopt or reject the innovation and then to facilitate further adoption or address rejection.

- Management Situation: To provide decisionmakers with an avenue to move on to the Implementation stage – signing on to the innovation implementation program or trial/pilot program.
- Audience: Within your defined audience in previous stages, this could be individual members of the community, opinion leaders, or whoever is at the point at which they are ready to make a decision whether to adopt or not.
- Message: The message is to communicate that there is a venue for the community to adopt or reject the innovation and voice their concerns or appreciation. The messages will vary in length and tone, but in general terms would be "By registering for this pilot program the community will commit to implementing the innovation."
- Media: This could be interpersonal communication with a verbal or written commitment to the decision or it could be a vote on adoption or not.
- Feedback: This will be the result of the decision made by the community.

This stage is particularly important as it not only determines if the innovation will be adopted and what adjustments need to be made to diffuse the innovation, but also is an indicator of political and community support.

Stage 4: Implementation – the objective is to assist the community in their successful implementation of the innovation.

- Management Situation: To communicate and demonstrate to the community how the innovation should be successfully implemented. To "hold their hands" through technical assistance and in the development of best practices for use.
- Audience: All members of the community who have agreed to adopt an innovation or all members who have agreed to participate in a pilot program.
- Message: The message will vary in length and tone but in general terms would be "we will assist members of this community in specific ways to ensure a successful and comfortable

implementation of this innovation." This message or talking point should be accompanied by actions (demonstrations, how-to sessions, etc.) of the change agent.
- Media: The message will be transmitted to the community primarily by interpersonal communication but also with some social media and mass-media-like teaching tools.
- Feedback: the use of mass surveys could work, but in-person interviews and focus group discussions work better as implementation becomes more technical as these formats tend to be more detailed and allow the interviewer to dig into the details more. As well, change agents could work with external affairs personnel to get feedback from their different audiences.

Step 5: Confirmation – the objective is to make sure the community is satisfied with the results of their decision to adopt the innovation and that they feel they have made the right decision.

- Management Situation: To communicate to the community that they made the correct decision to diffuse the innovation using evidence, incentives, and other tools.
- Audience: The community or community members and opinion leadership who have adopted the innovation.
- Message: The message will vary in length and tone but in general terms would be "based on the evidence before diffusion and after diffusion the community has successfully addressed their environmental problems and that they are recognized as providing improvements to their community." This message or talking point should be accompanied by actions (press releases, recognition plaques, etc.) from the change agent to affirm their decision to adopt.
- Feedback: Interpersonal and mass media communications with community members to ensure they feel they made the right decision. It is also important to ask, "would they recommend this innovation to other communities?" so that you can use that information in future diffusion programs. Continued assessment of the effects and use of the innovation are also critical so you can understand the ultimate impact of the innovation and how it can be improved.

Sustainability of the Innovation – the objective is to make sure the use of the innovation is sustained in the community.

- Management Situation: To provide a venue for community members to assess the diffusion and communicate with the change agent regarding status over time and facilitate the adaption of the innovation as needed.
- Audience: All community members and opinion leadership who have adopted the innovation.
- Message: The message will vary in length and tone but in general terms would be "the valued innovation this community has adopted requires periodic assessment to assure its value and the change agency is willing and able to assist." This message or talking point should be accompanied by recommendations (e.g., existing or new committee or board structure) and recognition (press releases, plaques, etc.) by the change agent.

- Feedback: Interpersonal communication to find out if the community members have created a venue for innovation assessment and assistance. As well, change agents could work with external affairs personnel to get feedback from their different audiences.

7.5 Conclusion

Environmental managers would benefit from adopting and adapting any of these plans – strategic planning, strategic communication planning, and diffusion of innovation planning – for their program, resource and political management. Simply put, planning is the skill of not only "thinking ahead" but also intentionally shaping future conditions – a critical requirement for managers. As well, plans should be created such that your successor or substitute can understand and execute them and be designed in a manner such that they empower your subordinates to execute your plan while assuring your superiors that you will successfully achieve your mission.

AUTHOR'S NOTE 7.1 Environmental managers not understanding diffusion

Environmental managers are not really delivering on program goals by believing that providing information equals diffusion of environmental innovations: simply injecting information is substandard implementation . . . *and* it might not be equitable.

I (M.L.) have always advocated that environmental managers must commit to managing the innovation/decision process of assisting communities to adopt (diffuse) environmental innovations. Simply put, diffusion of innovations (influencing behavior to adopt and maintain a better way of doing something) is a critical aspect of the technical assistance necessary for program delivery (implementation). Best management practices (BMPs) for preventing pollution as with Integrated Pest Management (IPM) in schools require this commitment to protect these communities (particularly children) from the risks from pests and pesticides. My experience has demonstrated that simply providing information, "injecting" – interestingly called the hypodermic needle model of diffusion by Rogers – into communities is not adequate. Further, that using a wholesale approach relying on injection and having the pest management industry (exterminators and pesticide manufacturers) be unbiased and competent change agents to public community "clients" does not result in the change required to diffuse and maintain a BMP innovation like IPM to protect human health and the environment.

While this wholesale approach of providing information and favoring a biased and often unrelated wholesaler is easier and less expensive for the environmental manager, a technical assistance program using a "two-step flow" to diffuse the innovation using change agents and a peer network sets up a relationship with an adopting community for

AUTHOR'S NOTE 7.1 (cont.)

creating awareness, persuading, demonstration, implementation, and confirmation – the innovation-decision process. It is my experience that face-to-face training (implementation) and recognition (confirmation) are more effective and efficient than simply making communities aware via websites and webinars. Using peers who are part of the social system (agricultural, industrial, public housing or schools, etc.) is superior to using contractors not of that system.

These concepts are the core of diffusion theory. Unfortunately, many change agents default to publishing information and disseminating it – injection. Not so much because they prefer that model but because a two-step flow model is not supported by the managing agency. Inadequate diffusion is a policy failure (see ideal policies and measuring policy effectiveness in Section 11.6), no matter how easy and cheap or how many "boxes" a bureaucrat can check off.

A short example: Responding to an emerging pesticide issue. If the Office of Pesticide Programs were to revive an effective program that provides a convertible infrastructure for diffusion of IPM, they could quickly address emerging issues such as possible detrimental effects from the herbicide glyphosate to human health or pollinators. Regardless of where one stands on the toxicity of glyphosate, I believe environmental managers have an obligation to address this. What resources do we have that would help communities adopt safer innovations to manage their weeds? I don't mean the statements from agencies and industry regarding toxicity. I mean influencing and maintaining effective IPM weed management that actually reduces the risks of pesticides to humans and the environment.

The mismanagement of change agents: The EPA and USDA often use universities, specifically their change agents (aka land grant Extension agents and specialists) to diffuse agricultural or IPM innovations. I have become disheartened particularly with their movement to the lowest common denominator of more "internet news letters," websites, webinars, and surveys. Though, I believe these instruments (particularly surveys for metrics) can be critical strategies for a strategic diffusion plan, I see too much effort and resources diverted from the "face-to-face" implementation required in a two-step flow process. The implementation phase of the "innovation/decision" process of diffusion has typically been short-shrift-ed by environmental managers who are not comfortable managing the resources of proper funding, effort, and political will. In my opinion, this poor understanding of how to manage diffusion has resulted in much "discontinuance" by communities who might otherwise sustain the diffusion of IPM. Agencies have justified the abandonment of the two-step flow innovation/decision process to the affected communities by switching to a "wholesale" implementation mode. In that, those communities requiring a behavioral change toward a BMP can make such a change by simply having a toolbox (websites and webinars) available to implement, which is not

AUTHOR'S NOTE 7.1 (cont.)

adequate and is not equitable. This wholesale approach favors industry practitioners who can invest in diffusion strategies such as training and demonstrations as a relative advantage to their bottom line whereas communities most in need whose practitioners are public employees are unable to benefit from such strategies implemented by legitimate change agents.

It has always been my assumption that agency resources should be directed at getting communities to reduce the risk of pests and pesticides in an equitable manner. When providing an educational venue, it should first and foremost be for IPM practitioners who are peers in the diffusing community. Sponsoring wholesale venues for the pest control industry exterminators who are not peer IPM practitioners, while not offering an opportunity for peer members (such as school custodial staff) to benefit from the innovation/decision process to diffuse IPM, is unethical. There should be a more retail approach to communities needing to diffuse behaviors other than applying pesticides, rather than a wholesale approach favoring an industry reliant on pesticides. I believe public agencies wishing to protect communities from the risks of pests and pesticides need to be accountable, as their responsibility is not to industry (pest management professionals or vendors) but to the public and the diffusion of verifiable IPM.

7.6 End of Chapter Questions

1. Why is strategic planning critical for environmental managers?
2. Why should an environmental manager know how to diffuse an innovation?
3. What three questions must be asked and answered in order to realistically implement a program?
4. What are the two basic questions a strategic plan should answer?
5. How do the values of an organization relate to its goals?
6. Using an environmental program of your choice please define the difference between a goal, objective, and a strategy.
7. Diagram the innovation/decision process.
8. What are the four basic elements in the diffusion of innovations?
9. Describe the five attributes of an innovation.
10. What are the five components of a strategic communication plan?
11. Define innovation.
12. Describe a pollution prevention innovation of your choice and how you, as an environmental manager charged with the implementation of that innovation, would diffuse that innovation to your intended audience.

8 Managing for Compliance and Performance: "Driving between the Ditches"

Skill: *Matching standards to performance.* The effective environmental manager must be aware of the mandates and professional standards that they and their programs are accountable to, as well as be able to demonstrate the performance and achievements of the programs they manage. Knowing your lateral operational limits and performing effectively and efficiently is not only critical to the career trajectory of the environmental manager but, more importantly, requisite for them to protect human health and the environment.

8.1 Introduction

This chapter is about accountability. While common legal "landmines" were discussed in Chapter 3, this chapter deals specifically with how the environmental manager can keep their program(s) compliant with the environmental responsibilities authorized or permitted to them as regulators and those that are regulated. As well, in this chapter the concepts of quality control and assurance, environmental management systems and program evaluation are discussed in terms of compliance and program performance.

8.2 Compliance Assurance

Simply stated, compliance is when an entity plays by the rules, and can prove that they are playing by the rules. If they do not play by the rules and they get caught, the consequences are intended to change their behavior and make them start playing by the rules again. Sometimes the consequences are disincentives such as fines or budget withholdings (i.e., not releasing or providing funding until compliance is restored), and other times they can be in the form of firing people, discontinuing programs, and cessation of the permitting process. Non-compliance related to environmental protection means that regulators are failing to protect human health and the environment and that entities permitted to pollute are willfully or accidentally disregarding human health and the environment.

8.2.1 Goals for Obtaining Compliance

Non-compliance is generally in reference to a local, state, or federal mandate or environmental law that an entity is expected to conform to – that is, the regulated entity is not upholding the standard they are required to. The US EPA, for example, has 44 compliance monitoring programs authorized by seven statutes including the Clean Air Act, Clean Water Act, and other similar, major environmental Acts (US EPA 2013b). The goals of these programs are: (1) Assessing and documenting compliance with permits and regulations; (2) Supporting the enforcement process through evidence collection; (3) Monitoring compliance with enforcement orders and decrees; (4) Creating deterrence; (5) Providing feedback on implementation challenges to permit and rule writers. How each agency or private entity pursues or performs their compliance is different, and it is up to the environmental manager to know and act in accordance with the practices and rules of their organization.

8.2.2 Compliance Actions

Most actions related to non-compliance are dealt with informally by simply communicating violations and solutions with the offending party or using voluntary disclosure policies or violation letters as the goal is to provide a path toward compliance (US EPA 2013a). As we discussed in Chapter 5, regulatory agencies that are not providing the appropriate monitoring and law enforcement are overseen by whatever agency or governing authority authorizes them. Thus, regulatory agencies must also maintain their own compliance in addition to ensuring the compliance of the entities they regulate. Again, compliance oversight and enforcement is usually a matter of communication toward mutual resolution but occasionally it can result in a formal withdrawal of authority from or restricting of an agency.

REAL-WORLD EXAMPLE 8.1 Environmental compliance assurance programs

Roger Ferland has practiced in the areas of environmental and natural resources law in both the public and private sector since 1975. I recently asked Roger, "What is the greatest area of weakness that you have observed in most companies' environmental compliance assurance programs?" The following is his response:

"In itself, the question raises two issues initially worth commenting upon. The question refers to compliance assurance programs, not the more politically correct "environmental management systems." The latter is a concept that embraces everything from sustainable product development to waste minimization. While these are all admirable concepts, most companies simply lack the resources (or need) to adopt a full-blown environmental management system with all the ISO-mandated "bells and whistles." At a minimum,

REAL-WORLD EXAMPLE 8.1 (cont.)

however, every company should have a systematic, structured mechanism for ensuring that it is consistently in compliance with environmental laws and that when non-compliance occurs, it is dealt with immediately and, if required to do so, the non-compliance is reported to the appropriate regulatory officials. A second preliminary issue is implicit in the term "compliance assurance program." By calling it a program, it is assumed that there are both written procedures covering its implementation and a single person assigned by the company to be accountable for that implementation. It is critical that the accountable person have the training, knowledge, and experience required to make the program work. In the past, companies assigned environmental compliance to a facilities manager or worker safety professional without training them or hiring the environmental expertise below them to carry out the assignment. In most companies that has changed, but not in all. Obviously, even the best compliance program is worthless if it is not implemented by people who know what they are doing. That being said, a consistent problem that I have observed with even the most sophisticated, well-staffed compliance assurance program is complacency. Complacency in this context translates to failure to observe the obvious compliance problem because it is assumed that the "system" will invariably catch the problem. In other words, it is assumed that proactively looking for problems is unnecessary. This sort of breakdown is most likely to occur in complex regulatory programs like RCRA or the Clean Air Act in which a compliance problem can arise and fester for months before it is discovered. For example, the increased use of CEMs or other continuous direct measurement devices to determine continuous compliance has meant that a problem with software or an equipment malfunction may go undetected until a particular event triggers an inquiry, only to then discover that the emission levels relied upon for compliance assurance were not accurately being measured for days, weeks or even months.

"The antidote to complacency is the periodic compliance audit. Audits have gotten a bad rap recently because regulatory agencies have either denigrated their value or been unwilling to protect audit results from public disclosure. Moreover, the cost of audits is usually the first thing to go when there's corporate belt-tightening. However, an audit, even an internally conducted audit, provides the fresh look at a plant's environmental performance that simply cannot be achieved by even the best compliance assurance program. Of course, the quality of the audit is only as good as the people that conduct it and many companies lack the internal sources to conduct their own audits. Nevertheless, even the cost of using outside professionals to perform the audit is relatively small when compared to the risk of allowing long-term non-compliance to remain undetected only to be found during an inspection by a regulatory agency or reported to that agency by a whistle-blowing employee. I would close by noting that audits should be viewed as an

REAL-WORLD EXAMPLE 8.1 (cont.)

enhancement to a company's compliance assurance program. They are not a substitute for such a program. Moreover, periodic audits make a good program better but will not save a dysfunctional program."

Dianne Frydrych (2004)

Formal compliance activities regarding regulated entities are resource-intensive as they require funding, time, and political will to effectively manage. Imagine the costs and resulting perceptions of shutting down the top employer in your jurisdiction. Not surprisingly, one of the enforcement tools environmental managers use is a "compliance agreement" or "agreed order." A compliance agreement is an agreement between facilities or entities whose activities may yield environmental impact and the regulators charged with monitoring that entity, that states what the expected standards are and how enforcement will be conducted if the standards are not met (US EPA 2013d). Some of the other commonly used formal tools are a Notice of Violation (NOV), civil penalties (fines), voiding operating permits, banning a registered product, and under the specific circumstances, incarceration if knowingly violating the applicable standards. For the environmental manager discovered to be non-compliant, the consequences range from reassignment to firing, or to increased oversight from superiors that would likely restrict or hinder their program, resource and political management capacity and their credibility.

8.2.3 Programmatic Tools for Monitoring and Managing Compliance

There are two basic types of programs that are used by entities who are required to uphold their responsibilities to environmental laws that address compliance, i.e., entities ranging from industrial production facilities that are regulated, to state-level environmental protection agencies that regulate pollution producers. One is a compliance assurance program. Compliance assurance programs specifically address regulatory program compliance such that regulated entities comply with the mandates of their permits or the applicable environmental laws. Compliance assurance programs are thus proctored by environmental managers and pollution producers alike to ensure all entities are in compliance with their mandates (just as the title, compliance assurance, suggests). The other program type is an environmental management system (EMS). Environmental management systems "embrace everything from sustainable product development to waste minimization" to assess and manage the production, and reduction, of pollution emissions throughout the processes of the entity under management (Frydrych 2004: 20). Both compliance assurance programs and EMSs are programmatic tools, meaning that they are not a physical item but rather a developed, formal process for monitoring and managing pollution production and other factors related to activities of the polluting entity.

8.2.4 Policy Procedures for a Compliance Assurance Program

In general, a compliance assurance program requires an environmental manager to apply written policy procedures for its implementation; to have training in, knowledge of, and experience related to specific environmental regulations; to conduct periodic audits; and to have a commitment from their administration that they are authorized to take action (Frydrych 2004). Specifically, for the successful implementation of a compliance assurance program, an environmental manager will require that:

- Line managers monitor compliance with environmental laws
- The organization's environmental policies are integrated into the daily work environment
- Self-auditing procedures are diffused in the organization
- Employees are trained to comply with environmental laws
- Incentives for complying are offered to those employees
- Employees who violate environmental policies are disciplined
- Improving compliance is a continually stressed goal
- Alternative approaches – such as minimizing wastes, preventing pollution, and involving employees in agency environmental decisionmaking – are adopted (Woodrow 1994)

For the US EPA, their compliance monitoring includes:

- formulation and implementation of compliance monitoring strategies
- on-site compliance monitoring: compliance inspections, evaluations, and investigations (including review of permits, data, and other documentation)
- off-site compliance monitoring: data collection, review, reporting, program coordination, oversight, and support
- inspector training, credentialing, and support to assure compliance by all relevant and regulated entities (US EPA 2013d)

There are several strategies the environmental manager might use to implement this program and to ensure that they and their employees can effectively carry out a compliance assurance program. Strategies for this might include ensuring that: affected employees keep up to date on the "state of the field of environmental management" through professional improvement/education; environmental assessments are done periodically regarding facilities they are responsible for; and effective, functional relationships and communications are established and maintained with their federal, state, and local regulatory communities (O'Leary et al. 1999).

8.2.5 Beyond Compliance

If during a compliance assurance evaluation or program the environmental manager finds that their organization is significantly out of compliance, but not immediately remediable (that is, there is regulatory room to attain compliance without formally mandated actions), they must: disclose (what is called a self-disclosure) the violation to the proper authorities;

cooperate with any investigation; implement a remedial program according to a formal or informal compliance agreement regarding fines, procedures, and timeline; and, in some cases, create and conduct a Supplemental Environmental Project (SEP) which would be considered "beyond compliance." A SEP is where an alleged violator may propose to undertake a project to provide tangible environmental or public health benefits to the affected community or environment, that is closely related to the violation being resolved but goes beyond what is required under federal, state, or local laws (US EPA 2013c). In following these guidelines, just as a driver ticketed for speeding can take a defense driving course to avoid a financial penalty or a ticket on their record, the organization can lessen or avoid serious penalties for non-compliance by undertaking an alternative action that directly benefits the community potentially harmed by their non-compliance.

When an organization finds itself not in compliance and wants not just to do what is legal but to "do the right thing," they may choose to go "beyond compliance" to attain compliance and make amends for their transgression. The EPA and state environmental protection agencies have a number of partnerships and programs that go beyond compliance (US EPA 2013a). These partnerships, leveraging and connecting tri-sectoral environmental managers, normally provide technical assistance and education from the agency to the regulated entity but also can include resources, such as grants, as partnership activity incentives. Some examples of these partnerships include Energy Star, Energy Star in the Industry Sector, Landfill Methane Outreach Program, Coalbed Methane Outreach Program Natural Gas STAR, agriculture-based programs, AgSTAR, Ruminant Livestock Efficiency Program, and High Global Warming Potential (GWP). As well, there are Stewardship Programs and Stewardship Initiatives like WasteWise, US EPA Region 1 Mercury Challenge, Water Alliance for Voluntary Efficiency (WAVE), Pesticide Environmental Stewardship Program (PESP), and EPA Region 3's Waste Minimization Program.

Stewardship programs are similar to beyond compliance programs in that they incentivize and facilitate the reduction of pollution by emitters, but participants in stewardship programs are often volunteers that join the program without necessarily having been found non-compliant, whereas beyond compliance participants usually have committed an offense first. The main difference between a stewardship program and beyond compliance programs is the scope of participants, as stewardship programs can and often do include pollution producers and non-producers that want to participate in the program, such as an academic institution interested in reducing their environmental impact. The stewardship program can guide the institution's choices toward sustainability while also giving them formal recognition for their efforts.

8.3 Environmental Management Systems

The second type of program used by environmental managers to address regulatory compliance, and even more so to address maintaining organizational or industrial standards, is an EMS.

an EMS formalizes procedures within an organization, provides accountability for continual improvement, and requires regular audits, the results of which must be communicated to management. The difference between an EMS Audit and a Regulatory Compliance Audit lies in the criteria used for evaluation. The criterion for an EMS audit is conformance to the standard itself [i.e., the standard the pollution producer establishes as their requirement which could be more but no less restrictive than the mandated standard], whereas the criteria for a Regulatory Compliance Audit are environmental regulations and other legal requirements. (Palasz 2010: 5)

Again, a compliance assurance program addresses regulatory compliance (i.e., compliance with mandated standards), and an EMS addresses standards compliance (i.e., compliance with standards of performance that may exceed but are no less than the mandated standards). Those standards are often related to the International Organization for Standardization (ISO) 14001 for the purposes of improving the organization's environmental performance. The three C's of ISO 14001EMS Auditing are:

The first C, consistency, relates to how well the procedure or process of the EMS relates to the others. In other words, do objectives and targets reflect the policy commitments? Are personnel trained on the correct legal and other requirements? The other two C's of EMS Auditing are Conformance and Continual Improvement. Conformance relates to addressing each of the requirements of the standard. Finally, Continual Improvement requires that the system lead to improvements in the system [or process] itself as well as with environmental performance. A system that has all the prerequisite procedures, but remains static, is not in conformance. (Palasz 2010: 4)

A formal process for developing an EMS that can be applied in private and public sector environmental management often looks like the following:

1. A policy statement declaring the senior management's commitment to the environment
2. A planning process identifying environmental aspects, legal requirements and goals with action plans to achieve those goals and targets
3. An implementation strategy including the roles and responsibilities for management and ensuring employees are trained, documentation is maintained, and processes for internal and external communication are established
4. Monitoring and tracking key activities and performance including compliance, remediation, and prevention of problems
5. A management evaluation as to whether the system is suitable for satisfying the needs of the organization (Stapleton, Glover, and Davis 2001)

To facilitate organizational decisionmaking regarding implementing an EMS, some helpful tools include: process diagrams outlining the parts and flow of an EMS; quantifiable goals for production and performance consistent with the organization's strategic plan and standards; reliable environmental data gathering and dissemination; and risk assessment tools for current and future operations (Matthews, Christini, and Hendrickson 2004). These tools can be employed in an EMS to help participants understand the EMS assessment process, provide guiding benchmarks to ensure the usefulness and the direction of the EMS program, promote the use and functioning of the program, and, most importantly, inform decisionmaking. Each entity using an EMS will develop and tailor their EMS to their specific organizational needs and standards.

8.3.1 TQM to LEAN and beyond

From a public management perspective, the EMS is part of the Total Quality Management (TQM) method developed by Deming after World War II (Deming Institute 2021). Deming's TQM sought to systematize how an organization seeks self-improvement through shared vision, pursuits, and leadership that promoted team efforts and synergy to create efficiency. It has since morphed into Total Quality Environmental Management (TQEM), Six Sigma, Lean Management, and Lean Plus, to name but a few examples of contemporary versions of TQM-based management systems. These methods all have the goals of effective, efficient goal achievement while continually improving production outcomes. There are slight tweaks in each method but in general they are about yielding better management. Unfortunately, sometimes these methods can be used as an excuse to do "more with less" rather than investing in the production quality and creativity of an organization which were key aims of Deming's TQM (Oakland, Oakland, and Turner 2020).

A requirement of TQM or similar management methods is that management shares power with their employees, and this can be a problem in many social systems and work cultures (O'Leary et al. 1999). Power sharing or empowerment is often talked about, sometimes feigned, but rarely actually done in many organizations, but employees who feel empowered do often substantially outperform employees who do not feel empowered (Andrade, Mendes, and Lourenço 2017). Thus, the environmental manager must truly assess the organization and organizational culture they are working in to find effective ways to empower their colleagues and subordinates.

Regardless of whether the environmental manager is working within compliance assurance programs or an EMS, ultimately they will need to provide quality control and quality assurance regarding their program management. Quality control (QC) is everything *you* do to make certain that your project is performing "up to specifications." Quality assurance (QA) is everything you have *someone else* do to assure you that your QC is being done "according to specifications or 'specs'." In general, QA is process-focused whereas QC is product-oriented; QA is therefore more about preventing defects or issues while QC is about detecting defects or issues that have occurred (Mitra 2016).

While many organizations conduct their own QA/QC work, some organizations choose to outsource this work to a "third party," such as a consulting firm that specializes in environmental regulatory compliance (this is especially true for private entities, but public entities do outsource certain functions to third parties as well, as we discuss in Chapter 11). There are many such third parties that are hired, or associations with paid memberships, that are able and willing to assure an organization is complying with regulatory rules or other desired specifications, group standards, and organization policies and goals. For the environmental manager, in general, this is a contract function (as discussed in Chapter 11). Regarding QA/

QC the most important variable is how a program is evaluated by the third party. Is this evaluation conducted by "checking a box," which really then becomes a self-evaluation, or are there in-person site visits and audits conducted by an independent assessor? What are the metrics being employed and how are they best measured – i.e., frequency, depth, and breadth of measurement? In alignment with the TQM focus of empowerment, the critical phrase every environmental manager must remember is "trust but verify." Whether conducting QA/QC internally or outsourcing it to a third party, the responsible environmental manager must empower and trust the team conducting QA/QC but also find effective ways to verify their findings.

Most all organizations have a QA/QC process in place to ensure they are attaining and improving their standards. And for public sector managers, program compliance and performance reporting is mandatory through the Federal Managers Financial Integrity Act of 1982 and the Government Performance and Results Act (GPRA) which were enacted for increased accountability.

8.4 Program Evaluation

Whether finding out if a program is compliant with the mandates it is accountable to, or tracking the progress of a program toward its pursued objectives, goals, and mission, program evaluation is required to measure and assess the effectiveness and efficiency of a pursuit. Program evaluation is the "systematic examination of a program to provide information on the full range of its short- and long-term effects on citizens" (Starling 2005). In other words, is the program delivering? To assess this, the environmental manager must ensure that the evaluation of the program is tuned to the strategic plan for the program, i.e., what you select and measure as indicators of effectiveness best capture your intended effects or impacts. As with the metrics of any program, the environmental manager should always remember, "you cannot manage what you do not measure" or more correctly quoted from Peter Drucker, "If you can't measure it, you can't improve it." This quote is not to imply that measurement is everything in managing. But when it comes to evaluation, one must have an eye on program delivery *and* improvement.

8.4.1 Five Basic Steps for Program Evaluation

There are five basic steps for program evaluation:

1. Find out the goals of the program – let us say that the main goal of an air quality program is to reduce air pollution by requiring more polluters to obtain and comply with operating permits.
2. Translate the goals to measurable indicators of goal achievement: How many air quality permits were issued in a specific time interval and what was the compliance rate with

permits that were in effect? What was the air quality of the area as a result of the permits issued? And how did this air quality effect human health and the environment?

3. Collect data on indicators – measure the air quality in immediate and surrounding areas of the permit holders and have a control site separate from the permit holders' areas to compare air quality against. Additionally, track human and environmental health effects from air pollution in the permitted area.
4. Analyze the data (longitudinal or cross-sectional) – compare the data across the appropriate variables, for example from the baseline air quality to resulting air quality, air quality across the geographic region, and air quality by type of pollutant and pollution emissions permitted.
5. Report the findings – remembering that this is part of resource and political management and requires effective communication (as discussed in Chapter 6):
   A. State the goals – in terms the target audience will understand and relate to, state what the goals of the air quality permitting program were initially.
   B. Report the status (data and analysis) and what it means to the goals – state the raw and interpreted findings in a message that is tailored to the target audience, i.e., in a format most appropriate to them to maximize understanding (as discussed in Section 7.3 in Chapter 7). Normally these messages will include visual aids to ease understanding of the target audience, as well as indicators of the products and effects of the program.
   C. State how things did or did not improve and recommendations – honest and open discussion of the performance indicators is required to share what good things happened but also where shortfalls were identified and how they can be improved.

One of the ongoing debates and shifts in program evaluation is what is and/or should be measured. Often the terms outputs, outcomes, and impacts are used to characterize different results of a program and indicate the effectiveness of a program (Royse, Thyer, and Padgett 2015). An output is the product or services that result as a function of a program. For example, an environmental management program might invest in environmental remediation such that four toxic sites are decontaminated. The output is the four sites. An outcome is the short- or medium-term effects of a program resulting from the output. For example, by remediating the four contaminated sites, environmental risk to surrounding communities may have been reduced by X percent. The percent reduction in risk is an outcome. The enduring result of remediating the four sites and primary and secondary effects are the impact of the remediation program. For example, the improved community health, well-being, and overall quality of life due to the reduced environmental risk is the ultimate impact of the program.

Most program evaluation tends toward measuring outputs, rather than outcomes and impact, both due to ease of measurement and because of their tangibleness (i.e., they can more easily be observed, pointed to, and understood). But it is the ultimate impacts that a program seeks to achieve that matter most and need to be accounted for, creating a discontinuity between what is most often measured and what really matters (Royse,

Thyer, and Padgett 2015). Thus, when designing a program of evaluation, the environmental manager must consider what they really want to know, what they can effectively measure, and what might be overlooked or not accounted for in the process. No evaluation program will capture all the effects achieved or problems created. But the design of an evaluation program deserves intentional and defensible planning or else the resulting product will be at best not useful and at worst misleading for future environmental management decisionmaking.

8.5 Conclusion

As the saying goes, "inspect what you expect." The effective environmental manager must proactively inspect the implementation and use of any management program to maximize performance and to ensure accountability. Inspection and quality assurance methods tend to have a negative reputation. But inspection should not be viewed as a punitive measure or due to a lack of trust, but rather as an opportunity for the environmental manager to understand the processes employed, to build relationships with folks doing the work, and to identify points of improvement throughout the broader processes. If a process is failing, the environmental manager is ultimately responsible, especially if such a failure could have been prevented by simple, proactive checks. It is always easier and more efficient to prevent than to repair.

8.6 End of Chapter Questions

1. What happens when an environmental manager does not Match Standards to Performance?
2. What are the minimum requirements for an effective compliance program?
3. What are <u>three</u> steps environmental managers can take to keep their staff from being negatively impacted by legal trends?
4. According to O'Leary, what are three activities you should implement if you get caught being out of compliance?
5. What is the fundamental problem organizations have implementing environmental management systems (EMS)? (Hint: Think what is the major impediment to implementing TQEM.)
6. What are the five basic components suggested for an EMS program?
7. What is QA and QC and how does it differentiate in terms of who implements it?
8. What are the five basic components of program evaluation?
9. What four basic items must be included in a program report?

9 Managing the Experts

One finger cannot lift a pebble.

Hopi proverb

Skill: *Leadership.* Integrating and applying previous skills while taking calculated risks so as to inspire. Just as the effective environmental manager has to manage programs, resource distribution, and many other factors, so too they have to manage the experts they supervise, contract, or otherwise engage. The experts employed by the environmental manager may themselves be managers but most often they are charged with a specific function that they either have substantial experience doing, have a relevant academic or professional degree or certification for, or often both. The experts often know what the right thing is or at least the right process by which things ought to be done. It is up to the environmental manager to lead them to doing it and to manage the contexts such that their efforts are facilitated – i.e., to enable them by reducing distractions and barriers.

9.1 Introduction

This chapter is about managing what is probably your most valuable resource – the people that work for you. We are not experts in human resource (HR) management, which is a specific field of work related to but very different from environmental management, though it will be discussed. Effectively managing experts, high-functioning individuals with specific, well-honed skills, requires not just management but also leadership. As such, this chapter begins with a discussion of leadership. "Management is leadership" (Drucker 1974). But it is helpful to think of leadership as the art of engaging, motivating, and empowering people and organization(s) toward a mutually desired endstate, whereas management is the science of how to get to that endstate – i.e., how to distribute resources, build a strategic plan, analyze the problem, etc. Some term these the soft skill and hard skills, respectively, but we disagree with this labeling as there is nothing soft or easy about being an effective leader. Leadership takes a blend of moral courage, commitment, empathy, and the ability to listen and decide; there is nothing "soft" about it.

Leadership typically occurs within an "organization," which is defined as "a stable system of individuals who work together to achieve common goals through a hierarchy of ranks and a division of labor" (Rogers 1995: 404). There have been volumes of material produced regarding leadership over the centuries, and leadership coaching and training is as strong an industry today as it has ever been, with about $366 billion spent by businesses globally on leadership training in 2020 (TI 2020). Perhaps one of the most quoted organizational leadership scholars is Peter Drucker who said, "Management is doing things right; leadership is doing the right things" (Drucker 2009). In this chapter we will discuss what the basic roles of the leader are, who they lead in terms of environmental management, and some concepts and skills for doing so.

REAL-WORLD EXAMPLE 9.1 Rules for meetings

A theme in popular culture emphasizes the loneliness of the governmental decisionmaker, the "man in the arena" in Theodore Roosevelt's famous quote. For those in government, however, loneliness is rarely a problem. Almost every decision, indeed every deliberation in government, occurs as part of a meeting.

Few meetings carry the import of presidential summits, or decision meetings of the National Security Council or the Federal Reserve Board. Nevertheless, meetings are an important part of every government official's life at any level.

Some meetings are a dreadful waste of time. There is value in setting aside calendar time without meetings, to give time for reflection or to attend to routine emails or phone calls. Peter Drucker's admonition about the importance of managing one's time remains as true today as when he published The Effective Executive in 1966.

That said, effective meetings are essential – to share information, to understand others' views, and to build cohesion around organizational priorities. For this reason, there is a voluminous literature on how to conduct effective meetings.

I found it useful to have three "Rules for Meetings." I borrowed these rules shamelessly from trusted mentors, and wrote them on a whiteboard in my office, so that we could all see them and remind ourselves to follow them.

Rule 1: Define Success

Ron Brand, my first governmental boss, a wonderful coach and teacher (and Drucker acolyte), began meetings by asking, "What is success?" This question works in two ways. The first is to clarify why the meeting is happening in the first place and what it is supposed to accomplish. The corollary question was, "If we succeed, how will we know?"

REAL-WORLD EXAMPLE 9.1 (cont.)

The second, even more fundamental value of the question is to challenge the group to consider the objective of the enterprise under discussion: what would constitute a successful outcome, and how could we tell if the enterprise achieved it?

It is fine to have meetings whose purpose is solely to build relationships, as long as it is clear that relationship-building constitutes success. "Relationships before transactions" is an important maxim for the successful leader.

Rule 2: Respect Each Other

This rule distilled a key principle from Getting to Yes, by Roger Fisher and William Ury, "separate the people from the problem." Meetings are not just about facts and figures. We bring our whole selves to meetings, and meetings encompass the whole range of human emotions. All of us can recall meetings where we remember little of their content but everything about how they made us feel.

Meetings are most likely to succeed if everyone knows that they are there for a reason. There is no better way to sabotage a meeting than to display disrespectful behavior. Note that there is a difference between passionate advocacy and disrespect; the former is aimed at the problem, the latter is aimed at the person. Mutual respect makes it far more likely that at the end of the day, meeting members will retain and build professional relationships.

Rule 3: Everyone Has the Right to Ask Why

Bill Ruckelshaus, the first and fifth administrator of EPA and a principled hero of the Watergate scandal, was famous at EPA for holding large meetings whenever an important issue had to be decided. Lee Thomas, who succeeded Ruckelshaus and followed similar leadership principles, has noted that this has many advantages. It enables meeting participants to see how a leader's mind works, how he or she considers multiple points of view, and, when a conclusion is ultimately reached, the reasoning behind it. It improves transparency in the organization and enables those whose views do not prevail to know that their thoughts have been considered.

This can be time-consuming. In some situations, decisions have to be reached quickly without the opportunity for such involvement. But if the support of many followers is essential to the effective execution of a decision, a wise leader will seek their input in advance whenever possible. It is always true, as Bill Ruckelshaus himself said, that "the right to be heard is not the right to be heeded." The converse is that empowering team members to be heard makes it far less likely that leaders will make bad decisions because a key piece of information is missing or, worse, withheld.

Stan Meiburg,
Acting Deputy Administrator, US EPA, 2014–17

9.2 Define Reality and Give Hope: Duties of a Leader

"A leader's role is to define reality and then give hope" (Napoleon Bonaparte). Defining reality is knowing what the situation is in the context of what you are managing – what we have termed situational analysis (Chapter 2) that is then built into a strategic plan (Chapter 7). Environmental managers need to know the policy, science, and social contexts that impact situations they manage. Below are some basic duties of a leader that will enable them to define reality and give hope (or better yet, a plan):

- **Have a vision** – The first thing a subordinate will ask their leader is, "what do we do?" Clearly the leader must have a vision of what is happening (the reality of the situation) and a possible solution for the challenge. The key parts of a leader's vision are an endstate – what conditions, accomplishments, or impacts are you trying to achieve – and a purpose – the why of doing whatever you are doing. With these two items leaders can empower subordinates to accomplish their shared mission.
- **Have a plan** – Again, this is where reality begins to translate into "hope" for a solution to the situation, where the leader provides a direction to a goal and plan for how to achieve it. But hope alone is not enough. Whether providing a simple tasking statement or a full strategic plan, a leader owes their subordinate(s) a task, purpose, and desired endstate.
- **Recognize dangers (and resistors)** – A leader's job is to facilitate and empower subordinates to "safely" achieve their goals. A leader must, therefore, do their due diligence to understand and overcome barriers to goal achievement, particularly the dangers or constraints to implementation and those entities that will provide resistance to implementation. In other words, they need to create a path to success for them or help them to make the path.
- **Recognize windows of opportunity** – A leader must know when to seize opportunity. As we will discuss in Chapter 12, a leader is a policy entrepreneur who has a plan ready to use when a window of opportunity opens. A leader must thus be forward-thinking while learning from the past.
- **Catch spears** – A "spear catcher" is someone who stands in front of their troops and is willing to sacrifice themselves to "catch" the spears thrown at them, be it the media, the community, or their higher authorities. A leader is always responsible for the actions of their organization and subordinates, and catching spears is part of that responsibility. This leadership inspires. This concept also applies to the idea of "not leading from behind" or hiding behind your subordinates to allow them to catch spears and protect yourself. Again, this is why we strongly disagree that leadership is a "soft skill."
- **Inspire excellence** – "Actions speak louder than words" … but "words matter" to use a couple of clichés. Clearly, a leader needs to act like a leader and to lead by example. One of General Colin Powell's (2012) rules for leadership for inspiring excellence is "Check Small Things!" Clearly, a leader needs to act as they want the people they lead to act, as they will be emulated.
- **Prioritize accountability** – Another of General Powell's rules for leadership: "Have a Vision! Be Demanding! – Effective leaders do not accept poor performance and mediocre

results. They hold people accountable for their performance" (Powell 2012). Leadership requires the moral courage to hold people accountable, and to hold yourself accountable, even when it would be easy to look the other way. Holding people accountable does not have to be demeaning, "cancelling," or other disparaging options that seem more commonplace today; it is a professional responsibility that can be handled in a professional manner (notably, this does not mean that public awareness or openness of accountability actions is not required in certain contexts).

- **Know how to compete** – Policymaking and implementation is a competition of agenda alternatives and execution both within an organization and between organizations – while on the same "team" most programs, agencies, and organizations are competing with each other even when seeking the same ultimate endstate. The environmental manager needs to understand where competing policy alternatives are coming from and how they are supported, i.e., who is supporting them, why they are being supported, and what strings might be attached. Environmental managers will have to compete with peer programs and managers and adjacent agencies and organizations alike; to do so requires professional competition that puts the ultimate mission, protecting human health and the environment, at center always.

In the context of an environmental organization, a leader must know who they are leading (Barnes, Graham, and Konisky 2021; O'Leary et al. 1999). Most often people who enter the environmental management profession seek to protect human health and the environment, and see their work as a form of service, and rightfully so. Knowing that a sense of service is at the core of your subordinates' efforts is critical for the environmental manager because it is what motivates them and, when nurtured and grown, can be instrumental in their success day in and day out.

9.3 What Is Service?

In his 1941 State of the Union speech just prior to America's entry into World War II, President Franklin D. Roosevelt spoke of the Four Freedoms: "freedom of speech," in that everyone has the right to speak "truth to power"; "freedom of worship," where everyone has the right to their own beliefs; "freedom from want," where humans should not be deprived of the necessities of life; and "freedom from fear," in that humans should not have to fear aggression or other harmful actions from others. Service comes in various forms. It can be sought in the public, private, and non-profit sectors alike. A key and necessary tenet is that "service in the spirit of democracy demands an unqualified commitment to the common good" (Gawthrop 1998: xxiii). Environmental management, when centered on protecting human health and the environment, is unequivocally a form of service that seeks to provide people with Roosevelt's Four Freedoms, but especially freedom from want and freedom from fear as it relates to the environment people live in and rely on for their everyday lives.

9.4 Leadership in Environmental Management

While the description of service we have presented may seem loftier than many in the environmental profession are used to, environmental managers should consider that those they lead have similar values, or else work to instill and grow them. These values are intrinsic to the protection of the rights of environmental justice, the right to know what one is being exposed to, environmental or natural resource ethics, the ability to drink safe water and breathe clean air, and to believe that one has, and one's children have, a future that is not only sustainable but enjoyable in terms of what the Earth has to offer. That is, the values of service and a sense of service are essential to the provision and protection of a healthy environment for human and non-human flourishing. If the environmental manager does not understand the values or ethics of their subordinates, or if their values and ethics are not aligned, it will be difficult to manage them (O'Leary 2020).

Knowing your people, understanding and growing their values and ethics, and enabling them to fulfill their sense of service is the duty of the effective environmental manager as a leader. It is no easy task given the variety of individuals that an environmental manager will lead and manage, from the policy wonk to a bench scientist. To describe the relationship needs and leadership and management skills involved in effectively leading subordinates of varying professional backgrounds, we will focus on one group that environmental managers will often engage with: technical personnel or "techies."

9.4.1 What Is a "Techie"?

As discussed in Chapter 1, there are many professions that fill the ranks of environmental managers. That said, the skills that apply to much of the program management for protecting human health and the environment are of a technical nature (Barnes, Graham, and Konisky 2021). Indeed, some "charter members" (employees at the creation of the agency in 1970) of the EPA state that the "Agency is an agency of scientists" (A. J. Barnes, pers. comm. 2010). That said, the EPA, like many state-level and other environmental protection organizations, is staffed by policy analysts, auditors, environmental managers, and environmental scientists in equal measure, and has most often been led by somebody with a degree in law.

Ideally, most of those developing standards for permits or registration, implementing permitting, monitoring for compliance, conducting enforcement and technical assistance should understand policy but must have a mastery of whatever aspect of program implementation they are working with – air shed modelling, water quality assessment, hazardous waste disposal or management, standards compliance, technology transfer, etc. Most agencies and organizations hiring environmental managers desire an individual who understands management, policy, and science. Why? Because they need managers who understand complex and volatile environmental issues and legal trends and how to co-produce ethical environmental solutions, and who are able to implement solutions that recognize these issues and are produced accordingly (Barnes, Graham, and Konisky 2021; O'Leary et al. 1999).

In contrast to an environmental manager, a "techie" is an individual who is essential to achieving program goals but falls into a different stratum in the hierarchy of the organization and does not necessarily concern themselves with all the additional contexts that a manager does. For example, a water quality analyst techie does not have to worry about the environmental justice contexts or ecosystem management frame associated with the sample they are analyzing for potential pollutants, they can focus solely on the analysis protocol.

It is interesting to note that in many agencies and organizations techies are valued (compensated) as much as, if not more than, the managers. An example of this is that the professional engineer (PE) is possibly the most sought-after hire in an environmental agency and thereby can expect commensurate compensation. So, why don't most managers obtain degrees in engineering, chemistry, or genetics? Simply put, it is often about math and market entry and exit (Kennedy, Hefferon, and Funk 2018). I (M.L.) often ask my students why they did not become techies and the answer is that they were not comfortable with math. As well, I ask students if they have family members who are techies, and many do have them. When asked what it is like having a debate with one of those family members, most answer that it is difficult because their techie loved ones debate based on facts in a linear argument without much room for interpretation.

Besides a disinterest in quantitative methods, what can also make market entry into a more technical profession difficult is the amount of specific training and levels of certification often required of those professions – and once in it can be difficult to transition to a different career. Management principles hold across many professions and industries, but analytical protocols for soil sample analysis, for example, are specific to a technical job and are less transferable to alternative professions. Many readers can likely relate to one side or the other, preferring an emphasis on the technical and the need to be consistent or more variation and transportability. It is a difference in personalities and worldview (Kennedy, Hefferon, and Funk 2018; O'Leary et al. 1999).

9.4.2 Managing Techies

How do environmental managers work together with techies to achieve goals when there is a difference in hierarchy and worldview (i.e., a difference in frame of reference)? First and foremost, a leader must identify and overcome certain barriers to communication (six of the eight described in Chapter 6) with techies, and this begins with understanding the techie culture (Garnett 1992). Each barrier presents unique implications for managing and leading techies (just as they will when applied to any other specific job):

- Language – Not unlike many professions, techies have their own language, and they like it. It makes them feel special, and mastery of it, learning the terms and their meaning, is an achievement. The manager needs to understand that, when they are on a techie's turf, it is helpful and a positive gesture to attempt to speak their language (which means first taking the time to learn their language and work). Otherwise, just like when some Americans go to another country and expect everyone to speak English, a barrier to communication is erected (aka the "ugly American"). One does not have to be fluent to make an attempt, and it is often appreciated even if stumbled through. For example, when I (M.L.) worked at the Arizona Department of Environmental Quality (ADEQ), the head of the air quality office, Nancy Wrona, was very accomplished and had a Master of Public Affairs in Environmental Management. She did not like math and was not a techie but endeared herself to her techies by learning their language and demonstrating an effort to use it properly. Further, Nancy was adept at decoding that language for the lay public and politicians.

- Prejudice – Of course this concept is about cultural differences. The environmental manager should not think of techies in terms of "nerds" or "geeks" or other pejorative terms that are sometimes associated with this group, but rather as equally valuable members of the organization, which is exactly what they are. Conversely, techies are usually very intelligent, and normally they know it. The removal of prejudice and development of a functional and effective relationship will come through mutual respect and understanding sought and promoted by the environmental manager; the onus is on the manager as leader to make this happen.
- Hierarchy – This is an intimidating variable in most management situations and can be very similar to prejudice when leveraged negatively. Hierarchy is inherent in nearly all organizations and has important functional purposes, but it is not a tool to power-check others with but rather one to empower people from – or at least that is what the effective environmental manager should do. Techies want to solve technical challenges, and realize they need resources to focus on that task. Ensuring that techies can do this is the manager or administrator's number one job in this relationship. For example, Nancy Wrona would frequently conduct political management in the form of testifying in front of the state legislature's allocation committee. These testimonies could be contentious and uncomfortable. But not only could she translate the language of the techies into understandable and relevant concepts, she would also not back down when confronted regarding those facts. She was considered a "spear catcher" (a characteristic we described earlier) and a hero by the techies for representing them well, and they appreciated and accepted her leadership as a result.
- Distance/proximity – Part of leadership is creating an atmosphere where members of the organization who have diverse and specialized abilities can work together. While these individuals may have separate units or offices in the organization, the manager must be a conduit and transmitter for communication and an anchor for messaging between all of the members of the unit they manage to reduce distance and promote cohesion. Thus, management by walking around and conducting face-to-face engagement, and especially joining them in some of their work in the lab, in the field, or wherever it takes place, is one of the best ways to overcome this barrier when communicating with techies. Reduce distance through engagement.
- Distraction – this is a communication barrier in almost any management situation. Like with providing resources so that techies can do their jobs, the manager needs to identify and address distractions as they occur in the workplace. Whether it is

issues between co-workers, non-beneficial required training, or other time-, energy-, and resource-consuming distractions, reducing them will not only improve the efficiency of the workplace but also promote employee satisfaction in day-to-day work. It is on the manager to block for their subordinates and keep the distractions to a minimum.

- Frame of reference – Techies deservedly have been recognized for solving technical challenges but sometimes do not relate that objective to the goals of the agency. The manager can overcome this barrier by relating the objective, for instance of developing more accurate and usable standards for air-quality permitting to the goal of better air quality resulting in fewer health problems. This is a barrier which can often occur when the employee does not understand the political implications of the pursuance of their objectives.

Information overload and faulty communication abilities (the other two of the eight barriers to communication not included in the above list), while they are common barriers to communication with techies, are not specific to the techie and can be overcome as described in Chapter 6.

Scientists, engineers, and other technical personnel want to accomplish their objectives and expect their managers to advocate for them (O'Leary 2020). The ideal environmental manager will have the:

> "Critical" attributes [of] flexibility, interpersonal skills, communication skills, vision, enthusiasm, and persistence. "Important," as well, are company interests, outside contacts, organizational skills, teamwork values, and inside contacts. A "desirable" characteristic is technical expertise. Finally, a "necessary" ability is paperwork skills. (Koning 1988)

Thus, techies are willing to be managed by non-techies but want their managers to have the management skills to allow them to achieve their objectives and do not necessarily expect them to be technical peers (Drucker 1974; Koning 1988). Unfortunately, many environmental managers do not take the time and effort to overcome the barriers of communication between them and their more technically oriented personnel, resulting in a less efficient and less productive working unit. The ultimate impact is less protection of human health and the environment than could be optimally achieved, a result that should be considered unacceptable.

Often, whether leading and managing technical personnel or any other professionals, the shortcomings of the effective environmental manager can be rectified through training, experience, or other avenues. However, there are instances where the shortcomings are due to nefarious or other ill intent of the manager that leads to the intentional neglect of their duties. In these instances, it is important that action be taken, but it must be ethical, strategic, and well-crafted action in order to address and rectify the issue while also not "burning down the house." In contemporary culture, with social media and other public avenues available, it is best to approach these issues with tact, care, and empathy, even when addressing a clear injustice through dissent.

9.5 Leading from Below: The Ethics of Dissent and the Three C's

"The ethics of dissent" is a phrase used by Rosemary O'Leary in her 2014 text, *The Ethics of Dissent: Managing Guerrilla Government.* Dissent from current leadership and the direction of an agency or organization is not necessarily dissent from the agency or organization itself and, in fact, can be done in an effort to maintain its integrity and mission (O'Leary 2020). For example, during two different EPA administrations (Administrator Gorsuch, 1981–83 and Administrators Pruitt and Wheeler, 2017–21) many within EPA felt the agency was functionally crippled in terms of employees not being able to conduct regulatory and non-regulatory programs. As a result a cohort of employees did not follow the deregulation trend and worked to uphold as many programs and efforts as possible to slow the shift (Barnes, Graham, and Konisky 2021; A. J. Barnes, pers. comm. 2021). These dissenters did not conduct illegal acts but did take action that would have been considered not in alignment with the leadership they served under. Understanding and learning from their experiences and other perspectives on how the environmental manager should respond to unethical behavior by administrations or others in positions of authority is a complex but necessary task for the environmental manager in training.

Dissent usually arises when one or more of the three C's has emerged. The three C's are co-optation, collusion, and corruption:

- Co-optation: Winning over through assimilation into an established group or culture by means of establishing connections based on similarities in interests, background, ethics, values, or other characteristics (Merriam-Webster 2021b). This ethical concept might best be explained with Maslow's (1943) hierarchy of needs where after a person feels safe and secure they desire a social system and recognition before they are self-actualized. In short, it becomes easier to assimilate into a group if they have the same or similar background (e.g., education and training in environmental science or engineering). For instance, a pesticide manufacturer technical representative with a budget to promote pesticide use and sales, and a land grant university entomologist who is a public servant with a moderate income and funding, might have the same training and "speak" the same language. One has a frame of reference of furthering the interests of the pesticide company and the other to publish research on the effect and impact of pesticides. The "tech rep" may offer to provide funding for research regarding their new pesticide and the university researcher can accept it, as long as they disclose it upon publication. Over time, the two become professional partners and maybe friends designing research protocols, providing updates (over meals provided by the tech rep), presenting the research at sometimes exotic locations (paid for by the pesticide company(s)), and publishing the results.
- As this relationship develops, the researcher can become less objective and/or be used to provide cherry-picked results to receptive audiences (farmers, pest control advisors, the scientific community, or policymakers) through presentations and possibly published articles (Barnes, Graham, and Konisky 2021; Mascarenhas, Ferreira, and Marques 2018; O'Leary et al. 1999). This co-optation to provide biased evidence intentionally

or unintentionally for the purposes of a group or culture is not illegal per se but it is unethical. Scientists believe they are always objective, they are not (Latour 1987; Ziman 1996). Co-option as a relationship can occur, affecting many steps (standard setting, permitting, monitoring, enforcement, and technical assistance). As with the other two C's, Collusion and Corruption, the environmental manager must establish and enforce policies that ensure this is not occurring. Transparency and disclosure policies must be standardized and enforced with auditing processes established to help prevent the emergence of co-optation.

- Collusion: "Secret cooperation for an illegal or dishonest purpose" (Merriam-Webster 2021a). Collusion, in most cases, is illegal. For instance, if two or more parties whether private, public, or NGOs, decide that an action is beneficial to them but do not disclose to other affected entities how that action was decided on (e.g., it is decided behind closed doors without disclosure to all rightful stakeholders), they have colluded. This has occurred with industry health standards or product registration results (Freudenberg 2018; Grant 2012; Huff 2007; Shelley, Ogedegbe, and Elbel 2014) and possibly with some past EPA administrations (Cohen 2018). One technique to avoid collusion is for the environmental manager or administration to know the relationship between existing partners regarding environmental decisions (see Author's Note 11.1 in Chapter 11).
- Corruption: Dishonest or illegal behavior especially by powerful people (such as government officials); inducement to wrong by improper or unlawful means (such as bribery) (Merriam-Webster 2021c). This is the most serious ethical violation of the three C's and the most criminal, and it is often clearer-cut in determination – i.e., easier to demonstrate or evidence. This relationship requires a "quid pro quo" where one party benefits another party for providing something involving a betrayal of the public interest or "common good" – i.e., bribery; for instance, if an EPA administrator is renting a residence at a below market rate provided by a representative from a fossil fuel company (Worland 2018).

One type of corruption that often flies under the radar is when a job is provided, or made in trade, for a regulatory action. There are rules in effect limiting (by moratorium from months to years) the taking of and/or working on a job in a company that a regulator has been involved with, regulating the "revolving door" between regulator and industry worker positions (we discuss this further below). Not all transgressions looks potentially like corruption, but the environmental manager must be vigilant in detecting it and weeding out potential instances.

9.5.1 Guerrilla Government

Guerrilla government is "when [government] employees work – sometimes quietly, sometimes not – against the wishes of their superiors" (O'Leary 2020: ix). Guerrillas can be found in any organization be it public, private, or non-profit, but most research on these individuals has been primarily in the public sector. Why would environmental professionals become guerrillas? In short, they do it when one or more of the three C's has emerged and/or when they are not allowed to abide by their code of ethics or when they feel the integrity and mission of the agency or organization that they work for has been compromised. Imagine the public servant who finds that their administration will not allow them to protect human health and the environment or, worse, is using them to inflict harm on human health and the environment. Waldo (1980) discusses this in terms of a map of ethical obligations; however, one need look no further than examples in the Code of Ethics of the American Society for Public Administration to begin to understand how public servants could feel violated during these EPA administrations:

1. Serve the Public Interest – Serving the public interest includes co-production, transparency, compassion, and science-guided decisionmaking. Under both the Gorsuch and the Pruitt/Wheeler EPA administrations, literally hundreds of scientifically developed rules to protect human health and the environment were overturned and scientific underpinnings were ignored (i.e., not funded or defunded for regulatory monitoring) and effectually removed (Carper 2018; Dillon et al. 2018; Popovich, Albeck-Ripka, and Pierre-Louis 2020; Woolhandler et al. 2021). The assessed results are substantial increases in risk to human and environmental health both in the US and globally (Woolhandler et al. 2021).
2. Respect the Constitution and the Law – Public servants are required to respect and uphold the Constitution and the federal, state, and local laws of the US, especially those laws they are charged with regulating and enforcing. Even at the highest levels of environmental management, there are recurring failures to uphold this standard. For example, the Pruitt/Wheeler acceptance of the president's many executive orders rolling back environmental protections was found to be unlawful in many circumstances (Gilmer 2021) and at the state and municipal level recent high-visibility examples exist such as the Flint, Michigan water emergency (OIG 2018; US EPA 2021f). Sometimes administrators do wrong wittingly, other times they may truly deem their actions lawful, and the courts disagree, which is why the courts play such a critical environmental role by interpreting and enforcing the law.
3. Demonstrate Personal Integrity – Just like any profession, but maybe even more important to the environmental manager given their mandate to protect human health and the environment, is the requirement to demonstrate personal integrity, i.e., hold yourself accountable even when your organization may not. The environmental manager must have the moral courage to hold themselves and the organization they work for accountable even when it puts their career trajectory at risk, otherwise they are complicit in the actions they have detected and deemed wrong. A lack of transparency about, and then the inability to avoid at least the perception of co-optation, collusion and corruption, has sullied the reputation of environmental professionals as well as elected officials (Rich 2018).
4. Promote Ethical Organizations – Organization accountability and protecting the "common good" outstrips institutional loyalty. This is where the employee and the organization put "country before party" or "others before self" in terms of the organization or agency's mission and its actions. Often this value is expressed in terms of the

actions of the guerrilla such as whistle-blowing – whether through initiating an Inspector General Investigation at the federal level or a similar institutional process in private industry – or other flagging of illegal action to ensure the ethical action of an organization.

5. Strive for Professional Excellence – The effective environmental manager is exactly what we have been working to describe and train in this textbook. Such attributes of the environmental manager striving for excellence include perpetual learning, openness to change and growth, competency as a requirement, and self-accountability always. Most managers know that the competence of their organization reflects on them, personally and professionally, and thus diffuse these characteristics and strivings into the personnel they lead and manage. That said, particularly in large bureaucracies or organizations, complacency can settle in quickly and stunt progress and innovation toward professional excellence (Drucker 2009).

Ultimately the ethics regarding the culture of an organization pertain to relationships. Understanding that culture is to a community what character is to an individual, Aesop (circa 550 BC) said, "you can judge a man by the company he keeps." Rightly or wrongly, the company you keep often affects people's perceptions of you and your organization, especially if and when the interactions of that relationship are not open and transparent. Thus, the environmental manager must regulate the relationships they and their organization have and ensure they maintain openness in all of their dealings internally and externally. This is true for public service environmental managers and organizations.

What allows guerrilla government to be successful is the "combination of policy entrepreneurs and the politics of expertise" (Kingdon 2003: 20). Policy entrepreneurs are "advocates who are willing to invest their resources – time, energy, reputation, money – to promote a position in return for anticipated future gain in the form of material, purposive or solitary benefits" (Kingdon 2003: 179). These individuals might wield this skill as a professional and/or personal obligation, as they are often involved in the policymaking process. However, as guerrillas they are especially powerful in that they understand the politics of expertise – why and how things work. Through their involvement and knowledge, they can affect the everyday and long-term outputs of an agency or organization.

When policy entrepreneurs are also guerrillas due to an emerging issue,

> in worse cases, some are whistle-blowers but, in most instances, "believers" quietly, legally, and ethically overcome the barriers erected by political appointees and anti-environmental lobbyists. … They have sometimes done this partnering with agency critics (offending administration supporters in the process) and working on their days off. They take risks, and they fulfill their agency's mission to protect human health and the environment. (Lame 2005)

In other words, a guerrilla that has the tactical skill of a policy entrepreneur does not destroy their house but works to renovate it in accordance with the blueprints, i.e., the mandates, laws, and mission that guide the agency or organization.

When one decides to become a guerrilla, even for the most honorable of reasons, there are downsides because government and organizations alike have punished and still do punish risk-takers at times. Whistle-blowers seldom come out unscathed even though there are legal protections for them. Whistle-blowers are viewed by lawmakers, law enforcement, organizational leadership and some in administration as beneficial to accountability while others view them as akin to traitors. There is a process to whistle-blowing which requires going up the "chain of command" until it is clear that the correct actions are not being taken – i.e., your superiors are not actioning the issue you present and evidence to them. Then, the whistle-blower must attempt to use the legal system before "leaking" information to the press or a similar brash action.

Whether protected or not, once the whistle has been blown it usually spells the end of the relationship between the whistle-blower and their organization, though this is not always the case, especially depending on how the process was gone about and the position of the majority of the workers within the agency or organization on the issue at hand. Further, whistle-blowers often lose friends and colleagues who either are angry or embarrassed. Finally, the need to "make things right" can alienate family and friends, especially if they are part of the same organization or agency (Delk 2013). All of these considerations must be weighed and balanced before the environmental manager commits to an action as the whistle cannot be unblown (Delk 2013; O'Leary 2020).

Guerrillas sometimes do present problems for the environmental manager. Why? Because, according to O'Leary (2020), their resistance becomes part of the organizational or agency system in that guerrillas have certain resources or developed skills such as institutional longevity, insider information, contacts, and the ability to co-opt outside groups. With these resources often "Guerillas can do it to you in ways you'll never know" as it is difficult to gauge their productivity and they can "set you up" (O'Leary 2020). Unfortunately, while some guerrillas can be reasoned with, others might be unreasonable in terms of meeting their grievances, such that a negotiated outcome is unachievable. And many public organizations are not equipped to deal effectively with guerrilla government, making the achievement of a positive outcome even less likely.

The tensions regarding the match between worker values and administration values – for instance, political appointments loyal to elected officials versus line staff – are inherent and will never be fully resolved but they can be and are regularly renegotiated. As a point for the professional, sometimes a guerrilla might be termed a "loose cannon"; it is helpful to understand that loose cannons on ships of the sea have always been considered extremely hazardous in that they are very heavy and when unmoored can hurt the crew and damage the ship to the point of sinking it. Once a "bad" guerrilla has infected the culture of an organization, their disposition can quickly spread and substantially reduce the ability to protect human health and the environment beyond whatever their initial grievance is. So, on one end of the spectrum a guerrilla might be working to uphold the integrity and mission of an organization, but on the other a guerrilla might just be trying to bring everything down. For the environmental manager, when considering guerrilla actions and guerrillas it is important to differentiate the two – and everything in between. Often the "why" is the best differentiator; it indicates whether it is a needed action or the action of a disgruntled employee.

9.5.2 Managing Guerrillas

O'Leary (2020) has some suggestions for environmental managers such that they can manage the guerrilla. Most of the suggestions involve the concepts of ethical communication discussed in Chapter 6. O'Leary's (2020: 127) suggestions direct the manager to:

1. Create an organizational culture that accepts, welcomes and encourages candid dialogue and debate. Cultivate a questioning attitude by encouraging staff to challenge the assumptions and actions of the organization.
2. (Actively) listen to the people but also for the root of the dissent.
3. Understand the formal and informal organization (in terms of what are the values and norms of the organization and what are the values and norms of the people who comprise the organization).
4. Separate the people from the problem (in terms that the people are not the problem and might be able to co-produce a solution to the dissent).
5. Create multiple channels for dissent (as with strategic communication planning, the manager should provide multiple venues such as suggestion boxes, meetings, work, retreats, etc.).
6. Create dissent boundaries and know when to stop (or value the dissent but reserve the right to decide).

9.5.3 What Do Administrators Find Valuable about Guerrillas?

In the long run, guerrillas force leadership to think outside the box and to overcome groupthink where individuals in a group or organization begin to lose creativity and make sometimes irrational decisions because of their desire to conform (O'Leary et al. 1999). The challenges to management brought by guerrilla government, in a sense, regulate an agency, thereby forcing creativity and innovation which can result in increased efficiency and competitiveness. As well, these challenges can point out areas of complacency and complicity, particularly regarding following the law or constitution and codes of ethics. At the very least, the challenge of guerrilla government should force environmental managers to become better communicators. Managing guerrillas will not always go well and they can sometimes be quite destructive no matter how well-intentioned or not. But they will always exist and thus the environmental manager must always be ready to effectively lead and manage them.

9.5.4 Talking to the Troops

Aside from simply but intentionally communicating with subordinates to avoid dissent, there are a few areas of communication skills the environmental manager should apply when generally communicating with subordinates, peers, and superiors. As part of their management responsibilities the environmental manager must strategically plan their formal verbal communications with their superiors, peers, and subordinates. To review, they need to understand the management situation, who their audience is, craft their message(s), employ the media necessary, and obtain feedback (Garnett 1992). The management situation and audience are context-specific (as discussed in Chapter 6); however, the concepts we describe below will be helpful regarding crafting messages, sending those messages, and obtaining feedback in any context.

The environmental manager, understanding the importance of their mission and the need to be respectful of their superiors', peers', and subordinates' time constraints, should be aware of all of the barriers of communication but particularly information overload. That is why many communications with these audiences are called "briefings" – as in brief. A common complaint of environmental professionals is the number and length of meetings, and research continues to demonstrate that such complaints are right that meetings are overly used and less useful than we like to think (Geimer et al. 2015; Perlow, Hadley, and Eun 2017; Rogelberg et al. 2010). Both the public and one's colleagues appreciate the concept of "doing more and talking less." During a briefing an environmental manager should consider and incorporate the following six concepts, also known as elements of the "elevator speech," as they will make for more efficient and effective communication:

- Accuracy – be prepared with the most accurate and up-to-date information as you know it, and if you do not know it admit it with the intent of providing it as soon as possible.
- Include pertinent evidence – stick to the situation and the facts. Do not go off on a tangent.
- Don't prejudice – be objective with your opinions and admit bias if you have bias (and work to reduce, remove, or transform it).
- Anticipate superiors' arguments – know your audience's management style and positional responsibilities.
- State correlation and causes – use the message(s) to relate the situation to areas of accountability (issues and laws) and to the reasons (political or technical) that might cause the situation.
- Present upsides and downsides – no situation is likely to only ever have positive outcomes, and so the message should include both the expected upsides and downsides (i.e., the pros and cons) (Garnett 1992)

Face-to-face engagement is considered the most effective way to meet (Garnett 1992). Today's managers can accomplish this by well-planned in-person venues (digital videoconferencing platforms can be substituted when physically in-person meetings are infeasible, though the debate on the comparative effectiveness of this option is still without a clear answer). Either way, it is the manager's responsibility to set and enforce guidelines for conducting these meetings, for example, not allowing the use of smartphones or laptops during the meeting except to retrieve pertinent information as multi-tasking at meetings has been shown to substantially reduce their effectiveness (Cao et al. 2021). The most important step to holding an effective meeting is to set an agenda – and enforce it (Perlow, Hadley, and Eun 2017; Romney, Smith, and Okhuysen 2019; Tropman 2003). Many meetings can wander and never address the issue they are allegedly held for.

While open deliberation and creativity should be fostered and not suppressed, the meeting must also be managed so as not to snowball into just another breakroom discussion. Finally, at times it is difficult to obtain what the manager considers an important meeting with their superior. Along with implying the importance of the meeting to the superior or their executive assistant, having an appropriate relationship with that executive assistant is very helpful, i.e., networking with your superior's staff members can be just as important as connecting with that superior. Knowing and gaining the respect of your superior's "gatekeeper" is a time-honored strategy for being able to compete for the superior's time and attention.

9.6 Task Assignment

When communicating task assignment with peers or subordinates, common sense applies. One should try to remember the experience of having a professor who did not provide needed instructions for an assignment or a boss telling you to do something without context. Below are some helpful instructions Garnett (1992) suggests for task assignment:

1. Explain why this task is necessary in the context of mission – as we said earlier, the "why" is critical because it empowers the task-doer.
2. State the actual results expected at job completion – i.e., the desired endstate.
3. Describe how this assignment resembles or differs from previous tasks that the subordinate has worked on.
4. Provide the information essential to completing this task: why, who, when, what, where, and how. Only be as specific as necessary in order to empower the task-doer and allow them to creatively carry out the task when possible.
5. Check your subordinates' understanding of the crucial aspects of this task. In other words, elicit feedback to ensure a mutual understanding of the task.
6. Encourage your subordinates to rely on their resourcefulness but to ask questions if necessary.

9.7 Making Email Work for You

Email is currently the most common medium for communicating in the context of management with superiors, peers, and subordinates. This communication should be strategically planned but with a few specific peculiarities. Aside from informing and affecting attitudes and behavior, the email is a "paper trail" not only for reporting but to protect the environmental manager (i.e., it can CYA – "cover your ass," in the vernacular). This email should be clear regarding intent (the subject line), be well crafted with deliverables highlighted, and be copied to appropriate audiences including the sender. By "copying," the sender implies there will be follow-up, appropriate individuals in the chain of command are involved, and it helps accomplish CYA. Note that non-professional or non-work-related emails should not be sent by the environmental manager using their business email as these emails, particularly for public servants, can be requested using the Freedom of Information Act (as discussed in Chapter 6) process and made public. Emails also can be abused, that is, overused, and should not be considered a substitute for direct engagement with people.

9.8 What Is the Revolving Door? Advantages and Disadvantages

This chapter, aimed at addressing the management of environmental professionals, will conclude with the concept of the "revolving door." Most people have used an actual revolving door. It allows one to move safely, with little effort and waste of energy, from the outside to the inside of a structure and vice versa. The revolving door in the instance of career change and consequence allows one to move from the public sector to the private or NGO sector and vice versa, though in this section the concentration is on the public and private sectors' revolving door.

The revolving door is a common occurrence, particularly considering that public sector environmental professionals, in general, are not paid as well as those in the private sector and that the private sector needs to employ well-trained environmental professionals – so they hire environmental managers that the government paid to train. There are two major advantages to both sectors using the revolving door. First, change of sectors allows for an entry-level employee to gain experience of the environmental regulatory system in terms of permitting, monitoring, enforcement, and technical assistance; of how compliance and application works for those functions; and of how the bureaucracy works. Certainly, this process benefits both sectors in terms of understanding the volatile and complex environmental issues and fostering the ability to co-produce between the private and public sectors, i.e., they will have shared experience and knowledge to facilitate their relationship. Second, it also allows for a broader perspective regarding the frames of reference that each sector has. For instance, permit applicants are normally willing to comply with permitting standards and permit requirements, but permit writers may change in the process and/or regulations may change in the process as shared knowledge of the inner workings of industry or government adjusts the relevant policies and practices. These changes may be viewed as the norm inside the regulatory agency but are extremely frustrating to the applicant who justifiably wants a predictable process as opposed to a "moving target." The ability of either sector to understand the frame of reference of the other sector, regardless of whether it is a low-level employee or a senior-level appointee, opens the door for a better process and/or at least better communication.

Unfortunately, there are some serious disadvantages to the revolving door in that it, either in reality or perception, "opens the door" (pun intended) to the three C's:

- Co-optation – It is fairly easy for an employee to be co-opted, as professionally they should view those in the other sector (where they just came from through the revolving door) as entities willing to comply with environmental regulations to protect human health and the environment and not as enemies per se. As mentioned earlier, these individuals probably have a similar training and background and perhaps might even have been in the same sector at one point in their career. To some degree, depending on the extent of interaction, these two entities develop a professional relationship, and this should be encouraged as adversarial positioning often only detracts from a relationship. However, the innocent occasion could arise where the public employee is told their skills would be useful at the applicant's firm. Where co-optation could possibly exist is if that employee "fast-tracks" or "cuts corners" regarding standards for an application or case for the organization offering a position to them, to prove their efficiency at their work. To this point there is no quid pro quo or corruption. However, that would be considered, though difficult to prove, if the employee accepted a position with the applicant's firm. This co-optation can occur at all levels and the public is aware of that possibility when a public official rotates into a lucrative private sector position, as often happens.

- Collusion – When a regulator (seeking good favor for a future job) works hand-in-hand with the regulated to remove or disable the environmental regulations they are charged with enforcing, collusion has clearly occurred in pursuit of the revolving door. For example, say a political appointee from the private sector chemical industry, now in a position to influence the regulation of pesticides, secretly with industry representatives develops a scheme to continue the registration and use of a pesticide that had been slated for discontinuance. Unfortunately this recently occurred at the highest levels of environmental management in the US between EPA Administrator Scott Pruitt and Dow Chemical regarding regulations of chlorpyrifos (Biesecker 2021; Friedman 2019). Notably this happens at all levels, municipal to federal, and across all sectors, environmental to transportation to health services. Determining whether the revolving door is collusion or not can be quite difficult because such transience is much more common today and often is done without nefarious actions or intent.
- Corruption – Again this is relatively simple in terms of determining whether it has occurred or not, but is still hard to prove as there must be a quid pro quo, i.e., a determinable agreement of employment was made that led to some form of favoritism or otherwise improper benefit produced. Since most people moving in and out of different industries and sectors move into a related industry and sector, the environmental manager has to determine that a specific action was taken, prior to the employee departing, that directly favored their new employer and that goes against standard policy or practice. Importantly, for the environmental manager, this means making clear to the people they manage what is and is not acceptable in external, professional relationships, and monitoring closely any developments that seem to push the boundaries of the acceptable.

What can the environmental manager do to maximize the advantages and mitigate the disadvantages of the revolving door? The simple answer is to employ the management and communication strategies discussed in this chapter and others, as most people of service want to do the right thing. More specifically, the environmental manager should develop, communicate, and enforce guidelines and policies to influence, monitor, and regulate the activities of employees before and after they go through the revolving door in either direction. Knowing, talking to, and guiding the people you manage is a critical leadership function of the environmental manager; practicing these skills can stave off many of the issues that stem from the revolving door while maximizing the benefits of varied experience amongst employees. As mentioned, there are policies that prohibit agency employees from working on similar cases for the private sector once they join it for requisite periods of time. As well, there are policies with requirements for employees who are in the process of leaving federal and state service. The environmental manager must be aware of these policies and ensure their proper application.

9.9 Conclusion

Managing the experts, whether they are techies or guerrillas, is no easy task. That is why environmental management is primarily about people management. At the end of the day, the people the environmental manager leads and empowers are what will make the organization they are all a part of successful or not at protecting human health and the environment. Even the best measurements, policies, and plans cannot overcome poor people management.

9.10 End of Chapter Questions

1. What is effective and efficient leadership?
2. What is the basic role of a leader?
3. What are the basic duties of a leader?
4. What are the Four Freedoms?
5. What do you think are the most important aspects of managing technically oriented personnel?
6. Describe the benefits and liabilities of the "revolving door" in terms of government service and contracting.
7. Explain the three C's of ethics violations.
8. How would you prepare to communicate with your superior?
9. What is a guerrilla in terms of environmental agencies?
10. Why would an employee become a guerrilla?
11. What are five basic principles in the American Society for Public Administration's Code of Ethics?
12. What are five ways you would manage a guerrilla in your organization?
13. According to Garnett, what are six good suggestions for assigning tasks to your employees?
14. Name five of the basic duties of a leader that will enable them to "define reality and give hope."

10 Managing Others to Do Your Job: Contracting

Skill: *Balancing internal and external resources.* Environmental management is about people management. In the previous chapter we discussed managing employees that work directly for you, be they full-time or part-time employees. But often the environmental manager will also have to manage contract workers, meaning non-full-time employees that are often hired for a specific project or task and are not integrated into the functioning, culture, and fabric of an organization. Some contracts last weeks, while others last years, often filling a gap in the organization's capacities for that specific project. Contractors can be a great asset by augmenting the capabilities of an organization, but they can also become intractable to manage, resulting in a host of factors for the environmental manager to monitor and manage.

10.1 Introduction

This is the second chapter pertaining to resource management, specifically people as a resource. Contracting, sometimes known as outsourcing or privatizing, is when an organization or agency relegates program functions or procurement (obtaining services and goods) to entities outside of the organization. Contracting is accomplished via a formal agreement or contract with negotiated programmatic details for services expected in exchange for agreed (often monetary) compensation; whereas grant funding through partnerships with states or other sectors is less detailed and often carries fewer expectations of exchange after the initial funding or other resources are transferred to the grant recipient. Often environmental managers interact with outsourced entities in the form of **contract personnel**, **shared services** (federal centers consolidating government services), or **business process outsourcing** (BPO) such as payroll and other standard business functions. Thus, this chapter, in part, is about establishing accountability for the reasons to outsource and the performance of those the environmental manager is outsourcing work to.

Normally, except for shared services (again, where a group of contractors are consolidated as one hub providing services to multiple government agencies concurrently), it is the private and non-profit sectors that would be in a position to act as contractors, consultants, or partners. There are three basic types of outsourcing instruments:

- Procurement Contracts – for purchasing products and services (e.g., leasing space, printing, construction of facilities)
- Grants – where the grant recipient has complete control over development of the work
- Cooperative Agreements – legal instruments that facilitate the transfer of something of value from federal executive agencies to states, local governments, and private recipients for a public purpose or benefit (LLI 2021)

The expectation of contractors and consultants is that the public sector environmental manager manage appropriate program functions with them as partners (M. Michaelson, pers. comm. 2021) and that they co-produce the agreement and eventual outcomes of the project. The relationship is not always harmonious, especially depending on the contexts within which the contractors or consultants were brought in – i.e., sometimes the environmental manager is ordered to outsource something they would rather keep in-house – but much of this depends on how the environmental manager manages the relationship.

REAL-WORLD EXAMPLE 10.1 Consulting for environmental management

Government, non-profit, and private organizations all have business decisions they must make to support the delivery of their mission. These include, but are not limited to, budget, financial management, organizational structure and human resources, information technology, data management, and other support services decisions. The business decisions of organizations impact environmental managers and the work they deliver. However, environmental managers and the programs they deliver should be the priority influence on business decisions. Unfunded (or underfunded) mandates and new policies should first consider the needs of program delivery to effectively implement change.

As a public sector consultant, I often work with US federal agencies through various business decisions and organizational change to help gain efficiencies and improve mission effectiveness. Some examples of this type of work include a one-year timeline to reorganize all business functions that support the programs delivered by over 20,000 field staff; business process improvements that directly impact the way field staff obtain information technology support to deliver their mission; workload studies that summarize to leadership the work products and services of the field and the number of staff required to deliver those; or data management decisions for how program data are stored, analyzed, and reported to support government data-driven decisionmaking.

Environmental managers delivering programs often feel at the mercy of organizational decisions and change when not involved in the business decisionmaking. Successful organizational change, management, and policy decisions should, however, begin with

REAL-WORLD EXAMPLE 10.1 (cont.)

the needs and input of the field. You, as an environmental manger, will have the opportunity to influence these decisions. To influence management and policy decisions, I would urge you to:

1. Participate in working groups or meetings which request input from mission-delivery;
2. Communicate priorities, feedback, pain points, and concerns to leadership often, especially during any organizational change efforts;
3. Steward the use of your program data. You are the data owner of your program's data as the manager who is closest to the program with the most intimate knowledge of the information. As the owner of your data, you should be involved in how it is managed, shared, analyzed, and reported to ensure your data are used accurately;
4. Advocate for the technology your programs need. Work with your IT department to communicate the technical requirements you need to successfully execute your mission; and
5. Document and communicate each of your program's work products and services so decisionmakers understand the scope of your work prior to any organizational decisions.

Most importantly, as an environmental manager, you are the number one advocate of the programs you deliver. Ensure you communicate your needs and be as involved as possible in business decisions that could impact the work you do. The most successful business decisions will begin with you.

Roy Fillyaw, MPA-MSES, Public Sector Consultant

10.2 The Movement to "Contract Out"

In government organizations, while contractors have historically been used to provide goods and services, the current logic and reliance on contractors dates back to when the Clinton administration embraced the concepts from David Osborne and Ted Gaebler's (1993) book, *Reinventing Government*. This shift led to the National Partnership for Reinventing Government, originally the National Performance Review (NPR), which was an interagency task force to reform and reduce government waste (Kellough 1998). It coincided with the Government Performance and Results Act (GPRA) of 1993 (later updated to The Government Performance and Results Modernization Act of 2010) (Federal Register 2021). The impetus of NPR and GPRA was the national perception that government was too big and not accountable. The NPR had four goals: putting customers (the American taxpayers) first; cutting red tape; empowering employees to get results; and reducing government back to basics. Similarly, GPRA had six provisions:

1. improve the confidence of the American people in the capability of the federal government, by systematically holding federal agencies accountable for achieving program results;
2. initiate program performance reform with a series of pilot projects in setting program goals, measuring program performance against those goals, and reporting publicly on their progress;
3. improve federal program effectiveness and public accountability by promoting a new focus on results, service quality, and customer satisfaction;
4. help federal managers improve service delivery, by requiring that they plan for meeting program objectives and by providing them with information about program results and service quality;
5. improve congressional decisionmaking by providing more objective information on achieving statutory objectives, and on the relative effectiveness and efficiency of federal programs and spending; and
6. improve internal management of the federal government.

NPR is a bit like Total Quality Environmental Management (discussed in Chapter 8) in that it has been reinvented as a management concept over time (and multiple times) and is based on concepts commonly taught in schools of management. Consequently, managers became skeptical of it as doctrine:

> Successors to Clinton's reinvention "did their own thing to differentiate it from what happened before," says Donald Kettl, Dean of the School of Public Policy at the University of Maryland. "Unquestionably there has been a large amount of reform fatigue – 'so this is what they're calling it now, who knows what they'll call it next?'" he says. "That bit of cynicism is worsened by budget battles and shutdowns, so it's one more piece that makes federal managers weary." (Clark 2013)

Nonetheless, the idea of more effective and efficient government, or any organization for that matter, is always appealing. And a number of these goals have been achieved over time as the NPR initiative used GPRA to require programmatic evaluation and formal accountability (Clark 2013). Further, it allowed regulators to change their attitude toward those they regulate from "bad guys" to potential partners – a much more durable and functional relationship construct. In *Reinventing Government – Two Decades Later*, Charles Clark (2013) discusses a conversation with one of the senior managers implementing NPR in the Clinton administration:

> Bob Stone recalls being surprised that the NPR's meetings with chief executives at major corporations such as Dupont, Alcoa, General Electric and Disney showed that businesspeople were put off by the regulators more than the regulations themselves. "They didn't like being treated like crooks. Many in industry actually want to do good," he says, but some don't have sufficient engineering or legal staff to comply with regulations. "Agencies then took it upon themselves to educate" companies while reforming compliance processes. (Clark, 2013)

Though there have been some minor tweaks, NPR remains the driver for contracting out services as it has become a federal system norm (Clark 2013), establishing a model often though not always followed at state and local levels of government and environmental management.

This chapter hopefully will answer the questions: Has NPR been a success in terms of more efficient government through outsourcing? Why would the environmental manager contract out services? What are the advantages of contracting? What are the disadvantages? And what are the strategies the environmental manager should use to avoid the pitfalls of contracting?

The short answer as to whether the NPR initiative has been successful in reducing government waste with outsourcing is yes, but it needs to be balanced, and monitored (Kelman 1998). The American Federation of Government Employees (AFGE) bemoaned the fact that when Donald Trump became president in 2017 the federal workforce was in fact two workforces – federal employees and contractors – illustrating the tensions between public sector unions and contractor lobbyists (Bur 2020; Rein 2017). Not surprisingly, NPR has resulted in a substantial increase in contractors since its inception (Bur 2020). Indeed, contracting has been used as a "bait and switch" scheme to make government appear to be smaller or not growing, when in reality the actual number of employees and expenses increased.

Further, the focus over time from a resource management viewpoint is that compliance becomes the management issue instead of performance which was the goal – i.e., getting contractors to comply rather than utilizing contractors to maximize performance. So why has performance not improved more through increased contractor employment? Former administrator of the Office of Federal Procurement Policy Steve Kelman had several ideas as to why net improvements for this initiative have been difficult to ascertain:

> To put the idea into language familiar to federal officials and their contractors, constraints direct our attention to compliance, and goals direct our attention to performance. Constraints favor a legal mentality, and goals favor a management mentality.
>
> Constraints reflect important ethical ideals. Furthermore, when levels of corruption and cronyism in a procurement system are high, companies will be hesitant to bid for government contracts. That reduces the system's performance, especially because high-quality businesses will be the most hesitant to bid. More broadly, rampant corruption and cronyism can be devastating for the overall performance of an economy because potential entrepreneurs might invest energy into getting favors from government rather than running legitimate businesses.
>
> My answer is that improvements in the procurement culture coincided with a major decline in resources available for contract management. The culture was getting better, but contract management was getting worse so that we have ended up not noticing net improvements.
>
> In the interests of honesty and completeness, I should note that the reinvention that produced a procurement system more oriented toward performance was also the one that produced cutbacks in the procurement workforce, which prevented improvements in the system from being translated into better performance. If one reads the reinventing government report, one of the biggest sources of the $108 billion in savings it promised was a 272,000-employee cutback in the federal workforce, a number that often became the headline recommendation of the entire report. (Kelman 2017)

Thus, contracting needs to be monitored such that performance is the priority as opposed to compliance, and corruption does not inhibit the ability to obtain effective contractors. Without the resources to provide monitoring – e.g., an internal workforce to manage the contracting process – performance will suffer. Indeed, when the Affordable Care Act (aka Obamacare) was rolled out there were huge problems with the contracted BPOs to the point where the "Professional Services Council, the leading trade association for federal contractors, was complaining publicly that too many agencies lacked skilled workforce to manage the contracting process" (Pearlstein 2014).

The environmental manager needs to assure certain human resources are available to monitor the outsourcing of goods and services in their program. These specialists normally come in the form of a project officer (PO). A PO is "an agency program official who initiates a *procurement* action for other than a small *purchase*, evaluates contractor proposals, and/or, in the contract administration phase, acts as the technical representative of the Contracting *Officer* for monitoring contract performance after award" (US EPA 1991). Whereas, the contracting officer (CO) "is a person who can bind the Federal Government of the United States to a contract. Contracting Officers hold a warrant that allows them to negotiate on behalf of the United States Government. As the Government's agent, only COs may execute, modify, or terminate a contract" (FAI 2021).

10.3 Why Would the Environmental Manager Contract out Services?

The simple answer is so that they can be more efficient and effective – to reduce waste in government. However, O'Leary et al. (1999) suggest there are more specific and practical reasons:

- It is easier (more streamlined) for the manager in a bureaucratic system – because of a difficult government HR system, contracting often is easier and quicker than recruiting and maintaining or even firing a full-time employee (FTE). If "you gave them the choice of handpicking whomever they want and hiring them under an open-ended 'umbrella' contract with a private firm (with the power to fire them at a moment's notice), nine in 10 would hire the contractor" (Pearlstein 2014).
- It can save money by providing expertise when needed – rather than having an FTE that uses their expertise on a part-time basis, an administrative unit can save money by contracting an expert only when they are needed. Further, contract employees usually do not require or obtain benefits, which are significant expenses.
- It promotes flexibility – in a survey of senior federal managers by the Government Business Council the top reasons agencies outsource goods and services "are to leverage external knowledge and expertise, and to gain access to external technologies and business processes" (GBC 2015). The expertise desired is sometimes very specific or newly relevant, and to be able to obtain the services, whether human or software, by passing the

government bureaucracy for a narrowly defined but critical service should be considered both effective and responsive management.

- It can help maintain continuity – long-term support contractors to environmental agencies often retain staff for decades and can provide an institutional and regulatory "memory" as part of the regulatory development process, which often occurs over several years. For example, Mary Willett, an instructor of environmental management and former environmental services contractor at the Eastern Research Group (ERG), recalls, "my experience was that ERG (and staff at other contractors) were often the ones who stuck around for a long time and held the institutional memory. The turnover at EPA was much higher than at ERG and myself and others often were the ones who had 10–20+ years of experience in the regulatory development process. For example, ERG provided/es new EPA staff with Effluent Guidelines Development training (we called it 'Water School')" (M. Willett, pers. comm. 2021).
- It furthers social objectives – contracting can allow an environmental manager to prioritize small- and disadvantaged-business goals of the administration in their solicitations. This is a common practice implementing "affirmative action" regarding contractors that are veterans, women, and/or minorities.
- It gains political support – the reasons noted above serve as politically positive impetus for contracting. However, as mentioned earlier, contracting can shrink government or give the impression of shrinking government which often is considered politically correct. Further, there are very active and large government contracting lobbies at the federal and state levels that support politicians who support outsourcing. The environmental manager must always monitor for an "institutional anorexia" where shrinking the public sector workforce results in pathologies that might, at some point, cause harm to an organization.

10.4 The Advantages of Contracting

Of course, the reasons an environmental manager would contract out goods and services would be because there are advantages in doing so. Clearly, there is potential for more efficient use of resources, enhancing the protection of human health and the environment, and promoting the responsiveness of the agency. However, there are other advantages of contracting that are more applicable for program management, such as contracting can provide a more robust and timely measurement and monitoring of performance. As well, the manager can assure hiring the most qualified contractors by requiring competitive bidding and, finally, the normal program operations can be separated from the contractors, allowing for less disruption to that program (O'Leary et al. 1999).

Taking advantage of outsourcing is contingent on having the resources to manage contracts. Without having procurement professionals, and clear oversight of how contractors are performing through QA/QC and other evaluations, these advantages may not be realized (Clark 2013; Kelman 2017; Pearlstein 2014). The number of federal contractors has increased over the years across all sectors of government. From 1996 to 2017 the number of federal contractors rose from approximately 3 million to 4.1 million (Nguyen 2019), while the overall ratio of federal full-time employees to US citizens has actually significantly decreased over the last 50 years (Hill 2020).

10.5 Disadvantages of Contracting out Goods and Services

This section begins with the question, "should government be run like a business?" The answer is probably another book's worth of discussion. However, this question illustrates the dynamic between pro-government factions and libertarian, small government, free market factions. The arguments might be best left to economists and public administration researchers; however, the short answer to this question might be found in "The Tragedy of the Commons" where Hardin (1968) shows that without well-crafted regulations, situations can be taken advantage of by a few "bad apples." Perhaps a more enlightened question for the environmental manager is "can your agency withstand the negative consequences of contracting out goods and services if they arise?" Again, there are a number of potential disadvantages of contractors:

- Inability to hold contract employees accountable – While it might be easier in some ways to contract out personnel, the environmental manager can have less control over accountability concerns with a contractor, especially if there are concerns that they may or may not provide their best working conditions for their employees and their employees are not subject to the same personnel policies that government employees are (Joyce 2013). On the other hand, contractors understand that if their project manager is unhappy their contract could be in jeopardy. Thus, the project manager always has the obligation to provide timely feedback (M. Michaelson, pers. comm. 2021). There is certainly give and take and in the long term it is in the best interest of both sides to meet or exceed standards, but in the short term contractors may find cutting corners more beneficial for various reasons (e.g., reducing their own expenditures to increase their margin of profit from the contract).
- Contract personnel might cost more – According to a survey of senior federal managers, contract personnel can cost as much as 80 percent more than those performing similar functions as federal employees, which led these managers to list overpaying contractors as one of their biggest concerns (GBC 2015). That said, contractors provide a mechanism of flexibility, e.g., the discontinuance of a contract, that full-time employees, especially in government entities where they can be hard to release from employment (i.e., to fire), do not allow for.
- The three C's – If not monitored well, co-optation, collusion, and corruption can and do happen in contracting which is why, for example, the Federal Bureau of Investigation has an Office of Public Corruption that investigates matters concerning the federal government procurement, contracts, and federally funded programs. Always follow the money! The outsourcing of public sector needs for goods and services is a multibillion-dollar industry and, like any industry, has the potential for ethical and legal violations.

- Contractors can feel as though they are second-class citizens – Contractors are often outsiders to an organization and are not always well received or well integrated into their role at an agency. As such, they are generally treated differently by full-time employees and not always in a positive manner. It us up to the environmental manager to effectively manage how contractors are integrated, viewed, and treated by their employees.
- Loss of expertise and institutional memory – One of the most insidious problems with outsourcing personnel is that there is a loss of expertise and institutional memory. If a full-time employee is replaced by a contractor, in-house expertise and institutional memory may suffer if that contractor does not have the level of expertise of the person being replaced (Joyce 2013). It can also happen that the use of a contractor forestalls the hiring of a new full-time employee, further elongating the process of restoring institutional memory and the expertise capacity of the organization. Institutional memory is "the information held in employees' personal recollections and experiences that provides an understanding of the history and culture of an organization, especially the stories that explain the reasons behind certain decisions or procedures" (SAA 2021). That said, some contractor organizations have worked in partnership with an agency long enough that they have developed their own institutional memory such that the agency can draw upon and leverage that shared history, especially if the agency itself has a high employee turnover rate.
- This negative consequence to outsourcing affects not only the environmental manager but the organization. In some quarters the loss of institutional memory is considered "corporate amnesia" where each time an employee leaves an organization a part of the organization's memory is lost (Tinline 2016). In terms of leadership, this loss needs to be strategically avoided through networking, mentorship, and teambuilding such that new members understand the who, what, why, and where for situations that will arise. For instance, most kinds of organizational crises have happened in some form in the history of the organization (Ashkenas 2013). The organization should not have to reinvent the wheel to address the crisis.
- Hollow government – Probably the greatest disadvantage of outsourcing and what ultimately amounts to a threat to human health and the environment is the "hollowing out" of government. Hollow government (see Figure 10.1) is when an agency or administrative unit becomes disassociated from its functions, possibly losing control of those functions, resulting in the inability to be mission-oriented (e.g., protecting human health and the environment).

Of course, if correctly managed, contractors and partners (with their consultants) offer value to an environmental regulatory system. However, there are functions that should not be outsourced. At the federal level, these are Inherently governmental functions such as policymaking, regulating, and managing federal employees. Maintaining these functions must be performed exclusively by federal employees because it is "a function so intimately related to the public interest as to require performance by Federal Government employees," according to the Federal Activities Inventory Reform (FAIR) Act of 1998 and Office of

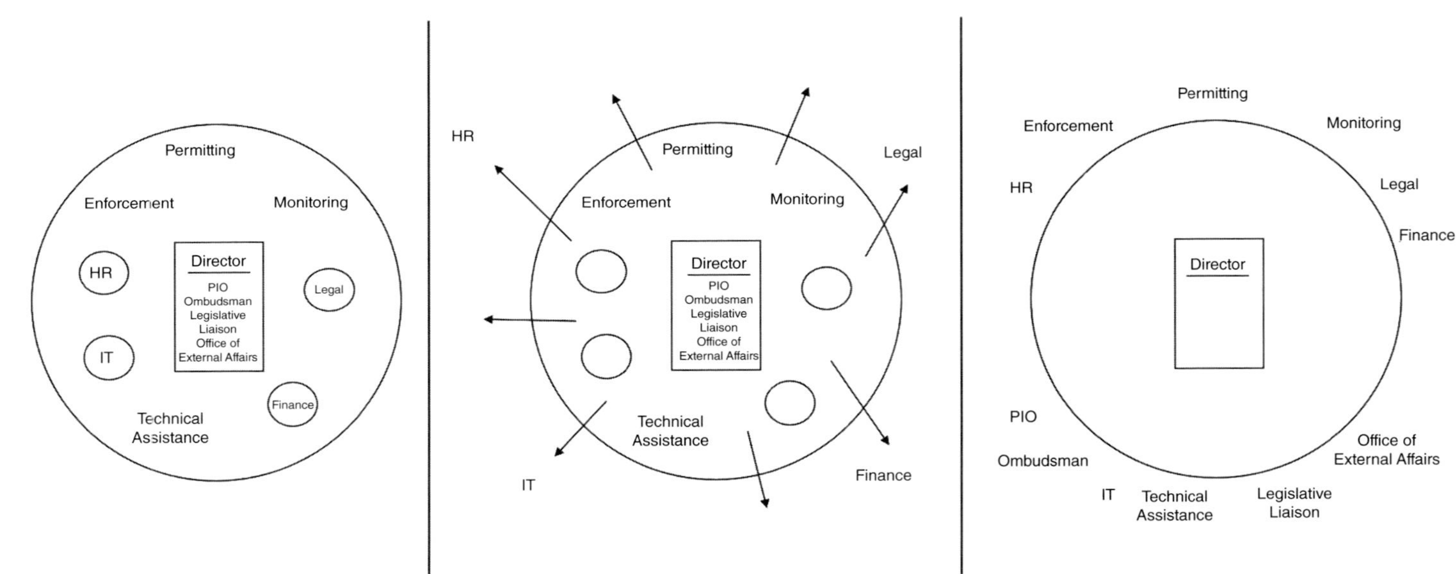

Figure 10.1 Hollow government.

Management and Budget (OMB) Circular A-76 (US EPA 2014i). This is not to say that external advisors (under the Federal Committee Advisory Act – FACA) should not advise federal regulators and policymakers but rather that external contractors or consultants should not create or influence regulations and policy. The idea that these regulations apply to state or local governments is unclear and varies by each entity. It can be observed that there is more reliance on outsourcing as it goes toward local levels and perhaps an over-reliance provides opportunities to divert from maintaining inherently governmental functions (Fernandez 2007).

REAL-WORLD EXAMPLE 10.2 From the other side

When you are an environmental contractor supporting an agency, here are some things that you care about the most:

Clear Statement of Work (SOW) in any solicitations – contractors respond to agency solicitations using overhead dollars, so writing a successful proposal is key. Contractors that don't win proposals don't tend to stay in business. The better the SOW, the easier it is to make sure that your proposal responds to what the agency is looking for and that the skills and experience you bring to the table will meet their needs. After a Request for Proposal (RFP) is issued, bidders have a chance to ask questions about the SOW – when in doubt, ask for clarification!

Clear directions from your project manager – government agency dollars are taxpayer dollars, so it is very important to perform the best work possible without wasting any resources. Good technical direction from your agency manager is therefore necessary and that requires good communication.

Building relationships with your clients – there are plenty of technical aspects to environmental contracting, but providing excellent technical service also requires developing good communication and a good working relationship with your agency manager. Start with active listening and keep communication happening throughout the process.

Being sure to document – when there is so much to do, you may be tempted to skip documenting your work and/or interactions with the client. Don't! Documentation is key to any contractor effort, both in terms of accountability and to show the path from SOW to final product.

Getting paid – contractors to governmental agencies are like any business, they need prompt payment of their invoices to pay their staff and stay in business.

Mary Willett, MPA-MSES, retired US EPA contractor

Thus, the idea that certain functions should not be carried out by contractors or consultants might be a legal requirement, but it is a critical management strategy as its compliance must be monitored particularly at the state and local levels where political pressures for privatization abound (Fernandez 2007).

How does hollow government work? It works because in the 1990s governments came to believe outsourcing made for more efficient and effective management. More important, today's environmental manager needs to understand how their agency looks in terms of a balanced organization. Figure 10.1 illustrates the worst case for hollow government at a state environmental agency.

10.6 Recommendations to the Environmental Manager for Contract Management

Upon reading this chapter, there are three basic and practical recommendations for the environmental manager who will be managing grants and contracts:

1. Outsource when the advantages can be realized while avoiding significant negative consequences.
2. The better and more detailed your Statement of Work (SOW) for a solicitation or grant request, the better your chances of getting good competitive bids and responses.
3. Always follow the money! As it is the metric for bad actors, relationships, and performance.

While there are many governmental requirements to take "the low bid" or RFP ... you get what you pay for and know what kind of contract you have. "Best value" contracting is where quality, the contractor's safety/health and past performance, take precedence over price. Or, the "lowest price technically acceptable" (LPTA) contract where the government determines its minimum acceptable technical requirements and attempts to award a contract to the lowest bid that meets the minimum requirements (TLB 2021).

REAL-WORLD EXAMPLE 10.3 Recommendations for contracting

Here are some specific recommendations from research and practitioner experts on government contracting (O'Leary et al. 1999; SPEA 2014).

Recommendations from O'Leary et al. (1999):

1. Critically examine circumstances where outsourcing is proposed
2. Be aware of the politics of the specific situation
3. Make sure competition is built in
4. Create a learning situation for rebidding
5. Check performance history – bad actors
6. Move incrementally when hiring contractors

REAL-WORLD EXAMPLE 10.3 (cont.)

7. Target opportunities for contractors and partners
8. Make them get insurance.
9. Use contractors to report non-compliance regarding the law and contract terms
10. Always try to buy outputs and outcomes rather than inputs
11. Move quickly when waste, fraud, and abuse are suspected, but within the boundaries of due process
12. Periodically review overhead costs
13. Be certain to investigate all laws involved
14. Be sure that your efforts to use outsourcing do not make the services for which you contract responsive to customer but not to citizens
15. Be alert that contractors and beneficiaries do not subtly – and not so subtly – begin defining what environmental management becomes.

Recommendations from "Government Outsourcing: A Practical Guide for State and Local Governments," an Indiana University, School of Public and Environmental Affairs panel report:

1. Determine initial motivations for outsourcing
2. Thoroughly assess the feasibility, potential costs, and potential benefits of the outsourcing initiative
3. Scope and plan the outsourcing project early on in the process
4. Develop clear, specific, and effective contract requirements
5. Encourage competition at different stages of the project
6. Select the "best" suppliers and partners
7. Spend adequate time negotiating and crafting the contract document
8. Assess contract performance both during and at the end of the contract
9. Minimize service disruptions and other difficulties associated with transitioning at the end of the contract

10.7 Conclusion

This chapter is an introduction to the skills required of specialized human resources that monitor contracts and grants and how the environmental manager might manage them. Further, as the reader might notice, the disadvantages of outsourcing often resemble what the advantages would be if they were not managed correctly. Contractors can cost more money, or they can save money; their inclusion can degrade institutional memory or, as long-term contractors, enhance it; they can be more accountable, but unscrupulous ones can "game the system"; and finally, they can add value to an agency or allow it to become disassociated from its functions of protecting human health and the environment. Either way, it should be understood by the environmental manager that while there are many advantages to outsourcing it is ultimately their responsibility to provide oversight, balance, and leadership in the context of what needs to be done, and who will do it, and that they have the resources to perform to your expectations.

10.8 End of Chapter Questions

1. What happens when an environmental manager fails to balance internal and external resources?
2. What are three basic types of outsourcing instruments?
3. What was the origin of outsourcing governmental goods and services?
4. Name three advantages of outsourcing government services.
5. Name two disadvantages of outsourcing government services.
6. What is a hollow government in terms of grants and contracts *and* what is the result of hollow government?
7. How would you manage a contractor?
8. Name three recommendations that would help an environmental manager regarding better contract management.
9. What is an "inherently governmental function"?

11 Understanding and Influencing Policy for Better Environmental Management

Skill: *Policy Entrepreneurship.* The effective environmental manager must also be a policy entrepreneur and be able to surf. Policy entrepreneurship allows the environmental manager to either innovate new policies for more effective execution of their role or new practices to better carry out the policies they are accountable to. Surfing is the process by which the manager understands and can navigate the contexts they work within to recognize opportunities and set themselves or their organization up to catch the wave.

11.1 Introduction

Policy is an idea which flows through all the ways in which we organize our life and a concept in the analysis of the process of government; but also, it is part of the process which it describes (Colebatch 2002). For the purposes of this chapter, policy can be defined in terms of what it should do. Policy provides guidelines for: (1) what action is to be taken; (2) who is to take it; and (3) whether these people have the capacity to take it. From that point of view, good policy, i.e., policy that is effective and efficient at guiding the achievement of its desired endstate, should result in successful implementation and practice.

This chapter will discuss the different policy instruments that environmental managers can use to protect human health and the environment and how they can become policy entrepreneurs. The first part of this chapter discusses the different regulatory and non-regulatory approaches analysts and policymakers should consider when developing strategies for "environmental improvement" (OP NCEE 2014). In particular, the traditional concept of command and control will be contrasted with challenge regulations. The second part of this chapter will illustrate the policy formulation process and why it is important for the environmental manager to understand it and to develop skills necessary to influence it.

11.2 US Environmental Policy: The Environmental Laws

To understand US environmental policy and the making of it, the federal environmental laws are the place to start (with the primary set of these discussed in Chapter 4). As you will read later in this chapter in Section 11.8, they are a compromise between alternative solutions to environmental problems, i.e., they are the chosen way to achieve a particular aim, among many other options, that was negotiated and agreed upon by elected officials. The compromises that are often in play in developing these laws are between providing protection for human health and the environment and the maintenance of our economy. The question that must be answered is can all risk to human health and the environment be eliminated? If Americans want to maintain economic well-being, then the answer is most likely no. That said, there is usually a bit, if not a lot, of room between all and nothing, and that is where our policies to protect human health and the environment generally settle. In formulating these laws, their intent was defined, standards for their provisional application were developed, and rules for implementation were devised. Environmental policy formulation is where the understanding and addressing of volatile and complex issues, the co-production of ethical solutions, and the implementation of sustainable ecosystem management come full circle.

Understanding that "science changes" and that circumstances – from administrations in charge to public preferences – also change, new policies or reformulated policies are and should be developed over time, adapting and improving their effectiveness at addressing an environmental problem. When environmental scientists better understand the effects and technical mitigation of a particular pollutant stream and situations change, normally a result of human activity (recall HIPPO in Chapter 4), this shift indicates the need for new or reformulated policies. Of course, these adjustments for improved organization and process will only occur if science is respected and will not be made if the science, or the contexts surrounding its development or implications, is rejected. The history of recognizing global climate change and the development (or the lack of development) of the policies to effectively address it is a good example of how scientific acceptance or rejection has impacted policy for protecting human health and the environment – in this case preventing the development of holistic and effective policy (Watts et al. 2015, 2017).

As well, circumstances regarding issues of accountability, ecosystem management, environmental justice, sustainable development, and unfunded mandates are impacted based on politics and public perception. Environmental justice, as a concept of environmental management, is a good example of how a specific framework of concern became an agenda item for policy. There was a shift in consideration regarding the idea that disenfranchised communities should not bear the brunt of environmental risk that resulted in the targeted identification of the unequal distribution of risk across the US, though much work to this end still remains today (Agyeman et al. 2016; Kojola and Pellow 2021; Shrader-Frechette 2002).

"The reality of environmental decision making is that Agency analysts are rarely in the position to select the economically efficient point of production when designing policy. This is partly because the level of abatement required to reduce a particular environmental problem is often determined legislatively, while the implementation of the policy to achieve such a goal is left to the Agency" (OP NCEE 2014: 4.1.1). Thus, the environmental policies environmental managers strive to implement encompass laws, rules, and regulations and they must employ a multitude of tools or instruments to effectively do so. But ultimately these mandates they are charged with permitting, monitoring, enforcing, and providing technical assistance for are policies developed by non-environmental managers that have been produced through balancing environmental science, public health, economic preferences, and social and political contexts. What does this mean for the environmental manager? It means that the policies given to them may not always be the policies they know to be best to effectively achieve their mission of protecting human health and the environment, which is in part why they need to be good policy entrepreneurs and surfers.

11.3 What Policy Tools Are Available for Environmental Managers?

At the federal environmental management level, policies are developed and analyzed through the EPA's Office of Policy. The OP guidelines include four basic approaches to environmental policymaking: (1) command-and-control regulations; (2) market-based incentives; (3) hybrid approaches; and (4) voluntary initiatives (OP NCEE 2014). According to the US EPA (1995: 2),

> An ideal environmental policy instrument would:
>
> 1. be cost-effective and fair,
> 2. place the least demands on government,
> 3. provide assurance to the public that environmental goals will be met,
> 4. use pollution prevention when possible,
> 5. consider environmental equity and justice issues,
> 6. be adaptable to change, and
> 7. encourage technology innovation and diffusion.
>
> Satisfying all seven of these criteria has seldom been possible in the past – and may be even more difficult in the future.

This prophetic 1995 statement puts the onus on contemporary and future environmental managers to understand what tools (or instruments) are available and how they are best applied. The EPA Office of Policy looks at environmental policies as regulatory and non-regulatory. Regulatory policies directly reduce pollution through the establishment of a standard while non-regulatory policies indirectly reduce pollution by means of incentives or other mechanisms to guide adaption away from pollution production (O'Leary et al. 1999).

REAL-WORLD EXAMPLE 11.1 The Clean Air Act and development of regulations

In addition to understanding the specific provisions of environmental laws in the development of regulations to carry out those laws, it is important that federal environmental managers understand specific requirements related to the regulatory process. This may sound like the bureaucratic part of an environmental manager's job, but it is vital to upholding and carrying out environmental laws.

The primary law that governs the regulatory process is the Administrative Procedures Act (APA) of 1946 (5 U.S.C. §551). This law ensures that the public is notified of proposed and final regulations by requiring that they are published in the Federal Register and provide opportunity for the public to provide written comment on a proposed rulemaking.

However, environmental laws can also have specific process requirements. For example, there are unique provisions built into the Clean Air Act (CAA) that ensure regulations that are developed under the law allow adequate public participation and transparency beyond what is already required by APA. This further ensures that CAA regulations are scientifically robust and legally defensible.

The Clean Air Act requires that all regulations be published in the Federal Register to provide the public the opportunity to request a public hearing (40 CFR § 51.102). Proposed regulations (in addition to other actions) must announce that opportunity for a public hearing no less than 15 days prior to the opportunity, and the public comment period must stay open for at least 30 days after the hearing. The Clean Air Act also requires that there be a public-facing docket that contains all records of any interactions with the Office of Management and Budget or reviewers during the interagency review process, technical support documents, and data upon which the rule is based, in addition to all other relevant supporting documents (CAA § 307(d)).

These procedural requirements allow the public to be actively involved and informed of the rulemaking process, which ensures that regulations developed under this law consider the latest science, implications for regulated entities, and public health concerns. In practice, this has proven itself because rules promulgated under the CAA that follow all proper processes typically hold up when challenged in court.

Jessie Mroz, MPA-MSES, is the current Tribal Coordinator for the Environmental Protection Agency's Office of Transportation and Air Quality, where she also oversees the process for all mobile source regulatory actions

**This work is not a product of the United States Government or the United States Environmental Protection Agency. The author is not doing this work in any governmental capacity. The views expressed are her own and do not necessarily represent those of the United States or the US EPA.*

11.3.1 Tools That Directly Limit Pollution

Tools that directly limit pollution are those that, through prescribed methods, have a predictable objective of lowering the emissions rate. These tools or "policy instruments" can be further divided as to whether they target single-source polluters such as entities that emit point-source pollution (smokestacks and discharge pipes) and multiple-source pollution or non-point-source (agricultural and stormwater runoff, mobile sources, etc.).

The command-and-control policy tool is considered the traditional regulatory instrument in that it "commands" with a prescription how much pollution a source can emit and "controls" (prescribes) how they will accomplish that (i.e., the equipment or process they must utilize in their production). The "command" relates to the harm-based standards which apply to a pollutant based on the maximization of a balance of social welfare and economic efficiency in the process of protecting human health and the environment. This protection of human health and the environment is more commonly expressed in our environmental laws in the form of:

- Design and technology standards – specify the technology and the production process that a source must use to control its pollution.
- Product bans and limitations – bans and restrictions on the manufacture, use, or disposal of harmful materials are a more extreme form of this traditional policy tool. "The most stringent form of prescriptive regulation is one in which the standard specifies zero allowable source-level emissions. For instance, EPA has completely banned or phased out the use or production of chlorofluorocarbons (CFCs) and certain pesticides. This approach to regulation is potentially useful in cases where the level of pollution that maximizes social welfare is at or near zero" (OP NCEE 2014: 4).

REAL-WORLD EXAMPLE 11.2 Command-and-control regulations: water and carbon dioxide

Environmental policy matters in the health of our planet, and a simple exercise based on some principles of physics you learned in middle or high school will make this clear.

In 1992 the US government required that showerheads release a maximum of 2.5 gallons per minute (gpm). Before that you could shower with a firehose if you could stand it. Moving from 3, 4 or 5 gpm to 2.5 seemed to be great progress. It was policy that got us there. Today there are very efficient showerheads to release at most 1 gpm and do so in a way that you would hardly notice that you've saved 1.5 gallons of water every minute.

If you think of the effect of this on climate change, the implications become stark. We'll assume a 10-minute shower by each member of a four-person family each of seven days of the week. The difference between a 1 gpm and a 2.5 gpm showerhead in the family shower translates into 15 gallons of H_2O for each shower for each of four people and for

REAL-WORLD EXAMPLE 11.2 (cont.)

each of the seven days of the week. The 1 gpm showerhead alone would save 420 gallons of water a week or nearly 22,000 gallons a year, from one household.

Now since these are showers, we tend to like them warm, so I will give you a way to think about the carbon cost of that 2.5 gpm showerhead that you refuse to swap out for a more efficient version.

If we assume that the mix of water for a shower is 50 percent cold water and 50 percent hot water, then the 15 gallons of water that is in "dispute" reflects 7.5 gallons of heated water and 7.5 gallons that comes directly from the well at wellhead temperature. So, what does it take to heat 7.5 gallons of water from, say, 10°C to 40°C so that they mix to a comfortable temperature?

From that high school physics class, you learned a simple formula that said a calorie is the amount of energy that it takes to heat one gram of H_2O one degree Celsius. Now if you knew how many grams there were in a gallon you would have some idea of the number of calories required to heat those 7.5 gallons the 30°C required for the warm shower.

It turns out that there are roughly 3,785 grams of H_2O in a gallon of water, so you have to increase the temperature of about 28,000 grams of water, and each gram 30°C. You next need to translate calories into a unit of British Thermal Units (BTU) so you can get to an estimate of the number of pounds of CO_2 per gallon of water, so each calorie is equal to about 0.004 BTU. 100,000 BTU is called a therm and each therm emits 12 lb of CO_2.

Something like $(28{,}000 \times 30 \times 0.004)/100{,}000 \times 12 = 0.04$ lb of CO_2 per 7.5 gallons of water.

This does not seem like a lot of CO_2 and hardly warrants swapping out showerheads. But in one family this turns out to be about 1.6 lb per day, or 585 lb per year. One family. If we have 100 million families in the US, this translates into about 58 billion lb of CO_2 annually because of inefficient showerheads. Old technology trying to accomplish what could be done more efficiently.

Of course, some of the assumptions could be wrong: not everyone showers every day, not everyone takes a 10-minute shower, and some people might already have low-flow showerheads. But we could reduce this carbon emission simply by making a new policy – what would be considered a command-and-control policy – that restricts the flow of a showerhead to something closer to 1 gpm. This would reduce water consumption as well as CO_2 emissions. Policy matters. It would not reduce emissions or consumption immediately because it cannot force compliance at the individual level. But over time old showerheads would be replaced by new, more efficient versions and consumptions and emissions would come down. This is much like CAFE standards for fuel efficiency on cars.

Pat Regan, Ph.D., former Professor at the Keough School of Global Affairs, University of Notre Dame

The advantages to command-and-control policies are that they can apply to specific pollutants and pollution production sites, they provide economic equity as they are not based on cost effectiveness but economic efficiency, and the results are more predictable for the regulators and regulated entities. The disadvantages are that there are no incentives to go beyond compliance (discussed in Chapter 9), their regulation is resource-intensive in terms of monitoring and assessment, and they are difficult to apply to non-point-source pollution.

Another tool that directly limits pollution, like command-and-control regulations, but that can effectively be applied to multi- or non-point sources of pollution as well as point sources is challenge regulations. Polluters regulated by these performance-based policies must still meet the harm-based standards commanded by the regulation but are further "challenged" to meet that performance-based standard with the most efficient technology the polluter deems appropriate. In other words, they are given the standard they must meet but, unlike command-and-control regulations, it is left up to them to figure out how to meet it. This challenge then acknowledges that the polluter might best understand how to reduce or eliminate a pollutant and it incentivizes the regulated entity to innovate, which is considered a desirable characteristic (Esty and Porter 2002; Porter 2011).

REAL-WORLD EXAMPLE 11.3 Environmental managers and environmental policymaking

The major pieces of environmental legislation – the Clean Air Act, the Clean Water Act, etc. – get a lot of attention, and without a doubt these regulatory policies are critical to managing our environment. But in Iowa, as is the case across much of the country, one of the largest policy influences on the landscape is the US Farm Bill. Reviewed by Congress roughly every five years, this is a sprawling piece of legislation that includes, amongst other things, authorizations for a host of voluntary conservation programs designed to incentivize, motivate, and assist farmers and agricultural landowners in minimizing the negative impacts of modern agriculture.

These conservation programs have acronym-laden names: the Environmental Quality Incentives Program (EQIP), the Conservation Stewardship Program (CSP), the Conservation Reserve Program (CRP), and the Agricultural Conservation Easement Program (ACEP) are probably the best-known. And while the bones of these programs are laid out in federal statute, there are also enormous opportunities to alter and improve the implementation of these programs by engaging with administrative policy such as rulemaking. Every state also has a state technical committee whose role is to advise the US Department of Agriculture (USDA) on the implementation of Farm Bill programs.

Engaging in administrative policy at any level can be tedious, time-consuming, technical and frustrating. It is also a critical tool for ensuring the government agencies are implementing programs and policies in alignment with both the letter and spirit of the

REAL-WORLD EXAMPLE 11.3 (cont.)

authorizing legislation. My day-to-day efforts to strengthen voluntary conservation programs involve working with farmers, agricultural stakeholders, the USDA and partner organizations to identify opportunities for and barriers to making the Farm Bill better. Inevitably, administrative policy of some shape or form dominates those conversations. Legislative policy – working with legislators and motivating grassroots action – is typically reserved for large-picture items: more funds, a new program, etc. The devil is in the details, and it is my work with administrative policies and procedures that truly impacts how programs, and therefore ultimately resources, are delivered to farmers and thus impact the landscape.

Taking part in administrative policy issues can be daunting. The learning curve is steep, and at times it seems like everyone knows more than you do. But it is a vital and necessary component to designing and implementing effective policies and programs that have their intended impacts on the landscape, and therefore something that all environmental managers must do.

Jorgen Rose, MPA-MSES, is program and policy coordinator for Practical Farmers of Iowa, a 501(c)(3) non-profit that works throughout Iowa and across the Midwest to equip farmers to build resilient farms and communities

11.3.2 Tools That Indirectly Limit Pollution

Tools that indirectly limit pollution are those that lower the "aggregate" pollution concentration in a specified area (such as the air quality within a city's limits) rather than the emission rates of individual polluters. Aggregate emissions will depend on the number of polluters and the output of each polluter. As either production rate or total size or number of pollution producers increases, so will aggregate emissions. Even when the standard is defined in terms of an emission level per polluting source, aggregate emissions will still be a function of the total number of polluters. Some examples of regulatory schemes to limit pollution by aggregate levels include:

- Cap-and-trade systems or emissions trading limit the total amount of pollution that can be produced and allow polluters to purchase and trade emissions allotments. "In a cap-and-trade system the government sets the level of aggregate emissions, emission allowances are distributed to polluters, and a market is established in which allowances may be bought or sold. The price of emission allowances is allowed to vary. Because different polluters incur different private abatement costs to control emissions, they are willing to pay different amounts for allowances. Therefore, a cap-and-trade system allows polluters who face high marginal abatement costs to purchase allowances from polluters with low marginal abatement costs, instead of installing expensive pollution control equipment or using more costly inputs. Cap-and-trade systems also differ from command-and-control regulations in that they aim to limit the aggregate emission level over a compliance period rather than establish an emissions rate" (OP NCEE 2014: 7).

- Paying to pollute through pollution fees or emissions taxes (tools that require a fee to be paid per unit of pollution produced) is another way to reduce aggregate emissions and incentivize pollution prevention. "Emissions taxes are exacted per unit of pollution emitted and induce a polluter to take into account the external cost of its emissions . . . As an example of how an emissions tax works, suppose that emissions of a toxic substance are subject to an environmental charge based on the damages the emissions cause. To avoid the emissions tax, polluters find the cheapest way to reduce pollution. This may involve a reduction in output, a change in inputs to production, the installation of pollution control equipment, or a process change that prevents the creation of pollution" (OP NCEE 2014: 9). Pollution fees or emissions taxes are thus a negative incentive in that they require the polluter to pay to pollute.
- Environmental subsidies have been used to incentivize pollution prevention by directly subsidizing (i.e., reducing the cost of) or providing grants for the purchase of new more efficient equipment, for the purchase of vehicles that minimize the use of fossil fuels, or for starting programs that reduce pollution, such as a city recycling program. "Subsidies paid by the government to firms or consumers for per unit reductions in pollution create the same abatement incentives as emission taxes or charges" (OP NCEE 2014: 5). Environmental subsidies are thus a positive incentive as they pay the polluter or somehow sponsor them to pollute less.

REAL-WORLD EXAMPLE 11.4 Pollution prevention grants from the US EPA

The Pollution Prevention Act of 1990 directed EPA to offer grants to state and Tribal programs to provide technical assistance to businesses to help them learn about, develop, and adopt pollution prevention approaches in their operations. Pollution prevention, also called P2 or source reduction, are practices that reduce or eliminate pollutants from entering a waste stream or the environment prior to recycling, treatment, or disposal. The idea is to prevent pollution from even being created so that it does not need to be controlled, mitigated, disposed, or cleaned up.

State and Tribal P2 Technical Assistance Providers receiving EPA P2 grants help businesses learn about, develop, and implement a range of P2 approaches which can help protect human health and the environment and often save businesses money. P2 practices include equipment or technology modifications, production process modifications, consideration of supply-chain impacts, reformulation or redesign of products, substitution

REAL-WORLD EXAMPLE 11.4 (cont.)

of raw materials and safer chemicals, and improvements in housekeeping, maintenance, worker training, or inventory control. Additionally, state and Tribal P2 Technical Assistance Provider grantees are required to document and widely share the P2 approaches they help develop or implement. That way, when P2 technical assistance providers help one business, other similar businesses have the potential to benefit from the lessons learned and potentially replicate the P2 actions.

Between 2011 and 2019, EPA issued 451 assistance grants totaling $48.8 million. The technical assistance from these grants helped American businesses identify, develop, and adopt P2 approaches resulting in the following cumulative benefits.

- $1.9 billion savings for businesses.
- 706 million pounds of hazardous materials reduced.
- 40.4 billion gallons of water saved.
- 16.9 million metric tons of greenhouse gases eliminated.
- 25.9 billion kilowatt-hours of energy savings.

Jenna Larkin MPA-MSES, Environmentally Preferable Purchasing (EPP) Program at the US Environmental Protection Agency

According to the EPA's Office of Policy, National Center for Environmental Economics (NCEE), environmental economists prefer cap-and-trade, pollution taxes and environmental subsidies (i.e., market-based policies as opposed to command-and-control or challenge regulations) because they are less resource-intensive, they require less programmatic input from regulators, and they promote technological innovation for better performance and efficiency (OP NCEE 2014). There are many other instruments or tools available to the environmental manager to help promote or create pollution reduction. Examples include:

- Information disclosure or reporting is an instrument that is commonly used in regulating pollution. Allowing affected communities and consumers to understand their risk exposure such as with the Toxic Release Inventory (TRI) or pesticide labels provides external pressure on polluters. "Requirements for information disclosure need not be tied explicitly to an emissions standard; however, such requirements are consistent with a standard-based approach because the information provided allows a community to easily understand the level of emissions and the polluters' level of compliance with existing standards or expectations. As is the case with market-based instruments, polluters still have the flexibility to respond to community pressure by reducing emissions in the cheapest way possible. The use of information disclosure or labeling rules has other advantages. When expensive emissions monitoring is required to collect such information, reporting requirements that switch the burden of proof for monitoring and reporting from the government to the firm might result in lower costs, because firms are often in a better position to monitor their own emissions. If accompanied by spot checks to ensure that monitoring equipment functions properly and that firms report results accurately, information disclosure can be an effective form of regulation. Without the appropriate monitoring, however, information disclosure might not result in an efficient outcome" (OP NCEE 2014: 11).

- Liability rules "are legal tools of environmental policy that can be used by the victim (or the victim's government) to force polluters to pay for environmental damages after they occur. These instruments serve two main purposes: (1) to create an economic incentive for firms to incorporate careful environmental management and the potential cost of environmental damages into their decision-making processes; and (2) to compensate victims when careful planning does not occur" (OP NCEE 2014: 13).

11.4 Non-regulatory Policies

As we have discussed in other chapters, regulatory program management is costly. Permitting, monitoring for compliance, and enforcement require resources that can be very expensive and difficult to obtain for the environmental manager. While technical assistance and education also require resources, they usually are less resource-intensive or costly because you are guiding someone else in their process (and thus it is their investment in all of the materials and other resources necessary for that process) as opposed to creating and managing your own. As such, promoting technical assistance and education programs are often a cost-efficient, non-regulatory (because these programs do not require the monitoring or enforcement of a regulation) way for the environmental manager to target and reduce pollution.

Policies or initiatives to prevent pollution through the transfer of technology are often promoted by environmental managers through partnerships. Sometimes these partnerships include grant programs such as the Pesticide Environmental Stewardship Program (PESP) that funds the diffusion of integrated pest management in areas the EPA feel will benefit from technical assistance (US EPA 2016a). In this case the "partnership" is formed around a policy of funding change agents for pollution prevention for pesticides. The EPA also has "centers of excellence" that provide an educational approach to pollution prevention by providing best practice training to state and city level environmental managers. There are also voluntary programs where polluters move "beyond compliance" such as with the popular Green Power Partnership (GPP) and Energy Star and provide added value and innovation to the pollution prevention efforts of regulators. Voluntary programs can use the following four general methods to achieve environmental improvements: (1) require firms or facilities to set specific environmental goals; (2) promote firm environmental awareness and encourage process change; (3) publicly recognize firm participation; and (4) use labeling to identify environmentally responsible products (OP NCEE 2014: 19).

These methods are not mutually exclusive, and most US voluntary programs use a combination of methods. All these programs and initiatives are examples of non-regulatory policies that the environmental manager can leverage to reduce pollution without or in addition to direct regulation.

11.5 Managing Available Policies

Like so many other management concepts and skills, policies – and their implementation – can be more effective if they are integrated. The environmental manager must determine what goals they want to achieve in terms of protecting human health and the environment but consider the system they are implementing those policies within or on (e.g., agrochemical industry system or a municipal transportation system) and the economic impact of that implementation. Often different policies are different sides of the same coin, like taxes and subsidies or information reporting and liability, and their design and implementation can and should be combined.

11.5.1 Guidelines for How the Environmental Manager Might Consider and Match Policy Instruments to Problems

There are four basic guidelines for how the environmental manager might consider and match policy instruments to problems to produce more efficient implementation processes and outcomes (O'Leary et al. 1999):

1. The more dispersed the sources of a problem are, the less likely it is that direct regulatory designs for monitoring and enforcement will be effective. For more dispersed pollution production management, challenge regulations and incentives programs tend to be more effective, for reasons we discussed previously.
2. The more dynamic a sector is, the greater the need for flexibility and innovation – as different industries and firms within an industry can have different, specific (and proprietary) technologies that address pollution. The environmental manager thus must recognize what best facilitates the achievement of emissions targets for those they regulate – as far as the policies they must uphold allow them to – and provision support or policies accordingly.
3. Problems that are local or regional in scope require local or regional response. For example, every US state is required to have a State Implementation Plan (SIP). This is because each state has different environmental, economic, and political factors that affect the utility and potential of the various pollution policies and regulatory approaches for their contexts.
4. Policymakers may combine various policy instruments to fashion effective strategies for responding to environmental problems more holistically and efficiently. As with most technological and behavioral solutions, an integrated approach is more effective.

REAL-WORLD EXAMPLE 11.5 One size doesn't fit all for regulatory purposes

As an environmental manager with the responsibility of being a government regulator, be wary of imposing "One Size Fits All" solutions to problems. Doing so can be counterproductive to achieving goals and instilling public confidence.

Take the example of California State regulators managing water conservation during a severe drought. Without question the state overall was suffering from extreme drought conditions. On May 5, 2015, the State Water Resources Control Board (SWRCB) adopted an Emergency Regulation mandating water agencies to significantly reduce water consumption.

Yorba Linda Water District (YLWD) was ordered to reduce water use by 36 percent or face state penalties up to $10,000 per day for each day out of compliance. Ironically YLWD was not locally experiencing the drought problem facing the rest of the state. The district and community had the foresight to invest in being drought resilient. They constructed capital facilities to store a minimum of five years' water supply in the event of drought. Investments in the Orange County Groundwater Replenishment System allowed purification of wastewater, returning it to potable standards, then using this water to recharge the aquifers, making it once again available for drinking water wells. Many state-of-the-art water conservation technologies were in use in the community, including leak detection, smart water meters and irrigation controllers, stormwater collection, and more.

Years of expensive drought resilience investment effectively protected YLWD and other Orange County water districts from suffering drought effects experienced by those who failed to make investments.

Proving the adage that "no good deed goes unpunished," imagine the shock to the YLWD community when regulators mandated a 36 percent water conservation mandate. No consideration or exception was allowed for those who made investments and were not in the same state of emergency.

When asked why no exception was made, the SWRCB stated "everyone must feel the pain."

The effect of this centralized environmental policy was counterproductive in many severe ways. After promising the community if they paid more on their monthly water bills, they would be drought-resilient, the State mandate discarded this promise and threatened fines for non-compliance. Public trust was destroyed and resulted in the YLWD Board of Directors being recalled. The 36 percent water reduction mandate and threat of fines for non-compliance required an instantaneous $25 per month water base fee increase per customer for the district to break even despite also using reserves. Customers had to use less water and pay more for it. The Yorba Linda Taxpayers Association followed with a

REAL-WORLD EXAMPLE 11.5 (cont.)

failed lawsuit to stop the rate increase, only increasing the financial obligation of customers for legal fees to defend against the lawsuit.

Ironically the revenue lost as a direct result of this centralized state policy was programmed for further water conservation measures which had to be scrapped. Environmental managers such as the YLWD management team and community should be held as the shining example for others to follow. Instead, a centralized emergency policy with no allowed exceptions destroyed public confidence in state water managers and wasted scarce public funds.

Marc Marcantonio, former General Manager, Yorba Linda Water District

11.6 How Does the Environmental Manager Evaluate the Effectiveness of a Policy?

In determining the effectiveness of a potential approach to the problem they are faced with, the environmental manager should consider the following factors and questions (OP NCEE 2014: 21):

- **Environmental effectiveness**: Does the policy instrument accomplish a measurable environmental goal? Does the policy instrument result in general environmental improvements or emission reductions? Does the approach induce firms to reduce emissions by greater amounts than they would have in the absence of the policy?
- **Economic efficiency**: Does the approach have the potential to achieve the most efficient outcome? Does the policy instrument reach the environmental goal at the lowest possible cost to firms and consumers?
- **Reductions in administrative, monitoring, and enforcement costs**: Does the approach reduce the administrative or other costs associated with the implementation of this program compared to other available options? How large are these cost savings compared to those afforded by other forms of regulation?
- **Environmental awareness and attitudinal changes**: While meeting goals, are firms educating themselves on the nature of the environmental problem and ways in which it can be mitigated? Does the promotion of firm participation or compliance affect consumers' environmental awareness or priorities and result in a demand for greater emissions reductions?
- **Inducement of innovation**: Does the policy instrument lead to innovation in abatement techniques that decrease the cost of compliance with environmental regulations over time?

11.7 The "Regulation Dilemma"

In general, environmental managers strive to implement policy tools that go beyond command-and-control regulations to facilitate innovation and increased performance, especially non-regulatory options. Some examples that are particularly attractive to policymakers are the concepts of partnerships and challenge regulations, where it is implied that regulated industries know best how to reduce their pollution and they deserve flexibility and incentives for regulatory compliance. To effectively utilize these options requires cooperation between environmental managers and pollution producers. The dilemma becomes that, while this cooperation offers many of the advantages of a good policy tool, both regulators and the regulated entity have incentives to game the system (Potoski and Prakash 2004).

Most often the cooperation required of polluters is in the form of moving beyond compliance when reducing pollution beyond the regulated levels is possible, i.e., that they should reduce pollution as much as they can and not just as much as they are required to (Potoski and Prakash 2004). However, even with challenge regulations there must be a level of trust between regulator and regulated regarding whether the technology a regulated entity has put forward as a means of meeting a prescribed emission rate or level does in fact work and is the most efficient option available. Examples of this might be witnessed in clean coal or carbon sequestration technologies where a pollution process is allowed to persist because mitigation technology sufficiently reduces the total pollution produced. However, there are instances when this trust is violated, such as with Volkswagen's recent vehicle pollution detection defeat technology scandal, that make this relationship a potentially difficult one to maintain (Schiermeier 2015). Thus, the overriding concept or dilemma which must be embedded in these types of programs is trust.

The purpose of regulatory relief, i.e., regulatory programs that incentivize or promote industry innovation rather than command-and-control regulations, especially for organizations and industries that have proven to be cooperative with environmental managers, is to provide flexibility to polluters and an assurance that regulators are meeting their goals without having a "policeman on every block" (Potoski and Prakash 2004). The intended outcome then becomes a "win–win" whose chances of success increase when parties have long-term, face-to-face interactions resulting in an institutionalized relationship (Axelrod 2009; Axelrod and Dion 1988; Hardin 1982). Unfortunately, even this long-term cooperation (and agreement) can be abused if one party or another has an overwhelming motive to cheat (Potoski and Prakash 2004). Often, this relationship changes with a change in management or administration, which can create a window of opportunity for an entity to exploit.

While not the only means to do so, there are three common strategies that the environmental managers can employ to avoid a lose–lose outcome (Potoski and Prakash 2004):

- The environmental manager can persuade regulated entities to join voluntary programs by demonstrating pollution prevention reduces pollution, production and insurance costs (Hart 1995; Porter 2011; Porter and van der Linde 1995; Schmidheny and Zorraquin

1996); enhances their reputation with consumers (Charter and Polonsky 2017); can stave off imminent, more stringent regulations (Salop and Scheffman 1983); and enhances their reputation with stockholders (Lozano, Carpenter, and Huisingh 2015).
- Provide incentives to partake in legitimate Environmental Management Systems programs where standards for compliance are evaluated and reported.
- Provide public participation and environmental mediation opportunities for environmental advocacy groups and industry to develop trust. Of course, this is easier proposed than accomplished as both groups not only have different worldviews but incentives to oppose one another.

This "nurturing" of relationships by the environmental manager can produce environmental benefits (Kettl 2002; Potoski and Prakash 2004) but only if they "trust but verify" their agreements and maintain credibility with appropriate enforcement responses if the agreements are breached.

11.8 Why Is Policy Formulation Important to the Environmental Manager?

The environmental manager needs to understand and develop skills for influencing policy for several important reasons: as leaders they are obligated to "know the reality" of existing and proposed policies such that they can better prepare their program to respond; to be able to have policy alternatives available when needed; and to know when "windows of opportunity" open so as to take advantage of opportunities to enhance the protection of human health and the environment.

Public policymaking in our governmental system in the US is composed of four fundamental, sequential parts: setting the agenda, choosing an alternative, authorizing a specific alternative, and implementing that alternative (Kingdon 2011). Policymaking, then, is a set of processes that are sequential and iterative, i.e., the process(es) repeat frequently as policymaking shifts and adapts to changing political, social, economic, environmental, and other relevant contexts (Kingdon 2011).

11.8.1 Setting the Agenda

The agenda, in the context of environmental policymaking, addresses problems that have policymakers' attention regarding environmental issues. Spillover of other real or perceived problems in the community – accountability, environmental justice, sustainable development, ecosystem management, and unfunded or underfunded mandates – could affect this agenda. These "problems" will change in priority and timing, based on what (or who) stimulates attention. The agenda then is the items that are assessed as important enough to require a policy action to address them. If there are many items on the agenda then they are usually prioritized in order of importance based on the needs and preferences of the policy decisionmakers.

Alternatives are "solutions" to those "problems." The specification of alternatives or choosing alternatives, generally, is to reformulate an existing policy, do nothing, or something in between. That said, new alternatives, rather than adapting existing alternatives, are sometimes what is called for. The results of this formulation or reformulation can be radical or incremental depending on the tolerance of those it might affect and those who would be required to implement it – in our case, environmental managers. In fact, an incremental approach to policymaking is not uncommon as existing policies are adapting to changing conditions while utilizing the frameworks and infrastructure of the existing policy to implement the adapted version (Kingdon 2011). The alternative that becomes authorized by decisionmakers depends on how those decisionmakers are influenced, which often depends most upon:

- who the participants or "policy entrepreneurs" are and their ability to influence;
- which alternative is most technically feasible and socially acceptable;
- and, which alternative can anticipate and/or overcome significant constraints (Kingdon 2011).

However, the best alternative might not see the light of day, i.e., be selected, unless a plethora of stars align, especially since "best" for one stakeholder may not be "best" for another stakeholder or for most of the stakeholders. The factors that affect agenda setting and alternatives specification are often generated by the participants who are active on the issue (the agenda item) and the process by which the agenda and alternative came into prominence, and who thus determine the direction and ultimate outcome of the process even where they are not the only stakeholders impacted by the outcome (Kingdon 2011).

Both talkers and doers in government live in a world of problems they aspire to fix. Whether existing policies (be they funded, underfunded, or unfunded) will effectively address identified problems, or the political decisions that govern whether those policies are successfully implemented, depends on who is forcing decisionmakers to do their job. Like many issues of environmental management, without clear and supervised accountability environmental management efforts often become ineffective, inefficient, or both.

11.8.2 The Participants

Typically, the participants or "players" that influence policy are either inside participants or outside participants (Kingdon 2011). Environmental policymakers and influencers inside government (inside participants) in the US include:

- Chief executives – such as a city mayor, state governor or the US president and their staff, all of whom hold public attention, have the ability to command institutional and organizational involvement and commitment regarding their environmental agendas.
- Civil servants – who, because of their protected status, offer longevity and institutional memory, subject expertise, and relationships to develop policy and political support. Further, these individuals often develop networks of information sharing to help inform

and shape policy decisionmaking based on their experience and professional, and sometimes personal, preferences.

- US Congress and their staff – are responsible for most of the overarching environmental policy in the US (Barnes, Graham, and Konisky 2021), have the legal authority to create laws, and often have more longevity than the chief executive. Further, like the chief executive, they have a high public profile and the ability to develop policy based on public perception and scientific data (Kingdon 2011).

Outside participants that can influence policy exist at many levels with varying degrees of influence. Often access provides influence and generally those that have the resources can gain access, e.g., individuals dedicated to obtaining and using access such as industry, trade group, or even environmental organization lobbyists. Outside participants usually consist of:

- Interest groups (businesses, NGOs, and government) – These groups are the squeaky wheels (involved and assertive entities) that influence agenda setting and alternative choice (Kingdon 2011). In general, these groups are composed of industry and trade organizations with substantial funding and interests in policy outcomes, whereas many environmental advocacy groups are not as well funded (Brulle 2018, 2020; Brulle et al. 2021). Further, after the pro-environmental impact of *Silent Spring*, industry and trade groups developed influencing strategies that became very effective in protecting their industries (Barnes, Graham, and Konisky 2021; Mayer 2017). While it is understandable and necessary that such stakeholders be involved in the policy decisionmaking process, evidence suggests that these groups developed robust ways of influencing and subverting environmental policy in favor of their production needs rather than the needs of human health and the environment (Brulle et al. 2021; Farrell 2016a, 2016b; Mayer 2017).
- Academic researchers and consultants – These individuals are often used to provide objective advice and testimony concerning policy (Kingdon 2011). This advising relationship to policymakers is relied on and has been formalized through the use of the Federal Advisory Committee Act (FACA), which created advisory committees such as the EPA's Office of Pesticide Programs' Pesticide Programs Dialog Committee (PPDC). FACAs are those advisory committees mandated to be made up of a balanced group of stakeholders including subject matter experts (e.g., research entomologists on the PPDC) and industry professionals. As well, of historical importance to the EPA is the Science Advisory Board (SAB), whose guidance and decisions impact environmental decisionmaking from the federal down to the city level. This board is critical to the development of EPA and the policies and rulemaking it develops. Historically, it has advised decisionmakers within the agency regarding the science relevant to a particular policy considering also the needs of industry and other stakeholders (Barnes, Graham, and Konisky 2021).
- Media – The media, in whatever form, can influence policy via how and what they decide to report on, as it has a strong effect on public opinion (Aerts and Cormier 2009; Kingdon 2011).
- Election-related interests – Interests that influence elections topics and outcomes, such as party officials and polling, have an indirect effect on policy as inside, elected participants

respond to voters or potential voters (Kingdon 2011). This impact of election-related interests on policymaking is, not surprisingly, particularly strong in the period before an election.

- Public opinion – Polling, interviews, and even public demonstrations influence policy, though typically more with regard to agenda setting than with alternatives (Kingdon 2011). The public, like interest groups, can become the squeaky wheels that force policymakers to pay attention to how they feel about environmental issues.

11.8.3 The Process

Kingdon (2011) describes three major policy formulation process streams that must "couple" or converge on a window of opportunity when setting the agenda and choosing an alternative:

1. Problem recognition or the "problem stream" in which problems capture the attention of decisionmakers. These can be new, previously unaddressed problems, like the discovery of poly- and perfluoroalkyl substances (PFAS; Ross 2019), where no specific policy exists, or the adjusting of existing policies that are not adequately addressing a problem such as particulate matter standards (Pope et al. 2020).
2. Formation or refinement of policy proposals – "policy stream" – where solutions (alternatives) to problems are developed. This stream is dependent on the understanding of what policies are currently in place and their impact on the problems that policymakers are paying attention to.
3. And the "political stream" in which factors such as public opinion, national mood and changes of administration allow alternatives to be noticed. In essence, this stream identifies and describes the politics of the problem and how it is currently being addressed. For instance, over time the politics and corresponding views of global climate change have fluctuated between natural weather patterns and a human-generated emergency, though tending much more toward the latter more recently as good science is winning out (Brulle 2020).

As with participants that influence policy, understanding the process of policy formulation and the streams that drive it requires the environmental manager to develop policy analytical skills. These skills are not just applicable to policy analysis but for determining what the "reality" is regarding what environmental problems exist or are emerging and the mood of stakeholders surrounding that reality. This requires the environmental manager to view the protection of human health and the environment from a bird's eye view, lift their head out of the everyday workings of their program, and not lose sight of the "forest for the trees" – where the forest is protecting human health and the environment, and the trees are the potentially distracting details of everyday program management. Professional attention and awareness are needed for not only environmental news, but also local, state, national, and internal organizational or institutional developments that can affect the environment. Frequent use of professional networks and executive briefings can aid in the development of this awareness when well constructed and disseminated appropriately.

The question then becomes, "If one understands the streams that need to converge such that an adequate alternative is possible, what can an environmental manager do to better protect human health and the environment?" They become policy entrepreneurs, "advocates who are willing to invest their resources – time, energy, reputation, money – to promote a position in return for anticipated future gain in the form of material, purposive or solitary benefits" (Kingdon 2003: 179). As stated at the beginning of this section, the environmental manager as a subject matter expert should have an implementable alternative "on the shelf" for policymakers (who find it comfortable to entertain alternatives that are well thought out and readily implementable) to choose and support.

Policy, like nature, loves a vacuum – that is, unimpeded space or a gap to fill. In other words, the environmental manager who knows the complex and volatile issues, is adept at co-producing ethical environmental solutions, and can deliver upon the goals of those solutions, better have an alternative at the ready because if they do not, another participant will. When the window of opportunity becomes available the successful environmental manager is ready to seize the opportunity and put forward their (or their team, organization, or institution's) alternative for implementation.

11.8.4 Windows of Opportunity

Assuming the environmental manager feels it appropriate to provide their alternative (and thus become an entrepreneur), when and how should they do it to best assure its acceptance? In the policy formulation process, the streams converge through a window of opportunity (Kingdon 2011; see Figure 11.1). This window opens when:

- Administration changes – How does the new administration's view on environmental issues align with that of the participants and the position of the outgoing administration? For example, the environmental regulatory preferences of the Trump administration were starkly different than those of the Biden administration with regard to almost every aspect of environmental protection (Barnes, Graham, and Konisky 2021). Likewise, shifts at the local and state levels can be just as dramatic, resulting in very different policy outcomes depending on what administration is installed currently.
- Changes in national (or regional or local) mood – As discussed earlier, media attention and public opinion are closely linked. There can be national inflection points – for example, after the publication of Carson's (1962) *Silent Spring* or the Flint, Michigan water crisis – that substantially change how an environmental issue is viewed and the amount of support for policy addressing that issue.
- New problems arise – Typically new problems are a crisis, i.e., a dangerous situation which has reached a point where a decision must be made (J. Godec, pers. comm. 2021). Paradoxically, environmental managers understand that what most often opens windows of opportunity is a crisis and the most urgent crisis is where risk (illness and death) is apparent and the crisis deserving of highest prioritization is one involving children. This

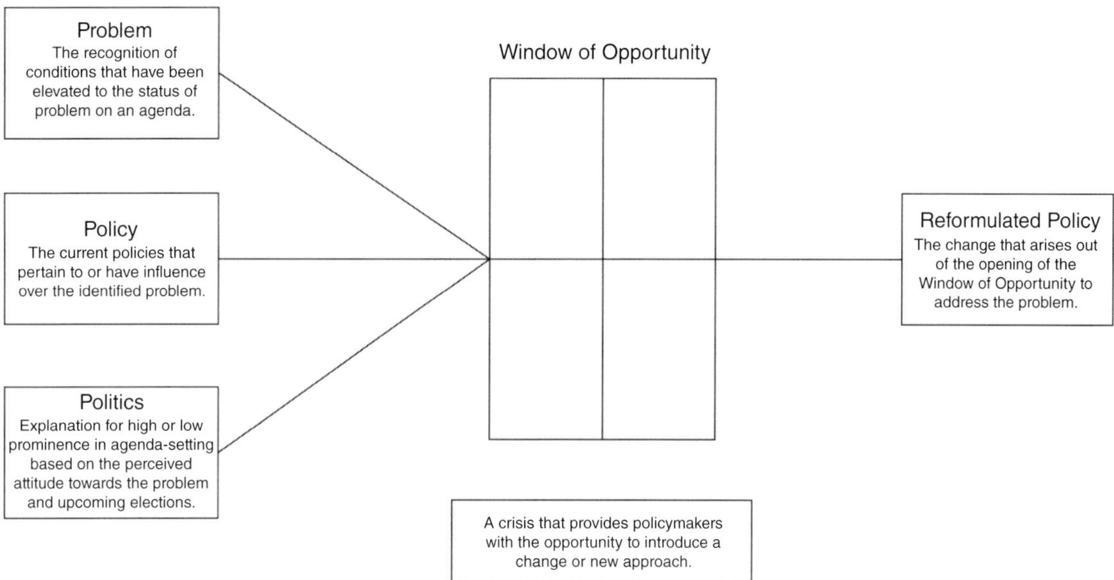

Figure 11.1 Policy formulation (Kingdon 2011).

progression appears to be "crude but true" with regard to media coverage and public opinion.

A window closes when:

- Participants feel they have addressed the problem and that the alternative they have chosen is indeed an improvement over current policy.
- Participants may fail to get action because of lack of political will, or the agenda is reset because of new priorities.
- Events opening the window pass from the scene (crisis vs. resources), again causing a reset of the agenda by policymakers.
- More personnel (policymakers or their staff) changes result in a reset of the agenda where what was considered a problem that should be addressed is no longer considered important.
- No technically or administratively feasible alternatives are available and the window of opportunity closes.

"The separate streams come together at critical times. A problem is recognized, a solution is available, the political climate makes the time right for change, and the constraints do not prohibit action" (Kingdon 2011: 88). These streams converge or couple through the proverbial window of opportunity when a new policy is ready to be implemented. If the window of opportunity for policy formulation is open, policymakers might just jump through it.

Whether they do so depends on whether a policy entrepreneur pushes them – and pushes them in the right way so that they take action.

Because they understand the process and can correctly analyze its components, including when opportunities present themselves, environmental managers must also become "surfers" (Kingdon 2011). Surfing, as actual surfers know, means paddling out to where waves begin to swell, waiting for the surfable wave, paddling toward the beach to match its momentum and "catching the wave." This skill requires the environmental manager who wants their alternative to be chosen to:

- Make sure the problem is on the agenda by providing data to assure that decisionmakers at local, state, or federal levels recognize that there is a problem concerning human health and the environment.
- Know when and where the problem, policy and political streams converge by anticipating the opening of a window of opportunity.
- Through their knowledge of what policy or change to an existing policy can improve the management of pesticides, have an alternative (in the form of a strategic plan outline) that is technically feasible and socially acceptable, and be in place to propose it to policymakers when they are presented with the facts regarding the risk of the currently used class of pesticides.
- Have in place and understand the political management necessary to develop the momentum to outcompete other alternatives.

11.9 Conclusion

If the environmental manager understands environmental policy tools and how they become useful instruments or concepts to devise implementable policies, they then can begin to become part of a process to successfully address and ameliorate problems affecting human health and the environment. However, it is when they develop skills to influence how those problems should be prioritized and addressed (agenda setting and alternative specification) that they can best fulfill the program goals and agency mission to which they have given allegiance.

AUTHOR'S NOTE 11.1 What can environmental managers do to have fewer regulations? The logic of pollution prevention

Environmental regulations usually result as a response to pollution. One way to have fewer regulations, of course, is for you or, better yet, your sector (industry or government) to establish a trustworthy relationship with regulators as a leader who proactively identifies environmental threats and consistently demonstrates existing

AUTHOR'S NOTE 11.1 (cont.)

rules adequately address pollution. Nonetheless, the "watch phrase" for all would be "trust but verify."

Perhaps more realistically, one of the most effective ways to escape being regulated is to prevent pollution that exceeds primary or secondary standards. Pollution is a manifestation of inefficient process and thus, except for agrichemicals and toxic substances, is largely made up of "byproducts." If industrial processes as sources of "byproducts" can be more efficient they will reduce waste, save money, and reduce regulatory reaction.

Quoting Gina McCarthy from her 2020 appearance on "The EPA at 50" webcast, at its inception "the EPA had the opportunity to really focus on pollution prevention ... that would drive new technologies in place that would significantly change our ability to protect health ... to focus on building that bridge between science and lots of common-sense, cost-effective ways to start moving the needle driving new technologies." Her presentation illustrated the three principles the EPA was founded on: Science, Pollution Prevention, and Adaptability, not only to protect human health and the environment but to provide for a sustainable and vibrant economy.

One example of this logic might be that, on a global level, auto makers who have increased the efficiency of their combustion engines because they were forced to innovate for more efficient engines due to regulation have become more competitive. In other words, if you are prohibited from a production process which might have an efficiency of 75 percent and innovate such that you increase efficiency you can be more competitive. Theoretically, an engine that is 100 percent efficient does not emit pollution and would therefore not require further regulation.

Another example is that instead of spending public school funding to apply pesticides in the school environment, which often are ineffective and dangerous, schools could implement Integrated Pest Management (IPM). In other words, school districts pay to apply sometimes unnecessary toxics. Pesticide use and cost is a result of pest presence or the perception of pest presence. By educating existing custodial, grounds and kitchen staff to monitor for pests and address conducive conditions (filth, uncontained food, excess water, and clutter) that attract and maintain pest populations they could substantially control or eliminate pests, thereby providing source reduction for pesticides. This would not only result in fewer or no costly pesticide applications (thereby eliminating pesticide use documentation and notification to the school community) but mitigate the negative effects of pests and pesticides, reducing absenteeism associated with asthma (particularly in inner city schools) and injuries resulting from pests (bites, stings, and uneven playing fields – gopher holes, etc.). Lower costs, more effective pest management and fewer regulations!

11.10 End of Chapter Questions

1. Why should an environmental manager become a policy entrepreneur and a surfer?
2. Give an example of an influential *inside* policy participant regarding environmental policy formulation.
3. Give an example of an influential *outside* policy participant regarding environmental policy formulation.
4. When do windows of opportunity open for the formulation of new alternatives (policies)?
5. Agrichemicals are continuing to contaminate the Chesapeake Bay under current rules regarding non-point-source pollution. Construct the Policy Formulation Diagram with specifics to this situation and *label* the policy formulation process for a reformulation of a policy which might be more effective that what is currently in place.
6. As an environmental manager, what do challenge regulations mean to you as opposed to command-and-control instruments? (Include: what the differences between these policy tools are, and the two managerial implications to choosing challenge regulations over command and control.)
7. What are the differences between policy tools that directly limit pollution and those tools that indirectly limit pollution?
8. What are the seven criteria for developing an "ideal policy instrument"?
9. How would an environmental manager evaluate the effectiveness of a policy?
10. What is the Regulation Dilemma (the trends that created it, etc.) and specifically what strategies would you implement to overcome this dilemma?
11. What are five ways government can induce firms to join voluntary programs involving non-regulatory policies?

12 Looking Forward

Skill: *Think forward to be most effective today.* Lessons and experience from the past should inform the decisions of today but those decisions must be made anticipating future conditions. The demand for the services and broad functionality of the environmental manager are only increasing. Thinking and planning for future contexts while considering the lessons, knowledge, and experiences of the past, all to produce the best practices for today, is essential for the effective environmental manager. Identifying, assessing, and orienting on future trends is critical to ensuring your program is operating with the most up-to-date and relevant technology and design; the environmental manager should embrace them and prepare for them.

12.1 Introduction

As stated throughout this textbook, science changes. It often does so because of new knowledge generated through the application of new technologies and methods. But science also changes because of new perspectives brought to bear by the scientists at work, just as management changes based on the manager. Currently there are major shifts and corresponding environmental management conversations ongoing due to the more broadly recognized concern for human-caused global environmental change. Human influence on the global ecosystem has never been stronger than it is today, and it is only growing (Rockström et al. 2009; Steffen, Broadgate et al. 2015; Steffen, Richardson et al. 2015; Steffen et al. 2018). While this change involves aspects separate from but related to environmental management – such as natural resource management, as we distinguished in Chapters 2 and 3 – they are all interrelated. That said, the risks that environmental managers focus on most, those stemming from toxic pollution emissions, are the single largest cause of human early mortality and morbidity globally today (Burnett et al. 2018; Fuller, Sandilya, and Hanrahan 2019; Landrigan et al. 2017). The challenges before environmental managers have never been greater.

The challenge to environmental managers is at all scales. Local environmental managers are facing new challenges, challenges that are often impacted in some way by the growing changes that regional and global environmental managers and policymakers are seeing increase as well. The interconnectedness of local to global ecosystems, and of humans more generally as a function of continued globalization, is becoming clearer and more pertinent to environmental managers at all scales. This trend is unlikely to change.

The effective environmental manager must stay current on the broad trends and changes within environmental management, paying particular attention to those areas that they work on most, be it contaminated site remediation, air pollution emissions reductions, or any other area of environmental management. Below we offer short notes on the state and direction of a few key topics ranging from changing technologies for the practice of environmental management to concepts for management that are on the rise. The list is not holistic nor are the descriptions of each exhaustive, but rather it is a general sampling of key areas of development for environmental management.

12.2 Examples of Persistent and Emerging Issues

There are many persistent and emerging environmental hazards that humans have created and continue to create (IPCC 2018, 2019a; Landrigan et al. 2017; Marcantonio, Field, and Regan 2019; Rockström et al. 2009; Steffen, Broadgate et al. 2015; Steffen, Richardson et al. 2015; Watts et al. 2019). From anthropogenic climate change caused by human-emitted greenhouse gases to toxic air pollution, humans are modifying the Earth system in ways that are creating greater and sometimes new risks to human health. We will mention just a few here to highlight some persistent and emerging risks because the list is long. Our point is to make clear the need for the expansion of environmental management both as a practice and a profession, i.e., we need more professional environmental managers but we also need more people to practice environmental management in their daily life.

12.2.1 Pesticides

Silent Spring brought an awareness that the misuse of pesticides was unbalancing the ecosystem and had become a threat to biodiversity (Carson 1962). The effects of new classes of insecticides, and neglect to clean up some of the original biocides such as DDT are again being brought to light regarding biodiversity (Gulland et al. 2020; Gunstone et al. 2021; Jensen 2019).

Because of decades of persistent marketing from current manufacturers and benign neglect from past manufacturers and regulators, pesticides have again jeopardized ecosystems, particularly aquatic ecosystems (Gulland et al. 2020; Jensen 2019; Lame 2005). Thanks to the policies and management of the EPA, environmental science regarding environmental risk assessment, environmental fate and toxicity has made quantum leaps since its inception in 1970 (Barnes, Graham and Konisky 2021). However, some policies regarding the registration and treatment of pesticides that become hazardous wastes have remained in effect due to industry influence, and need reformulation (Friedman 2020; Lame 2019). From a toxicological viewpoint, these policies should address the cumulative and synergistic effects of pesticides. Further, as with the global climate emergency, the simplistic equations favoring human health as a primary consideration must offer more consideration to what will happen to human health as our species loses other species such as pollinators, benthic organisms, soil organisms and marine mammals (Gunstone et al. 2021).

12.2.2 Plastics

The detection of plastics in the environment has occurred in almost every ecosystem sampled around the globe from marine to terrestrial environments (Boyle and Örmeci 2020; Choy et al. 2019; Huerta Lwanga et al. 2017). Microplastics and plastic particles have also been found in various species from seabirds to fish to the belly of whales (IUCN 2018). There are still substantial uncertainties about the effects of plastic pollution on animal, marine, and other biotic species, and on ecosystem functioning in general (Mendenhall 2018). But what is clear is that the governance challenges regarding the development, permitting, monitoring, and enforcement of any programs to manage plastic pollution are substantial given the broad human usage of plastics in various products and processes (Napper and Thompson 2020). Current estimates are that there is three times as much micro- and nanoplastic mass in the global ecosystem than all mammalian biomass combined, and correspondingly we are extensively and increasingly exceeding the Earth system's safe operating space for chemical pollution, with plastics being of one myriad forms (Persson et al. 2022)

12.2.3 PFAS: Per- and Polyfluoroalkyl Substances

Per- and polyfluoroalkyl substances are a group of man-made chemicals that have been manufactured and used in a variety of industries around the globe, including in the United States since the 1940s. These substances have been used in products ranging from food and water repellents on couches and car seats to non-stick cooking pans to flame-retardant uniforms and materials (US EPA 2021a). While their benefits to human processes are many, they are also "forever chemicals," i.e., they persist in the environment indefinitely. Just as they accumulate and persist in the environment, so too do they persist and accumulate in the human body (US CDC 2019; US EPA 2021a). As they do not breakdown and they are in many materials that humans consume and use, their pathways are into the human body from the environment (Benotti et al. 2020; Schaider et al. 2017; Sunderland et al. 2019; Whitehead et al. 2021). And PFAS are suspected of causing harm to human health, though to what extent is still under investigation and contested (Brown et al. 2020; Fenton et al. 2021; Pelch et al. 2019). Environmental managers will thus need to monitor developments in the science of PFAS and corresponding shifts in regulation and management in order to ensure their programs reflect the best practices for protecting human health and the environment.

12.2.4 Climate Change

The effect of human-caused climatic change on human health and the environment is substantial and growing. From increasing heatwaves (Mora et al. 2017; Raymond, Matthews, and Horton 2020) to coral reef die-off (Beck et al. 2018; Graham et al. 2014) to the interactions between different effects (Raymond et al. 2020), the changes are complex and myriad (IPCC 2018, 2019b). As we have discussed, many of the sources of greenhouse gases are the same as those emitting different forms of toxic pollution as well, especially air pollution. Yet the regulation of greenhouse gases, especially in the United States, has substantially less standing policy directed at managing them (Barnes, Graham, and Konisky 2021). For example, emissions of carbon dioxide and methane, the two greenhouse gases primarily driving anthropogenic climate change, are not regulated by the Clean Air Act or any other national-level policy currently. While the effects of climate change will be more acutely felt in other countries than the US, the US still faces substantial risks and has endured substantial human health and economic costs to date (Abatzoglou and Williams 2016; NOAA 2020b; Smith and Matthews 2015; USFR 2021; USGCRP 2018). While climate change spans the spectrum of environmental and natural resource management agencies, given the interconnection between greenhouse gas and toxic pollution sources, environmental managers are well placed to combat this growing threat and must be aware of its impacts on their current program, resource and political management.

12.3 Changing Technologies

The environmental manager can expect to see substantial development in the tools available for them to accomplish their mission regulating and protecting human health and the environment. For example, new wearable pollution detection devices that can sample the ambient air around you every few seconds are rapidly being developed and deployed (for a comparative example of two children in India and the toxic air exposures they face throughout their daily routine measured through wearable pollution detection devices see Wu et al. 2020). These devices, both worn and fixed in different locations, can be linked to provide a real-time spatialized map showing the distribution of pollution for more effective monitoring by managers and so that everyday citizens can be empowered through knowing their exposures. Mounting evidence shows that even short but acute exposures can have substantial impacts on a person's health and trajectory yet, especially regarding air pollution, measuring these exposures persistently can be difficult (Vermeulen et al. 2020).

REAL-WORLD EXAMPLE 12.1 Using real-time technology to address accountability and environmental justice

An important function of state and federal environmental agencies is the use of enforcement to identify and address non-permitted industrial releases. In communities that are located along the fence line of an industrial facility, off-site releases place people at higher risk of environmental health problems. So how does an environmental manager find and enforce on releases that are not reported and in the case of air emissions may not even be visible?

One approach is to take advantage of real-time monitoring technologies during industrial inspections or investigations of specific fence-line communities. For example, thermal imaging infrared technology can be used to "see" emissions that are currently not visible and often represent air pollution releases that are un- or under-reported. These releases often occur at industrial sites (e.g., chemical tank farms, petroleum refineries, or industrial facilities that pipe chemicals) located next to fence-line communities. As such, the use of this real-time technology by environmental managers can lead to more effective enforcement, addressing air pollution that disproportionately impacts communities that are at risk due to their location. The value in using technology to make visible that which is not easily seen addresses both accountability issues (when air releases are un- or under-reported) and environmental justice concerns.

Enforcement entities have used optical gas imaging FLIR™ cameras to visualize and record various types of emissions related to leak detection. To see what this looks like, visit www.youtube.com/watch?v=Sqg43EeBkY8.

Mary Willett, MPA/MSES, Faculty at the O'Neill School of Public and Environmental Affairs, Indiana University – Bloomington

Other technologies rapidly developing and changing how environmental management can be conducted, both by method and by cost reduction through improved sampling and analysis processes, include high-resolution mass spectrometry for lab and field detection of pollutants in different environmental media (Geer Wallace and McCord 2020; Kaufmann 2012; Vermeulen et al. 2020); remote-sensing technologies such as the National Aeronautics and Space Administration's soon-to-launch Tropospheric Emissions: Monitoring of Pollution (TEMPO) satellite constellation (NASA 2021); and developments in different geospatial software programs to integrate all of these different measures into a comprehensive dataset for analysis (Khan et al. 2019; Li, Batty, and Goodchild 2020; Rossetto et al. 2018). There are many other technologies that are being developed and deployed to aid in environmental management at local to global scales.

What does this mean for the environmental manager? First, the effective manager must be aware of what is being developed and what is available to them in their program and resource management to be put to use in permitting, monitoring, enforcement, and providing technical assistance. Knowing the landscape of available tools can help the environmental manager make more informed and efficient decisions regarding their program design and execution. New technologies may not always make sense to integrate into your program, or may not even be available, but it is important to know what's out there for the times when there is a better, more effective way to do the job.

As discussed in Chapter 9, environmental managers also need to know what new technologies are available and how they function so that they can effectively manage the experts who are using these tools. Not only is it important so that you can ensure that the experts you manage are using the best tools and techniques available (quality assurance), but it also facilitates your ability to support and lead those experts, i.e., you can speak their language and engage them about their work. New technology, no matter how complex or sometimes seemingly unnecessarily complex, should never be treated as a black box. In other words, the professional environmental manager takes the time to acquire at least the "manager's perspective and knowledge" of each new tool rather than resigning themselves to let others simply do it for them.

12.4 Integrative and Comprehensive Approaches

Increasingly, environmental management, especially through the use of the new technologies discussed above, is deploying more integrated and comprehensive approaches to tracking and measuring pollution hazards in local to regional ecosystems. For example, for water pollution management programs watershed-based programs that encompass the entire watershed, as opposed to just one tributary or section of a watershed, are increasingly deployed because they capture much more of the relevant impacts as opposed to just a slice (UN FAO 2017; Wolanski et al. 2003). The EPA has even developed an online portal called the Watershed Academy to provide resources for environmental managers and other interested persons to learn about watershed-scale environmental management (US EPA 2021l). A growing body of research is tracking and measuring the cumulative and interactive effects of different pollution streams (Alexeeff et al. 2012; GAHP 2020; Marcantonio, Field, and Regan 2020; Marcantonio, Javeline et al. 2021; Mora et al. 2018; Solomon et al. 2016). This work further evidences the need to take more comprehensive approaches to environmental management since, as we have said many times, pollution flows and does not recognize political, jurisdictional, or any other human-constructed boundaries.

As we have discussed throughout the book, particularly by way of ecosystem management as one of our five management issues, the integration of different monitoring, enforcement, and technical assistance programs to comprehensively address pollution management and reduction is needed. The environmental manager cannot look at pollution sources, types, and outcomes separately. And the technology available today makes these approaches much more manageable for the manager, i.e., as discussed above, new tools available to the environmental manager make taking a comprehensive approach to ecosystem management much more tractable. Not every program or management practice that the environmental manager will deploy will need to be comprehensive or integrative, but in assessing these instances the environmental manager should be prepared to explain why they did not take such an approach which, fortunately, is becoming more frequently deployed and the expected standard.

12.5 Evolving Management Practices

Similar to changes in the comprehensiveness and interactiveness of environmental management programs, new strategies for how environmental management processes and program development are done are changing, including deliberately planned experimentation and the integration of non-scientists. For example, the adaptive management approach is being deployed more frequently by environmental managers (Stankey, Clark, and Bormann 2005). Adaptive management is when the environmental manager deliberately includes experimentation in their program design to learn what works and what does not and to adapt their program along the way (Webb et al. 2018). Most environmental management is done using evidence-based environmental management which uses existing data and knowledge to develop a management program or policy; whereas adaptive management is initially developed using existing evidence but is designed to also generate new data in the process of deployment so as to adjust the program or plan during the implementation process (Gillson et al. 2019). Adaptive management allows the environmental manager to adjust to either changing or novel social or environmental contexts that they may face in their implementation process (Gillson et al. 2019). Often environmental managers have to do this anyway, but adaptive management takes a proactive approach to it instead of reacting when a program does not seem to be working.

Another changing management practice is the integration of persons who are not environmental managers or professionals into the "science of management." Often called citizen science or something similar, this process involves crowdsourcing or other forms of data gathering and analysis in support of an environmental management program (Ferrari et al. 2021; Hadj-Hammou et al. 2017). Leveraging citizen science can help managers to collect large amounts of data across a broader geographic scale and often at a low cost, especially when deployed with an app designed for the task (Luna et al. 2018).

Harnessing citizen science not only helps the managers collect data for their programmatic needs, but also provides a pathway for co-producing solutions of an issue and helps to better inform the participants of the environmental issue at hand (Irwin 2018). While there are many benefits to using citizen science, it also has its challenges, particularly regarding quality control – i.e., the data collected may not be as carefully or correctly reported as it would be if collected by a trained environmental manager or professional (Irwin 2018; Luna et al. 2018). And since the manager is still ultimately responsible for the quality and accuracy of the work, you must develop a means by which to review and ensure the quality of the work done by participants. Harnessing citizen science may not make sense in some, or potentially even most, cases depending on the role of the environmental manager, but when it does make sense, if implemented effectively, it can be a powerful, multi-beneficial tool.

12.6 Pollution Prevention: Education to Reduce Regulation

As environmental managers become aware that human behavior with regard to energy conservation, recycling, and toxic chemical use is a significant contributor to pollution, they must develop better ways to affect those behaviors. Technical assistance and education have been part of the program management for environmental managers since the inception of the EPA and state agencies. That said, those program functions often have not been given the consideration and resources that "change agents" could use and are often considered an afterthought from an administrative viewpoint (Lame 2005).

During the Civil War, land grant universities were established to bring agricultural education to the farmer. Then in 1914 the Smith–Lever Act established "cooperative extension" or the Cooperative Extension Service where the USDA and land grant universities provided, research and educational outreach to the agricultural sector (National Archives 2021). Perhaps it is now time for an "Environmental Extension Service" where the EPA and universities formally target the public and polluters with research and the requisite diffusion of pollution prevention technologies and behaviors using a well-resourced core of change agents? While the EPA already offers a substantial breadth of technical assistance options to state and city-level regulators, given the growing risk – and breadth and complexity of risk – to human health from human emissions the development of further environmental protection and management Extension services could prove to be useful and effective.

12.7 Conclusion

Environmental management, like any profession, is ever evolving. But the rate of evolution is increasing due to changes in management practices, the breadth of challenges to human health and the environment and changing technologies for measuring emissions as well as technologies for mitigating them. In case we have not said it enough already, it is clear that a critical part of the effective environmental manager's praxis is continued learning and training. This textbook is intended to serve as the foundation of the manager's training and to be a reference for them throughout their careers. But it is up to the environmental manager to continue to seek out professional development opportunities and to keep up with cutting-edge knowledge, techniques, and relevant policies and laws through peer-reviewed research and professional periodicals.

If the environmental manager ever stops learning, then they are no longer an effective environmental manager. There are certainly constants in the profession: management of people will always be the crux; program, resource and political management practices and corresponding permitting, monitoring, enforcement, and technical assistance provision are needed for effective management; and the key issues of accountability, environmental justice, ecosystem management, sustainable development, and funded and underfunded mandates will persist. But for the environmental manager to continue to find success and to most effectively protect human health and the environment, they must continue to learn and grow in their practice.

Case Study: The Case of Implementing a Pollution Prevention Program to Reduce the Risks of Pests and Pesticides in Children

This case study incorporates the different plans discussed in Chapter 7 and the diffusion of an environmental innovation. This is a case history demonstrating how change agents working with the EPA developed and implemented Integrated Pest Management (IPM) in schools in 41 states in 18 years and how, for more than a decade, School IPM became a national initiative. The case study first appeared in *A Worm in the Teacher's Apple: Protecting America's School Children from Pests and Pesticides* (Lame 2005).

In *Silent Spring* Carson (1962) was able to link the overuse of pesticides to ecological imbalances in the US. Many thought that she was simply anti-pesticide, when in fact she was against the misuse of pesticides. This damage to ecosystems continues as documented by the use of the latest classes of insecticides (neonicotinoids) being linked to the collapse of Japanese aquatic ecosystems (Yamamuro et al. 2019), and the legacy of DDT remains a threat to marine ecosystems in California (Gulland et al. 2020).

By the late 1970s, the pesticide industries learned the lesson of pre-empting public ownership of environmental health protection as in the past such actions resulted in environmentalism and the creation of the US EPA, further disrupting their aims deregulating pesticides. In the mid-1980s, pro-pesticide congressmen and the pesticide industry developed a strategy calculated to cripple regulatory agencies, and which by the mid-1990s began reducing program funding, calling for "studies" delaying the commonsense implementation of best management practices and providing rule-writing "assistance" by regulated trade groups to subvert or, at best, divert the priorities of the EPA, particularly with regard to pesticide reduction programs. This strategy flourished through the early 2000s, then became worse during the budget sequestration after the Great Recession. The pro-pesticide movement gained further prominence under the Trump administration during its "war on science" in the EPA and the appointment of chemical industry officials to that agency. In part, that administration's approach allowed children to be further exposed to pesticides slated for being discontinued (Lame 2019). Thus, the nation was approaching pre–*Silent Spring* pesticide dependence (Grullon 2013; Pimentel 1991), only this time with the industry aligned and prepared to counter all swells of popular environmental concern.

IPM as a technical assistance program was one area that survived deregulation and attacks on science because these programs were able to demonstrate their value to the American farmer. That said, IPM as an innovation was not well diffused in that community as agribusiness and the change agents associated with it presented little threat to the pesticide industry.

One area of IPM diffusion that did see increases in attention and funding was in the built environment, specifically where children could be exposed. Evidence of the risks to children exposed to pesticides had been building since the 1990s (Landrigan 2001; Landrigan, Garg, and Droller 2003; Landrigan et al. 2004) and in 2012 the American Academy of Pediatrics issued a statement about it. This included the findings that "Children encounter pesticides daily and have unique susceptibilities to their potential toxicity. Acute poisoning risks are clear and understanding of chronic health implications from both acute and chronic exposure are emerging. Epidemiologic evidence demonstrates associations between early life exposure to pesticides and pediatric cancers, decreased cognitive function, and behavioral problems" (Roberts and Karr 2012). As more evidence of the risks to children came out, the EPA seemed willing to support technical assistance/education programs protecting children. One such program implementing IPM in the school environment was supported by the Pesticide Environmental Stewardship Program (PESP) beginning in the late 1990s but was forgotten during the Trump administration. The case below illustrates the planning and implementation that resulted in the partial diffusion of IPM in the school community and the reduction of the risks associated with pests and pesticides in those communities.

A Window of Opportunity for IPM in the School Community

In the early 1990s a home in West Phoenix was over-sprayed (pesticides meant for crops accidentally applied elsewhere) by crop dusters and a mother swore her kids had developed chemical sensitivities. At that point, she also had some trust that her government would assist her with protecting her family and others from having future experiences of the same kind. The state lead agency for the Federal Insecticide, Fungicide, and Rodenticide Act (FIFRA), Arizona's Pesticide Review Board, could do little to help. Assuming they could find evidence, they had little ability to jerk someone's license for a single offense. Maybe a $100 fine. Strike one.

The Arizona Agricultural and Horticultural Commission (as the Arizona Department of Agriculture was then known) would not create, let alone enforce, regulations which would force farmers to limit pesticide use. These agencies are in every state. They are called the "state lead agency" (SLA) by their authorizing agency, the US EPA. FIFRA does not include language requiring a farmer to use alternative methods to prevent unnecessary but legal applications of a registered pesticide. More than once, I have heard agency officials state that it must be OK to use pesticides because the US EPA registers them. So, this family's farmer neighbors had the right to use as much pesticide as they wanted, and the SLA was not going to do anything about it. Strike two.

The mother in this story was not soft spoken. She got her story in the local papers and developed relationships with other mothers who believed their children were sick or in danger of becoming sick because of man-made toxins: air pollution, groundwater contamination, hazardous waste facilities, and pesticides. They went to the Arizona Department of Environmental Quality (ADEQ) for help. "Stop killing our babies!" In many states these agencies are under tremendous political pressure to permit pollution, keep the EPA "off our backs," and "make sure business continues, comes and grows for the creation of jobs (votes) in our state." The mother and her peers forced ADEQ to pay attention to what they were supposed to do but could not stop polluters from coming and operating if those applying for permits had the money and lawyers to jump through all the hoops. Strike three, these community members were outraged!

I was the ombudsperson for the ADEQ. The main function of an ombudsperson is to communicate with people outside the agency who raise an issue with the agency, analyze, and report that communication to decisionmakers in the agency. I was a "troubleshooter" appointed by the governor to work for the director of ADEQ. In 1991 I met the mother of the affected family, and asked what all ombudsmen should ask, "What can I do to help you?" After she told me I was probably just another "industry toadie," she said she heard I had been an entomologist for the University of Arizona and asked if I could help her stop her children's new school from spraying insecticides in the classrooms.

Months after finishing my graduate education in 1981 at The Auburn University I was able to move back home to Tempe, Arizona, where I was to begin my job as an IPM specialist for the University of Arizona's Entomology Department. Having a Cooperative Extension (way back we called it the Agricultural Extension Service) appointment, I was based out of the Maricopa County Extension Office, just next door to where I worked four years prior as a "field grunt" for the US Department of Agriculture's (USDA) Animal, Plant Health Inspection Service (APHIS).

In addition to housing the largest urban center in the state (Phoenix Metro), Maricopa County also generated the largest agricultural sales in the state. At that time Arizona was the third largest cotton producer in the US. Our irrigated farms produced some of the highest yields and best-quality cotton in the world. My job as the IPM specialist for the University of Arizona's Entomology Department was to demonstrate and coordinate an IPM program in cotton as farmers were experiencing severe insecticide resistance and needed assistance controlling their pests.

My office mate worked in urban entomology and would tell folks what conditions were causing their pests, and what they should do to prevent them. This seemed to be more fun than killing bugs on cotton to me. In fact, my emphasis in entomology as a graduate student was in the field of medical entomology: the study of human maladies caused by arthropods (bites, stings, diseases, and actual infestations such as maggots – yes, the subspecialty forensic entomology "bugs on dead people" falls into "med entomology"), and how one might manage those pests.

In a lot of ways, it was then that I learned that pest populations in and around buildings are influenced far more significantly by changing human behavior than killing them after a population explosion. I realized that "pest management is people management." Educating folks about their problems and convincing them that they can do something about them is a special skill. However, I also began to develop a real curiosity regarding just *how* one gets folks to adopt a new behavior. I remembered my Dad (an old-fashioned pediatrician) always telling me that it takes more than medicine, but also the "laying on of hands," to heal folks, to make them believe that they can be healed. I had no real idea how to do this. Entomologists are not trained to change people, but to study bugs back in the cotton fields. I was getting nowhere convincing farmers to adopt ways other than massive pesticide applications, to control their pests. They were not buying what I was selling – Integrated Pest Management.

Integrated Pest Management defined as an innovation to be adopted: IPM is a cluster of technologies (cultural, mechanical, biological, genetic, and chemical) which is an integrated application (based on biological information) designed to allow humans to compete with other species (pests).

In the practical sense, the concept of IPM goes to what environmental management professionals call pollution prevention. Concerned at the negative effects of hazardous waste on human health and the environment, environmental officials began discussing pollution in terms of waste. From this viewpoint, pesticides are a unique non-point-source pollutant. In that, they are intended to be applied to the environment, as opposed to other pollutants which are viewed as byproducts of civilization – or waste. Pesticide use resulting in unnecessary costs, pollution or adverse health effects is inherently a management/people problem. IPM, in fact, is one of the oldest pollution prevention programs (though not historically termed as such) for non-point-source pollution. IPM is recognized by the US EPA and USDA as a viable cluster of technologies which prevent non-point-source pollution and reduce pest management costs.

Even today, IPM has not fully diffused (been adopted by communities) in the agricultural or urban sectors (GAO-01-815, 2001, and Lowe et al. 2019). I began to understand while trying to work with my farmers in Arizona that it was unlikely to fully diffuse until IPM was, at least partially, diffused by the general population. And so, I began to realize that implementing IPM in the public schools of the United States would initiate (and possibly accomplish) the broader understanding, acceptance, and diffusion of IPM across all fields of pest management. Unfortunately, many land grant/Cooperative Extension institutions (the traditional implementers/change agents for IPM) either did not view the implementation of IPM in public schools as part of the Extension mission or were unwilling to jeopardize existing relationships regarding their Urban Pest Management programs of working with pesticide manufacturers and exterminators. Clearly, it was communication and the "laying on of hands" that was needed for the successful implementation of IPM, and this type of implementation seemed to be a lot more feasible in our schools than on our farms. When the concerned mothers of the school communities finally got around to asking, it presented a window of opportunity to push an IPM agenda in communities that had the potential to better diffuse IPM – beginning with schools.

What Our Children Deserve

The reason pests are managed in schools is that the school environment needs to be a safe learning environment. The presence of pests in the school building can represent both a threat to occupant health – through envenomation, allergic response, or disease transmission – and a distraction from the learning environment. How many of us remember the disruptive effect of the lone, foraging paper-wasp which happened to fly into our classroom on a warm spring afternoon, the mouse running across the floor, or perhaps the cockroach on the wall, or on the teacher's desk during a lecture? Further, the specter of "sick school syndrome," resulting in costly building closure, site remediation and negative public perception, has sensitized administrators to the effects of toxic materials used, stored or propagated (molds) in their schools. The routine application of pesticides in and around schools (landscape and turf) is, or should be, a concern of parents, teachers, and school administrations.

Ask any pediatric toxicologist. Ask your Congressmen who crafted the Food Quality Protection Act (FQPA). Children are not just little adults! Yes, they are smaller. But, their nervous systems are developing. The human nervous system is immensely complex and controls all conscious and autonomic (involuntary) functioning of our own body. Most of the pesticides we use are nerve poisons and biocides (able, in large enough concentrations, to kill any organism). For sure, many of the pesticides (organophosphates: e.g., diazinon, malathion, ortheneate) that generations of Americans were brought up on are the direct descendants of the nerve gasses of World War I and the Holocaust. Others (organochlorines: e.g., DDT and chlordane) are those poisons that Rachel Carson illustrated in *Silent Spring*, now proven to have killed or mutated whole species of invertebrates and vertebrates to the point of extinction.

Some of those pesticides are still in use, in our schools. While most of the pesticides used in our schools and homes today (synthetic pyrethroids, neonicotinoids, growth inhibitors, etc.) have active ingredients that seem to be less damaging to the mammalian nervous system, some, like pyrethroids, are confirmed to have endocrine-disrupting properties (Brander et al. 2016). The jury is still out on the effect of "endocrine disruptors" on humans. Perhaps of more immediate concern is how the inert ingredients, that the active ingredients are mixed with, affect us and our children (by proprietary law, the manufacturers do not have to disclose inert ingredients). Even pesticides made from plants can be considered toxic. One of the oldest insecticides that we used in this country was "Blackleaf 40" made from tobacco. It was banned years ago. As I mentioned before, pesticides have their place protecting us and allowing us to compete with pests. But why would we use any of these toxins unnecessarily?

The school community (students, parents, teachers, administrators, elected officials, and staff – maintenance, custodial, kitchen, clerical, and nursing) demands a safe learning environment. Tim Gibb, Al Fournier, and Fred Whitford demonstrated this (Fournier et al. 2003). The school community does not want pests in the school environment. On the other hand, a school community does not want to be exposed to toxins. That is a tough situation when you think about the traditional way of controlling pests in schools. By integrating other strategies, such as cultural control (sanitation) and mechanical control (exclusion), as the primary management strategies and relegating chemical control to an "as needed" strategy based on monitoring, schools can effectively address these demands (Gouge and Lame 2015).

What Indicates Schools Are Implementing a Verifiable IPM Program?

Minimum standards for the implementation of pest management in our schools?

- The school administration is aware of what their pest management program is.
- Those responsible for the cultural (sanitation) and mechanical (exclusion) components of IPM have been trained to incorporate them into existing job responsibilities.
- Those responsible for the chemical pesticide component of IPM are certified Pest Control Operators (with instructions to treat as needed and based on monitoring).

Is your school administration aware of what their pest management program is?

Call them and ask! Data from New York (Braband, Horn, and Sahr 2002) indicates that most school districts have pest management policies and most require inspections, monitoring, and record keeping. These are policies resulting from either state mandates or the threat of state mandates (as in Indiana). In Indiana, in Texas, in Michigan, if you ask them, they will say they have it covered, "they have a pest management plan." But are they implementing a verifiable IPM program?

A school district administrator knows how to deal with surveyors' enquiries regarding the use of pesticides in their schools. But, you call them and ask them. Ask them the specific questions I pose for "minimum standards for the implementation of pest management in schools" and let them know you are part of their school community and are watching.

Over the last 20 years we have asked schools (through their district offices) "what are you doing to control pests in your schools?" Most did not have a clue. They could not tell us what pests they had, what was being done to control them (including what, if any, pesticides were being used), or whether the individual applying pesticides held a license to control pests. Yes, today many school systems are required to have Material Data Sheets (MSDs) on site when applying pesticides, some are required to notify the school community regarding pesticide application, and a few are required to have an IPM plan on record.

If your management system demands pesticides, you will use them and you will pay for them. Pesticides really can work. They kill pests – if the applicator knows where they are and when they might be most susceptible! But you do not have to kill pests when you can prevent pests. This is why IPM is considered "source reduction." School districts I have worked with have consistently been able to reduce their pesticide use by at least 90 percent, and reduce pest complaints by at least 85 percent, without increasing costs over a three-year period (Gouge and Lame 2015). Finally, the required IPM plan is a great idea, but unfortunately an unfunded mandate by the states. Walk in and ask to see the plan and have the district facilities manager, superintendent, or your school's principal explain it to you. Most have it written for them by their contractor (i.e., the exterminator company) and have no idea what it is about. A manual your tax money paid for, it's great at gathering dust.

Are those responsible for the cultural (sanitation) and mechanical (exclusion) components of IPM trained to incorporate them into existing job responsibilities?

IPM does not have to cost more than traditional extermination programs. The IPM approach can be successful in the school environment because cultural (practices to reduce attracting pests) and mechanical (practices to exclude pests) strategies can be incorporated into existing custodial and maintenance activities. These include sanitation, energy conservation, building security, and infrastructure maintenance. Monitoring efficiency (verification of pest presence) is enhanced via the virtual full-time presence and perception of the school's human inhabitants. This strategy relies on an educational approach that creates awareness in all occupants of a proactive management strategy, versus the reactive strategy of chemical treatment. Thus, by incorporating IPM into existing school operations (sanitation, preventive maintenance, and classroom education) and partnering with a qualified Pest Management Professional as a consultant/educator (rather than as an "exterminator"), school districts can overcome their natural resistance to "adding pest management to an already full plate."

"Do What You Are Doing Now; Just Think Pests"

The way I get these overworked and underpaid members of the school community to even consider implementing IPM is to say, "Do what you're doing now; just think pests." What training program does your school district have that would allow custodians, maintenance workers, kitchen staff and teachers (all of whom have paid days for professional improvement programs) to become part of an IPM program? Such professional improvement programs have emerged for topics such as HIV, COVID-19 and school security (post-Columbine).

Custodians clean, they straighten up the classrooms, sanitize to prevent germs, repair or report broken windows, doors, etc. If they are provided with proper information, they can easily:

- reduce pest "harborage" (places where pests can eat, sleep and reproduce without being detected) by reducing classroom clutter
- clean up unsanitary conditions and spilled or unsealed food and water that attract and serve as resources for pests.

In short, they can "do what they are doing now" and prevent pests (including molds).

They are incorporating pest management into their jobs. It is the same for our friends in the kitchen. Is sanitary food prepared under sanitary conditions? Are they storing food in conditions that keep out germs? Does the food have bug or mouse poop in it? Sanitation and clutter control – those things are already their job! They just need to know what they can do to prevent bugs and mice.

Maintenance workers provide structural integrity to the school environment, so our children do not hurt themselves. In most school districts they maintain an energy management system that keeps cold air out in cold weather and hot air out in warm weather. These days they manage mold and other components of air quality in the schools with controlled airflow and water damage control. Whatever they are doing for those activities relates to excluding pests from the school facility or preventing pests via reducing damp habitat: weather-stripping, door brushes, caulking (actually, specific sealants dependent on the substrate), repairs to windows, water leaks, etc. – all exclude pests. If you can see daylight beneath an exterior door, cockroaches can enter the building. If you can fit a pencil under a door, a mouse can crawl through the same space. "Do what you are doing now; just think pests."

All members of the school community have the job of providing security for our children. Each has had some training to react to two-legged invaders: how to keep them out, detect them, and what to do if they are spotted. I ask administrations to let these people learn about the four-, six-, eight, and more-legged invaders they should keep out, detect, and react to. Who better to observe these invaders than the inhabitants of the schools? If education is the foundation for IPM, monitoring is the backbone of IPM. What pests are present? Where are they and are they numerous enough to constitute a real problem?

School districts are essentially centralized and linear in their management hierarchy. While district administrators agree to implement IPM, that commitment must be demonstrated by their willingness to communicate with principals, teachers, students, maintenance, custodial and kitchen staff. Whereas administrators and teachers normally are viewed as education professionals, the staff who are relied on to conduct cultural, mechanical and sometimes chemical control are not. Custodial staff are often long-term community members who take justifiable pride in their jobs. If communicated with as professionals and given control of pest management activities in their schools, they become more committed to the goals of the program.

It is interesting to note that custodians view teachers as responsible for attracting and harboring pests, and rightfully so. This viewpoint is not surprising when we consider that of all the individuals in the community, teachers are the only true, long-term human inhabitants of the school buildings. They with their possessions spend the most time in the facility. Teachers create an ideal habitat for pests via storing and consuming food products (for themselves, their students, and classroom pets), propagating and maintaining house plants and developing (sometimes over decades) a "refugia" of collectables best categorized as clutter.

Of course, students are second only to teachers in their ability to provide transportation of pests to the school and provide sustainable harborage in the facility (e.g., lunch bags, lockers, and even wheelchairs). It is my observation (M.L) that teachers understandably behave in the school somewhat differently than in their homes, in that they rely on custodial staff to provide "house cleaning" functions. I mean, wouldn't you modify your behavior (become a bit less fastidious) if you had a custodian come to your home every day? This attitude could be viewed as the major barrier to effective pest-related communication in this centralized system. The good news is that because of the educational culture inherent in the school community, all community members can be receptive to informational communications that affect the consequences to one's actions – a common classroom phrase.

Are those who are responsible for the chemical pesticide component of IPM certified Pest Control Operators (with instructions to treat as needed and based on monitoring)?

As already mentioned, most schools in the country still have exterminators, on contract, "inspecting" once a month. While it is true that many of these Pest Control Operators are not specifically contracted to treat the school monthly, it is my observation that they do. Here is how it works: Your school district facilities manager takes the "low bid" for pest management from a local exterminator, say $350 per school per year. For this, the exterminator fully treats the school once annually (normally in the summer – treating the baseboards of all rooms and often the external perimeter cracks and crevices). Then every month during the school year the applicator spends about 45 minutes to one hour (which is inadequate) "inspecting" the school. They call that monitoring? More often than not, they apply pesticides only where the teacher or head custodian tells them to.

In most of these low-bid contracts the exterminator can get paid for "additional treatment." If you look at the invoices to the school district, every time these technicians visit they end up treating some alleged potential problem site. This practice can double the yearly contract cost, but it does not show up as a line item on the districts' contractual budget sheet! So, they will tell you they have inspected, monitored and treated on an as-needed basis, but who knows?

Are these technicians really qualified to act as Pest Management Professionals? Almost never. If they are licensed or certified (by passing a test given by the SLA on the safe use of pesticides) Pesticide Applicators, does that mean they are trained and experienced in:

- Knowing the common pests for each particular school environment?
- Identifying each of those pests, as well as other less common pests?
- Identifying the conditions that attract or harbor (conducive conditions) each of those pests?
- Monitoring for the type of pest, specific areas in which they might be located, and extent of the pest infestation in those areas?
- Applying all technical options (integrating pest management techniques) in each situation?
- Communicating all of the above information regarding pest problems and solutions to the appropriate school community member?

Unfortunately, not necessarily. Many Pest Control Operators just pass the test. It is my experience that some of these folks are trained and experienced in IPM, but they do only what they have time for (which is damn hard under a low bid). Most states still allow Pest Control Technicians to "inspect" schools and apply pesticides. They work under the "supervision" of Pest Control Operators (who need not be on site). They might be trained to safely apply pesticides, are paid much less than Pest Control Operators, and often have little of the experience or training required to be a real Pest Management Professional. Often, this is the "McJob" of the pest control industry. Your school district administration should know and approve what real IPM is and who practices it. Ask them if they do!

Building an IPM Program

Reality Check for Parents and School Officials 1: IPM – Establishing the Basic Components

1. A management strategy based on communication/education supported by a committed school administration.

In fact, one of the major determinants for success when implementing IPM in schools is that the administration is doing this because "it is the right thing to do." But words are cheap. All major players in the school district feel like they already have too much on their plates. In this centralized system, leadership from the top is the key to commitment. These political animals (school board members and their district superintendents) know that policies without plans (strategies) are worthless.

Some of these folks will glad-hand you all the way out the door while waving their IPM policy for review. Find out if they have communicated their IPM policy, with a logical and reasonable strategy for achieving its goals, to school principals, custodial supervisors, food service managers, kitchen managers and maintenance supervisors sufficiently that these site-based managers can incorporate IPM into the existing functions of faculty and staff. Have they committed to a training or educational program for school staff and/or faculty? If not, the program will not work or, at best, will be less than effective.

2. A "partnership" of the school community (including concerned parents) with a qualified Pest Management Professional.

I will write this again and again, but the only way this type of program will work is if the pest management contractor (aka the exterminator) is converted to a "Diagnostician/ Educator." I have already described the minimum qualifications for the Pest Management Professional. If contractors are genuinely to conduct an IPM program for schools, they must be able to work with the people in those schools to deliver an effective program.

Custodians, kitchen staff, administrators, school nurses (ideally, teachers and maybe students) need to be educated to allow the Pest Management Professional to diagnose the pest problem and conditions causing it. These school inhabitants need to know what a Pest Vulnerable Area (PVA) is (e.g., kitchens, biology labs, teacher lounges, and custodial closets), and how to use and report the results of monitoring traps. They should be able to competently ask:

- Have you had any pest problems?
- What kind of pests are involved?
- Where are you noticing these pests?; and
- For how long?

These questions (and more) are key to the diagnosis and require access to and education of the school community, just as your medical doctor can make a more accurate and timely diagnosis when you are available to answer questions and you are informed about your health issues. Conversely, what good is a medical doctor (MD) who does not take the time (or a managed care organization that doesn't allow your MD the time) to educate you and communicate with you? That person is unprofessional, unqualified, or both!

The Pest Management Professional must be able to identify a receptive audience to remediate conducive conditions. This professional can't help prevent or control pests in the school if the school community will not provide cultural control with better sanitation (less clutter in storage areas and classrooms, cleanup after food in the classroom, fewer vending machines, improved cleaning techniques, etc.), and mechanical control with better exclusion (e.g., door sweeps, the proper sealant for foundation cracks and sealable food containers) any more than an MD can prevent or treat your disease if you are not taking care of yourself! These are management functions that the Pest Management Professional (or MD in the case of your health) can only recommend. It is up to the school officials to implement them. If the conditions that attract pests or allow access to pests are not managed, the chemical pesticides the professional might have to use will be of no more value than taking antibiotics for your respiratory infection will keep you from dying from cancer if you keep smoking.

3. An educational system which empowers the school inhabitants to eliminate or reduce the reasons for insects, rodents, and plants to become pests.

As I have mentioned, the school inhabitants ultimately are responsible for creating an environment that does not attract pests … along with learning:

- How to identify a pest (is it a praying mantis or a cockroach?);
- How many of those pests might constitute an infestation (was it a single adult cockroach or many, or were there immature cockroaches?); and
- How to report their findings to their school pest management coordinator or Pest Management Professional.

These school inhabitants (administration, staff, teachers, and students) should have some education through in-service training and/or periodic newsletters which will empower them to know what conditions might attract pests to the school environment. Understanding what attracts and sustains pest populations is enormously beneficial for the school community. Not only because of their work in the school environment but because they are all inherently responsible for their own living environment. Pest-conducive conditions can be pest-specific but most generally include improperly (accessible) stored human or pet food, spilled food or drinks, improper (accessible) waste management, filthy drains, pet feces, outside lights directly above external doors, water leaks, even warm temperatures.

4. An educational system which empowers school inhabitants to prevent pests from entering the school facility.

Once teachers, custodians, kitchen staff, nurses, and administrators learn how to minimize the school's attractiveness to pests, they can also learn what parts of their job responsibilities can, in fact, keep the vermin out. Is it any wonder that yellow jackets enter the school at the beginning of the school year when the soda pop syrup they love so much is in the open trash containers, placed directly outside the external doors that are left open to provide airflow at summer's end? What is attracting them and how are they getting in? It is all part of a cause-and-effect training program that occurs so often when schools encounter two-legged invaders.

5. An educational system that empowers the school staff to know when and how to remedy documented pest problems by integrating cultural, mechanical, and lowest-impact chemical control technologies.

Teachers can accurately report pest problems, reduce harborage in the form of classroom clutter and use sealable containers for their in-class refreshments and teaching materials such as pet food, macaroni art, etc. They can teach students not to accidentally bring household pests to school. Custodians should be empowered to report data from pest-monitoring traps and report conditions which would allow pests access to the school (poor weather-stripping, absent or broken screens, and leaking pipes) just as they would report any other maintenance problem. They should clean from the pest point of view rather than from the less-observant human view. Maintenance workers replace or repair doors, windows, and pipes as part of their building safety and energy management functions. Kitchen staff should use sealable containers for food, boxless storage to reduce clutter and cardboard (German cockroaches love cardboard – it provides both shelter and food), replace dark shelving and pallets with wire shelving and order supplies only from uncontaminated vendors. Administrators write (or sign) work orders for better equipment and maintenance response. They can write and enforce policies to reduce classroom clutter, cease doing business with sloppy and/or contaminated vendors (improper trash storage containers and pick-up, leaking vending machines, infested food supplies), replace dangerous chemicals with lower-impact baits and inform parents about these actions. It is all part of a well-managed school that already must satisfy OSHA, local health inspectors, parents, and the school board . . . but now by just thinking pests!

Reality Check for Parents and School Officials 2: IPM Is Not ...

1. IPM is not a term or job description added to that of an unwilling or unqualified individual (the school district must either employ or contract with a qualified Pest Management Professional).

Who is your school district's IPM coordinator? As I have mentioned, some of your school districts have them. Just what qualifies them to "coordinate" your pest management program? Are they people who were picked because the school superintendent needed to name someone according to the new policy, or did they volunteer? Why did they volunteer? Are they believers in IPM or just looking for a boost in pay? What is their training related to pest management and are they licensed? Do they really know what IPM is and how to practice it?

Nearly all pest control "specialists" I meet, whether contracted or in-house, tell me they are practicing IPM. Many are not licensed, and even if they are, too many could not tell the difference between a pharaoh ant and a pavement ant, a deer mouse and a house mouse, an organophosphate and pyrethroid insecticide. And they damn sure couldn't begin to discuss the proper monitoring trap placement for German cockroaches in school kitchens. Perhaps the most common and pervasive misinformation these guys lay out is that they are implementing an IPM program when, in fact, all they have done is switched from spraying baseboards to applying cockroach or ant baits. While I appreciate the change to lower-impact, probably more effective insecticides, this is substitution – not integration.

2. IPM is not a "low-bid" process subject to unqualified exterminators, or qualified Pest Management Professionals who will not be able to perform to professional standards.

I cannot say enough bad things about the requirement of schools, where our children reside for six hours a day (sometimes eight hours if they are part of extended day programs), that must accept the lowest bid to manage sometimes dangerous pests with sometimes dangerous pesticides. I have mentioned that, at the very least, this forces the Pest Management Professional to under-bid in order to win a contract. They become professionals who know how to diagnose and educate but cannot afford the time to do anything but revert to the 1960s pest management practices of scheduled spray applications. Of course, one could argue that these professionals should not take on jobs that will not allow them to practice up to their profession's standards.

The most common result of the low bid is that pest control companies normally will send underpaid and under-trained "technicians" to apply pesticides. Depending upon specific state laws this could be perfectly legal. I would recommend that this practice be outlawed in facilities that house children including homes. However, do not blame it all on the pest control industry. If technicians are told paying customers want calendar-based monthly sprays, then many companies will provide just that. In fact, in some instances school staff will force the technicians to spray. In talking to experienced technicians over the years I learned that they all have their own set of nightmare stories. Several report being told to spray kindergarten room carpets for head lice and hearing from decisionmaking school staff they will not pay exterminators if they "can't smell the poison" after their visit.

The most egregious results of the low-bid process are contractors or school employees who have been told to "get some bug juice and spray the school" by managers who just do not care. Reports are heard of individuals known to spray extremely toxic, agricultural-use or illegal products in classrooms because they didn't want to spend money on registered products or because that's what they did "down home" – spray into air ducts during school hours or pour gasoline into a nest of yellow jackets next to a school building. This negligence is a result of horrible management by the school administration.

The low-bid process does not have to be the way a pest management contract is secured. School districts can afford to have Pest Management Professionals conduct real programs if they require or allow them to partner with trained school employees for pest monitoring and pest-conducive condition inspection. Thus, the Pest Management Professional would not have to spend more time at the school facility, and the time they do spend would be as a diagnostician/educator or, when necessary, a low-impact pesticide applicator. Under these conditions, school districts can require a Request for Qualifications (RFQ) which would act as a filter for Requests for Proposals. School districts also could audit their current pest management programs to develop actual costs (remember the difference between contract costs and call-back application costs) and compare these to those of a quality program, then budget for an efficient, cost-effective IPM program. The Bloomington, Indiana, school system reduced its pests and pesticides by 90 percent while reducing costs by 35 percent.

3. IPM is not an "out of sight, out of mind" contractual function. The school community must be willing to communicate with its Pest Management Professional to facilitate long-term cultural controls (practices to reduce attracting pests) and mechanical controls (practices to exclude pests).

This point, of course, is related to the school district's and Pest Management Professional's willingness to partner on IPM. All the school districts that practiced traditional pest management relegated pest management to the "exterminator." The only time the administrator with control over that contract heard about it was either when the contract was up for renewal, when a school was being overrun by pests (at one school mice were literally dropping out of the ceiling onto students), or when the school had to be evacuated due to improper pesticide use. Those school business officials need to understand the importance of this type of service procurement. After all is said and done, IPM is a process (see Case Study Figure 1), not a miracle, and all actors must be a part of that process or it just won't work.

4. IPM is not a chemical pesticide program to prevent pests from entering schools.

In school buildings there is no such thing as a preventative chemical treatment for pests! It does not work, it is wasteful, it can be harmful, and it is a waste of school resources. In Las Vegas I had a national pest control company try to tell me it was practicing IPM by applying insecticide baits in and around the school kitchens annually. This was a required application with no regard to the necessary use of the chemical pesticide tool indicated by the presence or absence of pests. By now you know that this is what most schools and parents believe is needed and works – "spray and they won't come." Sometimes this works with our turf grass pests (particularly when we want our yards and football fields to resemble indoor green carpet), but for structural pest management the preventive application of chemical pesticides is normally misuse of expensive and hazardous toxins.

In structures, when used correctly, pesticides will kill pests in a given area as long as the pesticide is effective. Yes, there is a phenomenon called translocation among social insects (e.g., ants, bees, and termites) where the pesticide is carried back to the "colony," but the insecticide is controlling only an existing infestation, *not* preventing future infestations. Further, unlike the very persistent chlorinated hydrocarbons (DDT, chlordane, etc.),

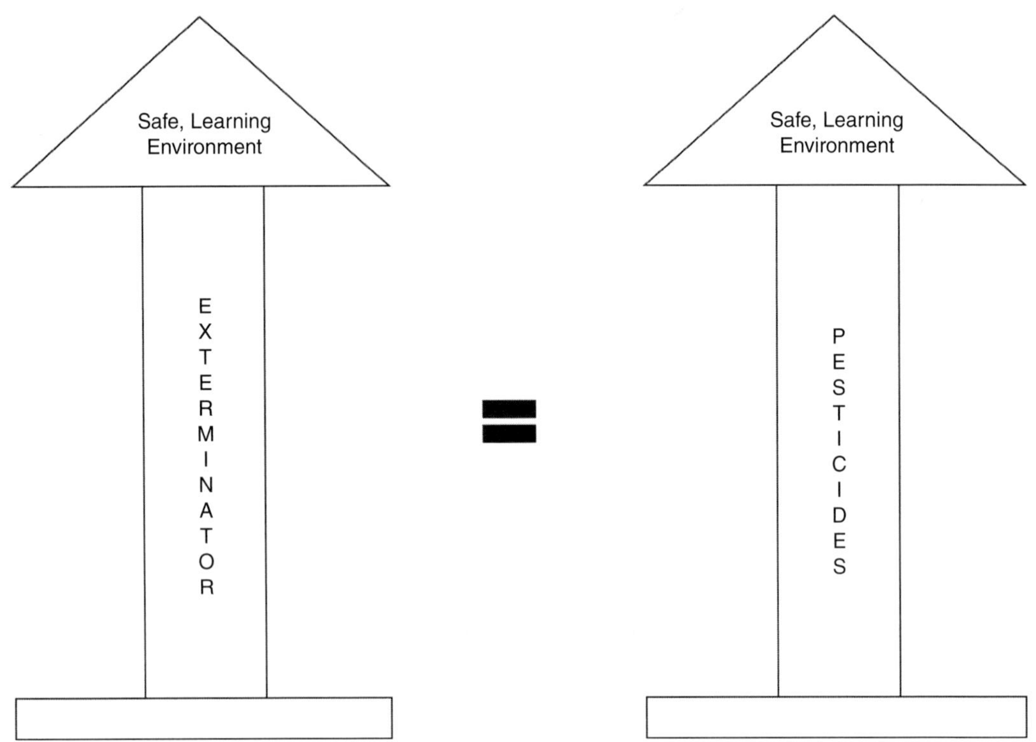

Case Study Figure 1 Traditional pest management model.

current insecticides are not very persistent. Depending on the environment (temperature, ultraviolet light, humidity) these toxic agents provide lethal effects for anywhere from minutes to weeks ... sometimes months in the case of very well-placed insecticide baits. Unfortunately, these baits have begun to fail due to increased resistance exhibited in some of our most common pest species. This, of course, is no surprise. As most scientists know, continuous exposure of adversity (toxins in this case) to organisms with a high generational turnover breeds resistance.

When living in Phoenix, I was always amused by customers (homeowners and school officials) who pay exterminators to spray the perimeter of homes and schools before total darkness. Daytime temperatures often exceed 110°F during a summer day, with unshaded ground temperature somewhere around 180°F, and the ground stays real hot into the night. This temperature deactivates most chemicals within minutes of application. Yeah, the pesticides will kill most pests they come in contact with, but during very hot months most pests are not exposing themselves to these killer temperatures! Timing and coverage are the keys to effective pesticide use. If the pesticide is not used when it is most effective on what it is supposed to kill, it is simply an act of unnecessary pollution and a waste of resourses. It may present unnecessary exposure to humans and non-target animals including beneficial insects and it promotes a false sense of security regarding dangerous pest management.

5. IPM is not a no-chemical pesticide option. Situations do exist in which chemical pesticides are necessary and can be used safely.

OK, now it is my turn to upset those who believe we can compete with pests for food, shelter, attention, and public health while living the suburban dream, without pesticides. E. O. Wilson (1987) very eloquently explains that humans would not be here without insects. They pollinate our food and sources of oxygen production, they decompose plants and animals to provide nourishing soil (acting as nature's recyclers), they provide medicine, clothes, shelter and (for most of the world's population) a significant part of their diet. Most people accidentally eat bugs! If you ask me, bugs aren't competing with us. They are just taking what is already theirs!

I appreciate bugs. I can live with bugs. I do not mind eating them, sharing my food with them, even giving them some of my bodily fluids. I am OK knowing that most humans have mites (follicle mites) living on them – all the time! I am OK knowing that most of us will, at one time or other, have head lice. My attitude is that when you have ectoparasites you are never alone (more bug humor), and I am serious when I say I would much rather become food for baby flies and beetles (maggots and grubs) than be pumped full of formaldehyde when I die. I can sleep with them and go to the bathroom with them – but can you? Be honest. How many people are that much into nature?

We all have different tolerance thresholds for pests. For some in my house the threshold for cockroaches and mice is zero, one or two for flies, three or four for ants. What is yours? Most of these thresholds have been established by the chemical companies so we use more chemicals – even deodorants for body odor tolerances. We of the Western world like to smell nice and not have bugs on us or around us. However, the bottom line is while there are some of us who can live with most pests, we will not allow ourselves to be killed or harmed by their bites, stings, and diseases they transmit. Thus, sometimes chemical pesticides become part of the IPM strategy. While the vast majority of pest problems lend themselves to preventive strategies, ongoing and explosive pest infestations sometimes dictate the inclusion of chemical or biological pesticides in the IPM program. Your schools should not tolerate massive German cockroach infestations due to the adverse health impacts on asthmatics sensitive to the cockroach allergens. Health codes establish certain pests to be minimally tolerated because of established impacts. In other instances, school districts may need to set their own tolerance standards. For example, schools rarely get honeybee colonies on site or near school ground. There are a lot of examples of when pesticides should be used for safety or public health reasons and to "knock down" huge pest populations that the lack of preventive strategies allowed to flourish. Once they are controlled, a full IPM program (particularly the education as to why these infestations occurred) can be implemented. But make no mistake, if we and our children are to maintain the quality of life we have come to demand, chemical pesticides will remain part of modern pest management. But also make no larger mistake: chemical pesticides should be used only as a last or only resort, based on necessity, with the lowest toxicity possible, applied assuring the least possible exposure, by a licensed, experienced pest manager and you have a right to know about it!

"Pest Management Is People Management": Planning for the Successful Implementation of IPM in Schools

This planning is for the diffusion of IPM by school communities, understanding that for humans to adopt something new, they must change their behavior, and therefore they need to communicate. Clearly, this communication is needed for the successful implementation of IPM in schools. Currently, communication regarding IPM in schools is relegated to either the pest control industry (which does not view its time spent on communication as reimbursable) or your school administration, which does not consider IPM worth communicating about – and it may not have the expertise to send the proper message(s) about pest management, in any case.

The school IPM model or plan introduced in this chapter is based on communication – more specifically, the communication subspecialty of diffusion. You will witness how a model of diffusion for school IPM demonstrated its potential in the US. In his 1995 book *Diffusion of Innovation*, Everett Rogers outlines how the change agent should manage diffusion (Rogers 1995). As I have mentioned, many of the change agents responsible for implementing IPM (Cooperative Extension, school administrators, pest management professionals, regulators, etc.) have little knowledge of the diffusion process (the process of adoption of an innovation by a community over time). It probably is not necessary that IPM programs be managed by diffusion experts. However, individuals who are managing IPM programs should have knowledge of the diffusion process and access to experts in this field.

IPM has some unusual features as an innovation. It is technical but could also be classified as both a soft innovation and a hard innovation. It is hard in the sense that most technologies that comprise IPM are either concrete procedures (such as pest monitoring or exclusion practices) or products (insecticides, weather-stripping). It is a soft innovation in that it has an intellectual component: IPM in practice is the plan or strategy of implementing the technologies in a coordinated way, using models and policies.

Important subcategories of innovations include Rogers' (1995) five characteristics: relative advantage, compatibility, complexity, trialability, and observability. Relative advantage, compatibility, trialability, and observability are considered to have a positive effect on diffusion while complexity has a negative effect on the rate of diffusion of an innovation. It is important to look at the properties of the IPM innovation itself to understand how the community might diffuse IPM. In general, the compatibility of IPM with the public school system is excellent as it pertains to education, sanitation, energy conservation, and structural maintenance. IPM is very trialable – it can be visually demonstrated. In this context, a three-school trialable pilot program was part of the district IPM implementation plan. The most positive of IPM's attributes is its trialability and observability – affirming for one school district superintendent that his decision to try IPM was right to the extent that he attempted to persuade another superintendent to try IPM. The relative advantage of IPM to schools in the past has not been viewed as an economical alternative to current practices of pest control – it was difficult to demonstrate the economic benefits of IPM over traditional controls. However, when the school district becomes more aware of its environmental health costs, IPM becomes viewed as having distinct relative advantage over traditional controls.

The complexity of the IPM innovation has a definite negative impact on the rate of adoption (Lambur et al. 1985). IPM is a technology cluster of integrated parts and this integration involves more planning. Further, utilizing technologies within the cluster requires planning more involved than scheduling monthly applications of an insecticide. IPM technologies require knowledge of biology of the insect complex (both pests and beneficials), the timing of special applications and the coordination of management operations (sanitation, maintenance, food preparation, and education) to integrate cultural and mechanical controls into one's operation.

The Change Agents

When we question which entities will effect the behavioral changes necessary for the successful implementation of school IPM, we arrive at the three key actors: local government, program managers, and technical experts – entomologists. Originally, the partnership that implemented the Monroe IPM Model consisted of a local government entity (the school district), a program manager contracted by the US EPA, an urban/industrial entomologist from the private sector, an Extension entomologist, and usually a representative from the state lead agency. The combination of these entities as change agents has had a synergistic effect on program development and implementation. This has resulted in an innovation or model based not only on the IPM concept, but which also mandates the incorporation of organizational management, policy development, and management communication. All partners were initiated by first demonstrating their commitment to a successful IPM program.

In most school districts in the United States the administration structure is highly centralized (a school board, superintendent and principals administrating faculty and staff). The centralized nature of this system is extremely important to implementation of policies and communication within the school community (Rogers 1995). The administrative decisionmakers (superintendent and assistant superintendent of extended services) examine the results and recommendations of the initial pest management assessment and establish a commitment to the program by moving away from traditional "exterminator" services.

Strategic Communication Planning for IPM Implementation

The change agents for this program agreed to use a technology transfer plan grounded in James Garnett's "strategic planning for communication" innovation. This plan requires describing the management situation/object, the message needed to accomplish the objectives, the audience(s) whose behavior would be influenced, the media tailored to the audience's stage of adoption, and a feedback mechanism for program evaluation purposes (Garnett 1992). This plan was then incorporated into a process described by Rogers (1995) as the innovation-decision process. This process traces or represents a sequence of actions and choices by which individuals or organizations (e.g., an adopting school community) evaluate an innovation to decide whether to incorporate it into their management practices (Rogers 1995). The innovation-decision process has five stages: knowledge, persuasion, decision, implementation, and confirmation.

The Management Situation

The management situation for any IPM change agent is to take the adopting community through the innovation-decision process (awareness, persuasion, decision, implementation, and confirmation) by presenting and demonstrating a program that allows schools to manage their pests while maintaining the safest learning environment more effectively. Again, a pest can be defined as an organism that competes with humans for food, water, shelter, or attention. The strategies for successfully competing with pests in the school environment present an unprecedented requirement and opportunity for environmental and behavioral management of the host community. This community consists of administrators, teachers, students, custodial/maintenance staff, kitchen staff, nurses, parent volunteers, and pest management contractors (see Case Study Figure 2). Activities in this environment include education, recreation, and the maintenance of human bodily functions (food and water intake and output). This environment and the pests associated with it are largely uniform with relatively minimal regional variations.

The Message

The IPM approach can be successful in the school environment because cultural and mechanical strategies can be incorporated into existing custodial and maintenance activities such as sanitation, energy conservation, building security, and infrastructure maintenance. Further, monitoring efficiency is enhanced via the virtually full-time presence and perception of the host community. This strategy depends on an educational approach that can create an awareness by all occupants of a proactive management strategy versus the reactive strategy of chemical treatment. Thus, by incorporating IPM (see Case Study Figure 3) into existing school operations (sanitation, maintenance, and classroom education) rather than adding pest control currently performed by contractors, school districts will be able to overcome their natural resistance to adding pest management to an already full plate.

Primary – "Integrated Pest Management will allow your school to reduce the number of pests while using fewer potentially harmful pesticides, and you can do so without significantly increasing your current job responsibilities."

Supportive Secondary(s) – "Nationwide, most school districts that have successfully implemented IPM say they did so because 'it was the right thing to do'."

The vast majority of what will be asked of you involves incorporating the knowledge of "what our pests are and why we have pests" into your existing job responsibilities of facility/food sanitation, energy conservation, building security and education.

Tools you will learn and use to prevent pests in your schools will also work for your home.

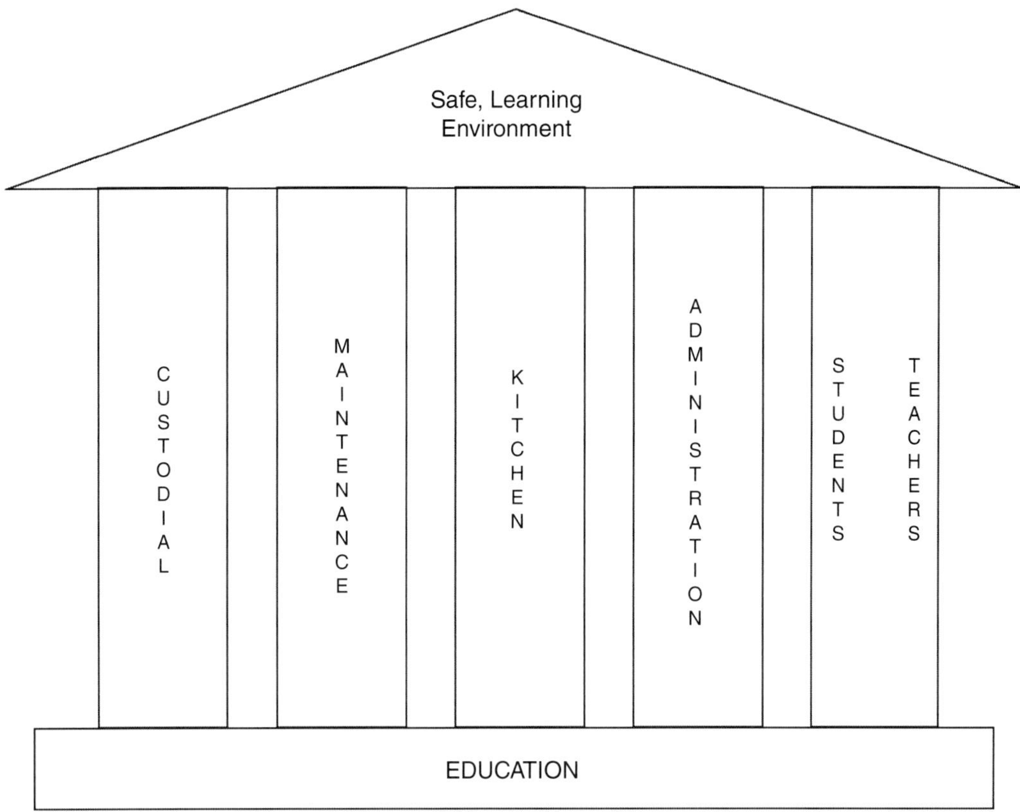

Case Study Figure 2 The current support system for public schools.

The Audience(s)

As discussed earlier, school districts are essentially centralized and linear in their management hierarchy. While district administrators might agree to implement IPM, their commitment to adopt it can be confirmed only by their willingness to communicate what this commitment means to the principals, teachers, students, maintenance, custodial and kitchen staff. Whereas administrators and teachers are normally viewed as education professionals, other staff who would be relied on to conduct cultural, mechanical, and sometimes chemical control are not. The custodial staff often are long-term community members who take pride in their jobs. By communicating with them as professionals and giving them control of pest management activities in their schools, they become more committed to the goals of the program.

I find it interesting that teachers are viewed by administration and staff as the audience primarily responsible for attracting and maintaining pests. After all, teachers are the only true, long-term human inhabitants of the school buildings. If one wants to find pests, look in the teacher's lounge or in the cupboard of the most beloved kindergarten teacher. Of course, students are second only to teachers in their abilities to provide transportation of pests to

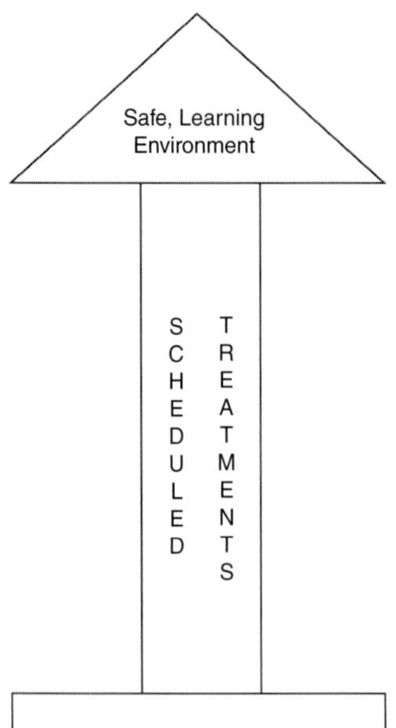

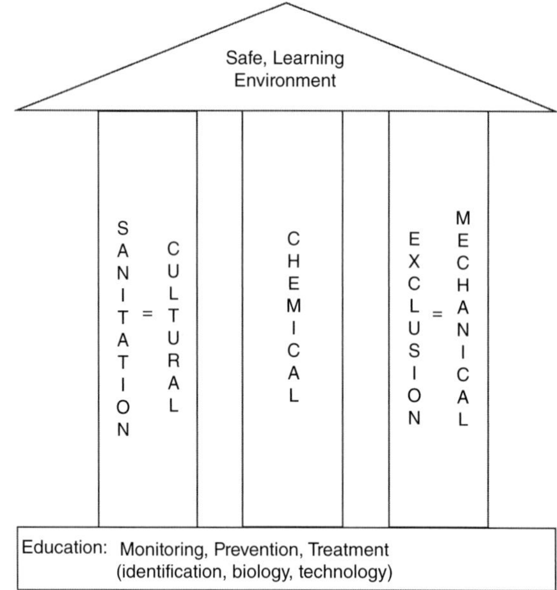

Case Study Figure 3 A shift to IPM.

and their maintenance in the facility (e.g., lunch bags and lockers). Both of these audiences understandably behave in the school somewhat differently than in their primary residences, in that they rely on custodial staff to provide sanitation functions. This frame of reference could be viewed as the primary barrier to communication in this centralized system. However, because of the educational culture inherent in the school community, all audiences are receptive to informational communications that affect the consequences to one's actions – a common classroom phrase.

The following subsections describe these stages of the innovation-decision process and demonstrate that, while the message communicated fundamentally remains the same, the media and specific audiences (school administrator, teacher, students/parents and kitchen, custodial, and maintenance staff) can change in the process.

The Awareness or Knowledge Stage

At this point the sender transmits the message, "Here is a new solution to a problem (old or new)," with the intent of communicating that the innovation (IPM) will meet the community's need. At this stage the primary audience could be the school administration or local change agent.

The Persuasion Stage

The stage at which the adopter unit develops an attitude toward innovation is the persuasion stage. Two specific messages should be sent to audiences: (1) "The presence of pests is a result of their ability to enter your facilities and subsist on the food, water and shelter that you provide for them"; and (2) "By incorporating your knowledge of pests in the performance of your job (sanitation, building maintenance, education, etc.), you can manage pests without increasing your workload." In other much more meaningful words, "Do what you are doing now – just think pests!"

The Decision Stage

At the decision stage, members of the community decide whether they will adopt the innovation for use in their program. At this point the school district is sending the message, "Yes, you have convinced me/us to try out your new solution." However, a more important exchange is taking place within the communication network of the adopting unit. This is the communication between early adopters (administrators) and later adopters (faculty and staff). Newsletters should be distributed to the schools to reinforce some of the issues addressed in training and presentations. The newsletters contain updates and information about the program and pests in schools. Entertaining facts about insects can be included as well as pest control hints. Testimonies from members of the school community should also be included, as these testimonies prove invaluable. Phrases like, "I've been watching for bugs," "It's just the right thing to do," "It's a no brainer," "I have less work because my teachers aren't complaining about bugs so much," and "We just don't see roaches anymore" decide the fate of the adoption.

The Implementation Stage

Implementers include the message, "Yes, you are using this innovation correctly" or "No, but here is how you can make it work properly." At this point, it is critical to avoid information overload. The implementer may recognize, and should accept, that adopters will utilize various coping methods including reinvention. Nurturing may best describe what the change agent must do at this stage.

The change agents need to conduct periodic visits during the next two years. These site visits include inspections for the presence of pests and conditions which could be conducive to pest infestations. Remedial measures should not only be suggested but observed and critiqued as well. However, visits also are conducted to facilitate communication: both to the change agents from the community and by the change agents to the community regarding pest status/problems, program status/problems. This type of communication extends beyond the actual visits. As with relationships between the Extension urban entomologist and the private sector exterminator, these entomologists must make themselves available for questions from the school community – most often from the IPM coordinator, but occasionally from teachers and staff.

Probably the most important phrase I use during this "handholding" stage is, "IPM is a process, not a miracle" (see Case Study Figure 4). While the adopters of school IPM have been sold on implementing recommendations for achieving real IPM, they often are

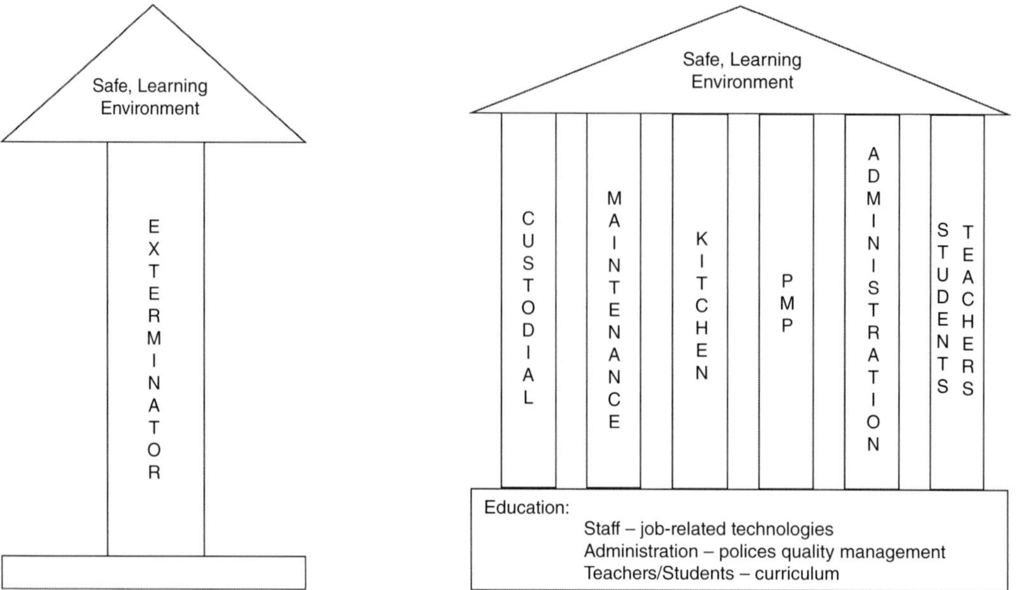

Case Study Figure 4 A shift to "partnering" for pest management.

uncertain at what pace such actions should occur. For example, achieving the suggested level of sanitation to really prevent pests from being attracted to or allowed to infest the school environment will take training, oversight, and time. Further, it is not reasonable to expect maintenance crews to replace all the door sweeps in a school, even in the first year of implementation. We ask school facility managers to address, with appropriate pest monitoring, situations that are most likely to become, or are, an infested area. We must remind them that IPM is an ongoing process as with their everyday energy, security, and sanitation procedures. This is critical communication and makes your audience more comfortable.

The Confirmation Stage

The implementers and community members should gather, and exchange information related to the success or failure of the innovation. Pats on the back using mass media or interpersonal communication are necessary.

The US EPA, the National Foundation for IPM education, and the IPM Institute of North America have recognized, with award plaques and certifications, successful program implementation by school districts that used the Monroe IPM Model. The plaques contain phrases such as "... for your successful efforts in protecting the children of ..." These plaques or letters of congratulation are presented to the supervisor of whoever made the decision to adopt. Further, we often are successful at obtaining news coverage for the award ceremony. A job well done is recognized by newsletters, local newspapers, and, when we are lucky, regional or national media.

When we consider both the nature of IPM as an innovation and the findings on adoption practices, two additional policy implications can be mentioned. The first of these relates to the idea that at least two types of information about IPM need to be disseminated. The first type deals with the nature of the innovation itself. This information describes IPM generally and addresses its use and possible benefits. The purpose of this type of information is to spread the idea and to influence potential adopters to give IPM a trial. From the perspective of the individual adopter, this type of information is important in getting started on an IPM plan or program. The second type of information regarding IPM deals with the mechanics of individual techniques and their implementation. This information is procedural and lays out not only what needs to be done to initiate a technique, but also what should be done to sustain its use.

From the perspective of developing a dissemination strategy for IPM, both types of information are critically important. Furthermore, since adoptions of an innovation occur at different rates in communities, it is important that both types of information be disseminated simultaneously. In this way, information is available to those who are at the decision stage of the diffusion process and additional information is available to sustain those who have already made adoption decisions. This is accomplished by change agents in two ways: First, ensure that both types of information are continually available, that is, built into the dissemination plan. For instance, promote IPM education to the general public (students and parents) in the context of general awareness of environmental health concerns. Second, be sure to disseminate each type of information. Unless detailed follow-up information is available with this type of innovation, there is a danger that the number of adoptions will begin to approximate the number of discontinuances.

The implication of understanding diffusion as a subspecialty of communication science is that change agents who implement IPM improve their overall communication skills. As change agents assemble their implementation plan with some emphasis on diffusion, the communication strategy becomes critical. Once these change agents have answered the questions: "What action has to be taken?" "Who is to take it?" and "Do these people have the capacity to do it?" (Starling 2005: 402), the agents, as communicators, then describe the management situation, the audience, and the preferred medium of communication (Garnett 1992). A plan involving a conscious observation of diffusion forces communication focus through process. The implementer/change agent necessarily identifies the specific management situation, audience, and media by the time the innovation-decision process stage occurs.

Finally, I should note that technological advances that involve the mass media and certain interpersonal media (e.g., interactive television, Internet, cellular telephones) allow more convenient communications between the change agents and the school community – but not necessarily better communication. With the large amount of information, one could send out regarding IPM, it becomes critical to avoid information overload.

This concern places the change agent in a position of needing to use more basic and effective forms of face-to-face communication to determine which, and how much, information is needed. In that, in order to converge on a mutual understanding, the change agent eventually must rely on the most basic interpersonal communication, including intonation, facial expression and body language. In my opinion, the most effective communication is usually face-to-face. Change agents should be prepared to be vulnerable when utilizing this more intimate form of contact. Those who implement IPM should not over-rely on remote training delivery, and technological devices used for, and sometimes in place of, interpersonal communication. As I tell my students, the best managers will make every attempt to "manage by walking around."

The Process for Success: "IPM Is a Process, Not a Miracle"

Over our Integrated Pest Management model's years of development there were many requests from change agents throughout the country for a description of the process one needs to implement what is now called the Monroe IPM Model. With the help of EPA and the National Foundation for IPM Education, we co-produced the process we used in persuading school communities to adopt IPM. About 15 change agents spent several days together agreeing on this process, its impact on the IPM community, how it should be improved, and how to name the model.

A Co-produced Strategic Plan

In preparation for this "co-production meeting," each step in the programs that we had implemented in the previous six years was described in the most effective way and the reason for each step explained. I can say that before I start any new pilot programs I now review these steps for myself and with my team. They work! When one includes other school districts throughout the country that have consciously incorporated substantial parts of this model into their implementation of IPM, we know that more than a million children have benefitted from reduced pest and pesticide exposure.

As discussed in Chapter 5, the successful implementation of any public program requires that three basic questions must be asked and answered: What action must be taken? Who will take that action? And do they have the resources to take the action (Starling 2005)? The Monroe IPM Model combines the process for answering these simple questions with the Rogers innovation-decision process. Below are the 13 steps or objectives with strategies (presented as a narrative) a good manager should utilize to assure the best chance of adopting a successful school IPM program:

Vision Statement: That all toxic chemicals in the US be used according to best management practices to protect human health and the environment.

Mission Statement: That IPM be adopted by the school communities in the US to protect children from the risks of pests and pesticides.

Values:

Humans are an integral part of the ecosystem.

Community members are responsible for the health of the school environment.
Decisions to address the risks of pests and pesticides must be science-based.
Goal: To reduce the presence of pests and the unnecessary use of pesticides in the school community by diffusing IPM.

Objective 1. Secure resources for program implementation. The first thing the IPM model program manager must do is agree with the funding agency on the goals of the program and what resources will be needed. Why do they want you to conduct this program, and what do they want to get out of it? Conversely, what do you want? I hate to say it, but this is negotiation, and you don't stand a chance of getting what you need if you don't know what you want.

For the most part, I have worked with the US EPA, Office of Pesticide Programs, to fund the implementation of this model. We agreed that we wanted to expand the IPM innovation – and how to accomplish this goal. By demonstrating real IPM to a school community, that school community can incorporate it into the school management, and be able to expand IPM to other school communities in the region. The EPA and their change agents had matching goals, maybe for more political reasons, but they see it as part of their general mission and values, specifically under the mandate of FIFRA. All parties had to be willing to compromise on the goals of expansion, program timing and media exposure. This co-production was a partnership and I treated them as such. I also expect to be treated as a partner, not as merely a grantee/contractor.

The Monroe IPM Model costs the funding agency about $40,000 to implement in an average-size school district. This includes assessments by Pest Management Professionals, administrative overhead (paperwork), recommendations, publicity, travel, and on-site inspections by the implementation team. Money is hard to come by for school IPM, and I resent any wasted dollar. This attitude fits pretty well with the requirements for seeking a grant from a government agency. I am not as good when it comes to quarterly reports. However, the Government Performance and Report Act does require this paperwork. I do it, but then I also expect to be able to present the results of our models to others around the country. I believe that government reports left on the shelf are a waste of taxpayer moneys.

Determine the best venue/community for implementation. I always work with the funding agency to strategically plan where the proposed program will have the most chance of success and impact for area-wide expansion. We need the most bang for our bucks. Are the facilities and grounds of the district well managed? Is this a progressive school district that has demonstrated a commitment to other innovative practices and partnerships, either with its facilities or academic excellence? Do the decisionmakers in the district interact with their peers in other districts? Are they leaders? These are questions that should be considered before you and your funding partner invest money, time, and reputation.

While it is nice to choose a run-down, pest-infested school district as representative of how the government wants to help disenfranchised populations, some school districts are run-down and pest-infested for reasons that are incompatible with demonstrating IPM. If a school district cannot manage the behavior of the school community, either because of funding or competence, then positively affecting the sanitation, maintenance, and security functions of the district to "do what you are doing now, just think bugs" could be very costly or fail. It is not a matter of old school buildings that have pests; it is how those schools are managed. IPM is a management system, and the most cost-effective way to demonstrate the IPM innovation is in a system that is currently well managed in terms of its sanitation, maintenance, and security.

Objective 2. Awareness stage: create awareness that IPM is an innovation that will benefit the school community. Make contact with local change agents and school district decisionmakers (the audience for the awareness and persuasion stages). Once you have an idea which school district will offer the most potential for pilot program success and possible regional expansion, you need to establish relationships with interested change agents who reside in the target area (e.g., entomologists, regulatory personnel, and children's health activists) and with the representative public school district and the model implementation team.

Recruiting an Extension specialist and an on-site IPM coordinator is not something I do via the Internet. I use my personal and professional network of friends and colleagues who might know someone and/or work in the target area that the funding agency and I have agreed will offer the most potential. Occasionally, the funding agency will have a local contact for me, but often I make my own contacts with the Extension or regulatory folks in the area through networking. It is my experience that without some type of "entry" it is difficult to get these folks to listen to you, let alone work with you. Entries that work best are actually knowing the players, knowing someone else who knows them, and/or mentioning money. I reach as high up as possible: agency directors, department chairs, etc. That way, when or if they have to put you in touch with the person most closely connected to school IPM, that person will know the boss wants something to happen.

School contacts are a little bit more difficult, and I often rely on Extension or regulatory staff to connect with a school administrator. Network – you might be surprised at how many folks you already know that are part of the public education system. Bob King, assistant principal at Lake Valley School on the Reservation, turned out to be the brother-in-law of my best friend in Phoenix. Bob ended up being a great help to me during implementation of the Navajo program. I have cold-called some school administrators, and usually have been successful obtaining an appointment to discuss the Monroe IPM Model, I usually try the superintendent's office first and always mention my university affiliation and the agency I am working with. Except for a few districts, I have always been treated well.

Objective 3. Persuasion stage: persuade the school community that they can safely adopt the IPM innovation. In all cases I bring examples of past programs and newspaper clippings with me. I provide evidence so as to obtain a verbal commitment (proactively providing the decision stage venue). Since it is a major waste of time to work with a school district that is not really committed, I am fairly assertive from the onset. I insist on communicating with the highest-ranking school official (preferably the superintendent) to persuade them to say to subordinates and the school community, "I need to try IPM in my schools." This step is obviously different than mere contact because it implies action. I usually walk away if they cannot introduce me to the person who has authority to make decisions about pest management in and around the school facilities. It is one thing for school officials to politely listen, and another for them to feel the need to commit resources … and I work very hard to impress them with this fact. They do not want to waste their time either. At this point I make sure they understand how they could benefit or lose when taking on a pilot program. Do I feel like I am wearing a white belt and white shoes while I am selling this used car? No, but I sometimes wonder if I'll be able to deliver the goods.

Once I am introduced to someone who has appropriate understanding of how facilities are managed (sanitation, maintenance, grounds, etc.) we can begin to cooperatively design a pilot program for the school district. When decisionmakers understand what the Monroe IPM Model is, they can designate schools that will best serve as pilots, based not only on pest pressures, but also on which school principals might be most receptive. They can assist in scheduling and facilitating training programs that are critical for our program, but they also have the authority to divert their staff from their normal duties. By having decisionmakers cooperatively tailor the model to their school district, they begin to own it.

There is risk for both the school district and the funding agency with these pilots, in areas such as time, reputation, and money. One way I try to mitigate the risk of adoption is by providing resources. Resources typically take the forms of:

(1) Personnel (nationally recognized trainers and assessors with experience at successfully implementing IPM in schools in other states);
(2) educational and implementation information and materials;
(3) monitoring stations and technical materials; and
(4) recognition (awards, publicity, etc.).

Most important: district decisionmakers must be made aware that their political risks are small (they are using a proven program to "do the right thing" for their community), that their actual costs are negligible because this is an externally funded program, and that we are willing and able to do almost all the work.

Objective 4. Decision stage: obtain a decision to adopt the innovation by the school community. Obtain a memorandum of understanding (MOU). This is a critical strategy. Obtaining an MOU among the school, local change agents (Extension and regulatory), the on-site IPM coordinator and the implementation team is not legally binding, but it ethically forces compliance. While I encourage inclusion of the pest control contractor in discussions for this step, I usually do not require the contractor's signature since that person is an "employee" of designated district decisionmakers (usually the facilities manager or assistant superintendent) who are putting their name on the dotted line.

I spell out everything I need to assess the pilot schools, the resources (personnel, time, and facilities) I need for training, the historical records we need regarding pest management, the authority to prohibit pesticide applications unless we agree, and the possibility of publicizing the program upon successful implementation. In turn, I let all program participants know the resources we will provide, how often, and when. I expect everyone to sign the MOU and use it if folks are not holding up their end of the agreement.

Objective 5. Implementation stage: show the community how to use the IPM innovation; conduct an assessment/audit of the school district. These strategies are usually conducted at the same time, normally during a three-day visit by the implementation team. The audit consists of two parts: (1) the technical assessment of pests, pest-conducive conditions, and pest management; and (2) the management assessment of historical and current costs of policies dealing with pest management. This is where the implementation team splits up and goes into the pilot schools. They complete forms for each of the physical and managerial parameters I have listed. This assessment or data is critical as baseline information for program evaluation and to develop preventive or remedial recommendations for program implementation. The report goes to all members of the implementation team and then to the school district. When the report does get to the designated decision-maker at the district, it should be presented and fully explained by the model program manager and on-site coordinator so corrective measures can be taken with pests and conducive conditions.

Train the trainers (early implementation stage – opinion leaders). This phase – education – is the beginning of actual IPM implementation in the pilot school. We invite (actually insist that) the pest control contractor, technicians, local change agents (Extension, regulatory, and local activist – if known), and the designated on-site coordinator to join us during our technical assessment. These people are responsible for implementing change in the system, so this walk-through training is extremely important. However, it is equally important to use this opportunity to develop a good working relationship with all parties involved. We become less teachers and more mentors for these individuals who we hope will become educators/diagnosticians. This networking is extremely valuable for real-time program troubleshooting. It develops a conduit for moving information we need via email, phone, and already-produced materials (fact sheets, newsletters, websites, etc.).

Train the school staff adopters (mid-implementation stage – peers). During the last phase of the assessment visit we conduct training for staff (custodial, kitchen, maintenance, etc.), and if possible faculty (including, if we are lucky, the principal), in each school pilot location. The subject matter is pests, pest-inducing situations, preventative measures, and possible remedial action. Most of all we explain what IPM is, what their current actions mean to pest management, and that we are asking them to "do what you are doing now, just think bugs." All the local trainers who just completed their own walk-through training with us are expected to attend. If we communicate effectively, these local trainers actually begin to teach at this point. The best way to learn is to teach, and staff begin to view the program as an opportunity for professional improvement. We have learned that this communication works best when we view all staff as professionals and genuinely treat them that way.

Objective 6. Monitor for pests, pesticide use, and conducive conditions for QA/QC (implementation stage – pest management). The on-site coordinator monitors schools at least monthly, and, if possible, other team members – particularly the pest control contractor – do so as well. They use a variety of methods (traps, staff interviews, visual inspection, etc.) to monitor for the presence of pests, pesticide use, and pest-conducive conditions. After discussion with the team pest management professionals, team members facilitate recommendations for preventative measures and possible remedial action at each pilot school. They should be accompanied by senior staff (head custodian, cook, etc.) of each school. If the principal tags along, that's gravy.

Objective 7. Introduce IPM to the whole community (late implementation and early confirmation stages). In the initial stages of the pilot program, usually only those who are responsible for the school facility are introduced to and educated in the ways of IPM. Soon, though, it becomes important for the remainder of the school community to be introduced to IPM. In this step we initiate a mass media awareness campaign for teachers, administrators, parents, and students – the untrained community. If possible, and with the permission of the district, we use local newspapers, but most definitely we use school newsletters and offer to present the program at faculty meetings, school board meetings and to parent–teacher organizations. This stage is critical in garnering support for continuation and expansion, and it also begins the IPM training process for this segment of the community.

Newsletters: Unlike the standard school newsletter, this medium is specific to pest management. Ideally, the whole school community is the audience, but normally the district administration finds it cumbersome and expensive to disseminate such information further than the school facility itself. Thus, it is distributed to faculty and staff and posted in appropriate areas of the schools. Some districts include information about their IPM program on their websites. The Monroe County Community School Corporation has developed a website dedicated to its IPM program in hopes that not only will it benefit their school community, but also other school communities throughout the country that wish to use the IPM innovation.

The "Pest Press" first defines IPM and what it means to the school community. It usually addresses a pest of the month, explains why the pest is attracted to the school and tells how to prevent it from infesting the school. It also can include the status of the IPM program, recognize folks who are helping make it work and present short "bug facts" about simple biology insects in general. The key to this newsletter seems to be that whatever information on pest management is provided, it should relate to what is going on in readers' homes. Bingo.

Objective 8. Conduct midterm evaluation (implementation and confirmation stages). Halfway through the pilot program the implementation team conducts a midterm evaluation. Basically, it is a reassessment of the pilot school program, pest problems, and pesticide use that compares IPM results to pre-program conditions, evaluates progress, and addresses any new problems. This is a fairly in-depth document that points out progress and problems, with recommendations (action items) to solve those problems. It needs to be disseminated promptly to all those who signed the MOU, then reported back to the school. More often than not, problems at this point are not pest infestations, but rather conducive conditions resulting from human behavior. This is where MOU agreements, which apportioned specific responsibilities to each team member, can be used to help get folks to hold up their end of the program management tasks.

Midterm adjustment meeting (mid to late implementation stage): In this step the IPM model program manager should communicate with local implementers to explain the midterm evaluation recommendations. Ideally this is a face-to-face meeting that explains results of the midterm evaluation and how the partners can adjust for program improvement. Program adjustments need to be implemented quickly and there should be a scheduled follow-up for each action item. The IPM program manager should elicit the school's response to issues brought up in the evaluation. One common issue we have noticed is the need to improve the work orders process so that facility repair (door sweeps, torn screens, structural cracks, etc.) can be expedited.

Handholding or nurturing is a natural step during the implementation process. This step is really the IPM program manager's follow-up to implement midterm recommendations by using a multimedia communication to continue education and overcome administrative hurdles. You must take them by the hand and walk them through the implementation process because local school implementers often are not sure what happens next. Fact is, IPM implementation to most of them is like a walk in the dark – scary. The more experienced team members need to be on call for more training, more one-on-one inspections, and more listening to find out why conducive conditions are not being remediated. Handholding is for all audiences, even the implementation team.

While it is a nurturing process, handholding also is a time for the district to analyze whether or not internalizing Pest Control Operator functions will work. The IPM program manager should introduce this concept to administrators. Based on the district's relationship with its contractor, and the contractor's ability and willingness to partner for real IPM, it might be time to wean the district from the contractor type of management. The district should reevaluate its PCO contract if there is one.

Objective 9. Integrate the Pest Management Professional into the model and specific IPM standards (implementation and confirmation stages – program sustainability). Once the relationship with the exterminator-turned-Pest Management-Professional is defined (internal, external, even a combination) the team should spend time working with and emphasizing real IPM with the PMP and the school district. I call this professionalization, whereby the district will:

- Recognize what a quality pest manager is and what a reasonable workload for that person should be.
- Standardize uniform training requirements and joint in-service training for any pest manager.
- Professionalize by upscaling qualifications and compensation to industry standards.
- Centralize by combining and internalizing pest management responsibilities focusing on the school as the unit of operation rather than the program area (food service, grounds, and facilities).

This entire process can be accomplished by ensuring that the school district and its Pest Management Professional are continuously exposed to professional meetings and publications about quality pest management, and that they be made aware of the opportunity for continuing education credits provided by Extension IPM specialists.

Objective 10. Confirm to the school community that they made the correct decision to adopt the IPM innovation (confirmation stage). Conduct a final evaluation of the program in terms of how the goals of the funding agency were met using a comparison of the baseline, midterm and end of program data. We compare costs, pests, pest-conducive conditions, pesticide use and exposure, and finally the attitudes of the adopting school community. We detail what action items from the midterm evaluation have been implemented or not, and why. Distribute the evaluation to all members of the implementation team and then to the school district. When the report does get to the designated decisionmaker in the district, it should be presented and fully explained by the model program manager and on-site coordinator. The report addresses the percent increase or decrease in costs, pests and pesticide use, program progress, and recommendations.

Objective 11. District Expansion (implementation to expand the diffusion). Upon a successful evaluation – demonstrated significant decrease in pests, pesticide use, and with no significant increase in cost – the IPM program manager then needs to propose a district-wide IPM program for the participating school district. The IPM program manager must provide an analysis of costs and timeline for implementation. They must be able to propose a strategic plan for this implementation, and if need be, propose possible sources of funding. I normally develop this proposal with my implementation team, including the state lead agency, the state's IPM Extension specialist and the US EPA Regional Office prior to presenting to the school district. A standard plan includes funding options, a strategy for expansion (larger districts usually need to implement progressively rather than all at once), personnel requirements (those acting as professional pest managers), training requirements and options, and even a public relations campaign. An additional strategy is to assemble a committee of parents who will move expansion forward. Parents have clout. Combined with a committee, their involvement can make a significant contribution.

Objective 12. Stabilize the diffusion of the innovation and prepare for regional expansion (confirmation stage and program sustainability). By recognizing the success of its IPM program, confirmation comes home to the school community and local implementers that they made the right decision in diffusing IPM. Rewards can take the form of award plaques, certificates, and positive media exposure. We usually send letters of congratulation to the school district and Extension or entomology department administrators with copies to deans, state agency directors, and sometimes the governor's office. These letters congratulate the school district and announce recognition of its efforts in an agreed-upon-in-advance press conference. The key words are, "… for your efforts to protect the children of the state of …" It works great!

In preparing for Area-Wide Expansion (implementation to expand the diffusion) your greatest resource is the successful program (as opinion leaders and peers) and the local team that worked with you. Use these peers (school facility managers, Pest Management Professional, Extension specialists, etc., from the successful program to get the next group on board. I suggest implementing the hometown and coalition models (see below) for expansion.

Objective 13. Program evaluation: prepare final compliance narrative for the funding agency, national press and publication for technical peers. Second Law of Thermodynamics – without the input of energy, all things tend toward a greater state of entropy (disorder). School districts need to maintain momentum to keep the IPM ball rolling or people will tend to return to pesticides and behaviors conducive to pests.

School districts and their Pest Management Professionals must be careful not to react to pest problems (that is, lazy), but to be more proactive. By doing so, they establish a schedule for assessing every school over a reasonable period of time, making sure pest presses are being distributed and that they are spending more time working with staff. Spraying pesticides is not acceptable without documentation and is not a preventive measure.

It is obvious that this strategic plan requires intensive communication and partnership. It is based on sound pest management as practiced by national experts, but also on the fundamentals of public management that address program implementation, politics, and resource acquisition and use. School districts in Alabama, Arizona, California, Florida, Indiana, Ohio, Oregon, Washington, Wyoming, and three Native American communities use this model to implement IPM. Policy adoption and faculty/staff education programs are in place in most districts. Most of the districts that have implemented the model have internalized pest management operations and no longer use regular contracted services. The average pesticide reduction has been 90 percent, with a similar reduction in pest problems. Each team that adopted the Monroe IPM Model has developed statewide education programs to develop model peer implementers for its state under different supported agreements with the state lead agency. Further, each team has participated in programs with, or for, its state's Professional Pest Management Association and/or Association of School Business Officials. Last, implementation teams reviewed the impacts and shortcomings, made recommendations for model improvement and expansion, and named the model.

This model was to be the Monroe IPM Model named after the Indiana county it was developed in. Because of the success of this model and subsequent recognition, all change agents not only recognized the importance of and required management techniques related to IPM in schools, but also that IPM could actually be implemented in schools without onerous mandates or fiscal upset. Further, these IPM implementers were now willing to consider similar programs for childcare facilities. They (particularly Cooperative Extension and school administrators) strongly felt they benefitted from positive exposure in the news and EPA recognition of their efforts by their superiors. It is always nice for your bosses to recognize there is importance to working on behalf of children.

The IPM implementers felt that complexity (13 objectives) is definitely a negative attribute of this model and of IPM (as a technology cluster) in general. Culprits included the required amounts and types of communication and nurturing. However, simplifying the process could decrease the abundance of positive attributes, particularly the ability of the model to be tested and observed by the adopting school district. Obviously, increased attention to communication should mitigate much of this complexity, as will the increase in numbers of demonstrations around the nation. Many of our implementers felt the model needs to be more considerate of day-to-day operations (including how work-orders are developed) at the schools, and the degree of staff turnover at schools (especially urban schools). For instance, negotiating time away from work for staff in schools has been problematic. In most cases, proper planning, with plenty of lead time, has introduced a greater degree of comfort to this issue. Newsletter production needs to be simplified for school administrators. A series of pre-written IPM newsletters is available on change agent agency websites to solve this problem.

Our change agent audience described problems associated with adopters pursuant to this model (obtaining past pest/pesticide use records for baseline data development, engaging state pest management associations and obtaining a more binding commitment for participating school districts), but the audience suggested few improvements for the model. All agreed that school principals are a key resource and should be co opted at a very early stage in the process. Our audience suggested involving pest control contractors to a greater degree; however, most felt that there had to be a genuine expression of cooperation from the contractors for this to occur. The Monroe IPM Model is a demand-driven model. I am convinced that professional service providers will respond to their customers when those customers are able to say (demand) what they want. Implementers of school IPM recognized that on a national level a number of management tools must be used in order to successfully diffuse IPM in schools nationwide. Those tools include information dissemination (websites, technical manuals), provider (pest control contractor) education and university research and verification/certification programs. However, when the implementers I worked with (during the development of this model to demonstrate IPM in schools) met, they proposed only two basic recommendations for how this model could be diffused nationally.

The hometown proposal: This proposal allows university entomology departments to access non-competitive funds for a departmental team to implement the Monroe IPM Model in the immediate vicinity of the university. This two-year grant requires the implementing entity to sign an MOU with the US EPA, state lead agency and the hometown school district. The second year of funding is contingent on successful implementation of first-year processes. To my knowledge, there is no existing model for this program. Perhaps Auburn University's is similar, but unfortunately most academics are not coordinated or motivated enough to cooperate on such a community program. That is too bad, seeing that even entomologists have children in public schools.

The coalition proposal: This proposal uses Monroe IPM Model adopters to implement statewide diffusion of IPM in schools, in combination with change agents and funding agencies. Essentially it is a project that uses the initial school district (after one complete year of district-wide implementation) and the implementation team to initiate the Monroe IPM Model for a coalition of school districts, as a springboard to statewide expansion. This requires that the regional or state IPM implementation manager form and chair a taskforce for the expansion of IPM in schools. The coalition should include school pilot program participants, administrators, professional pest manager(s), public health officials and children's health advocates in coalition with selected suburban rural and intercity school districts (one to two pilot schools/district). The administrative training is to be coordinated by the initial pilot program school district and faculty/staff trainings conducted jointly by peers and Extension specialists. Other aspects of this proposal follow the 13-objective model to achieve IPM diffusion in schools. However, several managerial aspects have been added so state program coordinators can sustain their program.

Dawn Gouge at the University of Arizona partnered with me to implement a coalition model in Arizona. I got my old employer – the Arizona Department of Environmental Quality – to talk Region 9, US EPA, Office of Pesticide Programs, into funding this as a children's environmental health initiative. Dawn talked Stan Peterson, Facility Manager with the Kyrene School District, into chairing our coalition partners. The coalition was forged among strategically picked school communities and various change agents including the Arizona Department of Environmental Quality, Arizona Department of Health, Arizona Structural Pest Control Commission, the University of Arizona Department of Entomology, the US EPA, and private sector Pest Management Professionals.

The Case of Implementing IPM on the Navajo Nation

In early 2001 the US EPA, Office of Pesticide Programs contacted me to see if we could develop a pilot program for IPM in the Bureau of Indian Affairs (BIA) school facilities on the Navajo Indian Reservation. It seemed that no one had ever looked at what the BIA was doing for pest management in Tribal schools. The very dangerous pathogen hantaviruses are a family of viruses spread mainly by rodents, and infection with any hantavirus can produce hantavirus pulmonary syndrome in humans. The Navajo Nation spans Arizona, New Mexico, and Utah, three of the four-corner states with the highest number of cases. At the beginning of 2016 it was clear that hantavirus disproportionately affected Native American communities, who accounted for 18 percent of hantavirus cases but only 1.7 percent of the US population. The case fatality rate in Arizona between 2001 and 2014 was 42 percent. No doubt this generated interest in how we might prevent this type of problem in Reservation schools. The US EPA had presented the results of our programs in Arizona to the BIA environmental folks and got an enthusiastic response. I guess it looked like a workable project, and they asked me to head out to Crown Point, New Mexico, (ground zero for the last hantavirus epidemic) and see if it could be done.

Now, many entomologists have traveled the world looking at bugs, but I am not one of them. I just never got around to it. Growing up partially in the Southwest gave me some exposure to Native Americans and, of course, I viewed their cultures as exotic compared to mine. Further, I had worked with many of the tribes both at Extension, and even more when I was with the Arizona Department of Environmental Quality where one of my job responsibilities was to be Tribal liaison to the 25 tribes in the state.

I knew that Tribal governments were way underfunded in the area of environmental protection. It was my experience that, while Indigenous people are often acutely aware of their ties to nature and can in many ways be considered the first environmentalists, they do not all have the resources to staff technical experts and build robust environmental programs. This is why the federal government (US EPA) implements most programs on their behalf, although they can be granted the authority and treated in a manner similar to states under certain provisions of environmental laws. While the government is supposed to be acting in the tribes' best interests, historically it has been convenient for our government, some industries, and certain unscrupulous Tribal leaders (yes, all cultures have them) to suppress environmental education and regulation or not take into consideration their unique relationship to the land or the way they use it. That allows tribes and their partners (e.g., participants in uranium mining, clear-cutting forests, hazardous/radioactive waste storage and disposal) to indulge in environmentally risky behavior and sale tactics that result in overuse of pesticides, particularly herbicides). In my environmental management class at Indiana University, I address these types of situations under the heading of "environmental justice." This issue is sometimes called "environmental racism," whereby a disenfranchised community is exposed to unequal distribution of environmental risk (O'Leary et al. 1999).

Except for a few very interesting attitudes (explained later in this section) regarding what we consider pests, folks in Indian Country feel about and deal with pests the same way the rest of America does. They do not want to live with them and are willing to do what they have to in order to compete with them. Let's be clear here. They watch the same commercials that we do and use pesticides just as so many other Americans do. Anyway, I sat in on a conference call with the feds in DC, the BIA's Eastern Navajo Agency school facilities people in Crown Point and the Navajo EPA pesticide regulators. I got an idea of what was going on there and who I would be working with, and shortly after I made my way out to the Navajo Reservation.

BIA has five "agencies" on the Navajo Nation (some in Arizona and some in New Mexico) which is the largest Reservation in the continental United States. My assessment began at the Eastern Navajo Agency Facility Offices in Crown Point, New Mexico, about 40 miles from Gallup. I met with a work group that consisted of representatives of Navajo EPA who were also members of the Tribe, BIA facility managers, and the contracted pest control operator. Three of the 18 schools managed by the BIA had been selected for this program initiation: Crown Point Community School, Lake Valley School, and Mariano Lake School. At this meeting, we discussed the conditions (which, by now, we had outlined in our MOU) necessary to implement a successful program. In particular, we outlined the need to have a committed school administration, an on-site program manager, the opportunity to conduct an audit (documenting current pest problems, pesticide use, and cost of pest management), and the training of facility personnel (regarding how to prevent pest infestations, recognize pest-conducive conditions and how to remediate those conditions). I found attendees at this meeting to be very receptive, particularly the facilities manager, Bob Villarreal, and his Environmental Specialist (a position unique to the Eastern Navajo Agency) Chad Bourgoin.

While the Navajo EPA representatives were very interested in implementing IPM, I noticed they were not overly willing to work with the BIA. This animosity goes back to the late 1800s and stems from the sovereign status of the tribe and its relationship with its US government "overseers" – the BIA. It is also the result of broken treaties and mistreatment by the federal government from the start of this country. Quite naturally, the tribes resent the authority that BIA commands on the Reservation. Conversely, BIA personnel have mandates to follow, and believe their input is needed to assure standards for education, health, etc. This is an old and ongoing mess over which the Department of the Interior (where the BIA is administratively located) and tribes throughout the US are deadlocked – particularly as it pertains to Indian "trust" monies. I learned there were "BIA Schools" and "Tribal Schools" and was beginning to realize I would not be able to work with both in the initial stages of implementation.

Crown Point Community School, Lake Valley School, and Mariano Lake School all were BIA schools, while Crown Point was actually in the town of Crown Point. Lake Valley and Mariano Lake are located in more traditional villages. Most BIA schools have dormitories where students live during the school week. This would become a particularly challenging aspect for our project when it came to managing pests. Also, it was a real learning experience with regard to communicating with a particular audience. Being aware of the big Indian School in Phoenix, I had always heard that Navajo, Hopi, Apache, Pima, and other Tribal children were forced out of their homes and villages and made to live in schools controlled by the government so they could learn the ways of the white man. These days the dormitory schools are located in or near villages of the student population and are very culturally oriented with regard to language, traditions, etc.

I was surprised that the schools and dorms were receptive to being developed as models. BIA personnel had warned me that these were "rough places" when it came to pest-conducive conditions. However, I found the facilities to be in good condition regarding both building repair and sanitation. After some of the schools I had seen in our previous pilot programs it would have been hard to call these schools "rough places." I attributed the better-than-expected conditions to above-average facility management and to the fact that the Indian Health Service (IHS) inspected the facilities (regarding sanitation) more frequently than environmental health inspectors visited non-Tribal public schools.

Evidently, IHS inspectors were very concerned about disease transmission. Hantavirus is very scary stuff. In the early 1990s dozens of people were killed by this rodent-borne disease (in the urine and feces of the deer mouse). Every year several more residents of the area contract the disease. Some die. Most now survive, but sometimes they are permanently disabled. Thus, the community is acutely aware of the disease as it is related to sanitation and rodent exclusion necessary for prevention. From our point of view, this awareness would aid in implementing additional cultural controls for pests other than mice. It is interesting that this area of the country also produces one or two cases of plague (as in the Black Plague) every year. Plague is a flea-borne disease associated with rodents. No matter who your audience, you can always get immediate attention when you mention plague.

Rodent prevention practices in the schools for exclusion and chemical baiting were inadequate. Entryways for deer mice were evident in all the schools. Many of the external doorways had ground-level gaps that allowed entry to mice. If you can fit a pencil beneath any door or crack in the foundation, a mouse can squeeze through (Corrigan 2001). Of greatest concern were portable classrooms built above ground. Structural openings for pipes, electrical conduit, etc. were unsealed and conducive to pest entry. Because of the threat of disease transmission, snap traps for deer mice are not used in Reservation schools. Therefore, control of mice that might enter the school facility was limited to poison baits. On the good side, the pest control contractor was not using the pelleted baits that could be translocated by mice, instead utilizing the larger biscuit-type bait. Unfortunately, these baits had been improperly distributed and were not containerized. They were not as effective as they should have been and also presented a potential danger to smaller children in the schools. Pretty green biscuits can be tempting to hungry five- and six-year-olds.

Following discussion with the work group and a cursory inspection of the pilot schools/dormitories, I found pest pressures to be low. Flying insects and spiders (bees, wasps, and house flies) were not a problem in terms of presence or tolerance, and cockroaches were non-existent. This region has a very dry climate and is at a high altitude that makes it tough on traditional structural pests. I did find one of the very large (6 to 10 inches long) centipedes common to the region and heard of the very occasional scorpion. Harvester ants and what residents called sugar ants were present and considered to be pests. Sugar ants entered facilities via cracks and crevices. Harvester ants were found only outside the buildings. Harvester ants or "pogos" (a shortened scientific name – from genus *Pogonomyrmex*) are very common in the desert Southwest. A large ant, they almost never sting, but they can and if they do it is an experience you will never forget. Years ago, when I was working on rangeland caterpillars in northeastern New Mexico I got one up my pant legs, and when she finally decided to sting me, I was driving down the freeway. In abject pain, I pulled over, dropped my pants, and scraped her off. My leg cramped for the rest of the day, and my always-compassionate colleagues questioned whether any kind-hearted motorists had offered to stop and help me with my pants.

Of real interest to us entomologist types were the more-than-occasional head lice and bedbug infestations. When visiting Fort Wingate High School I did get to visit a dorm room that had recently been treated for a massive infestation of bedbugs. I actually was able to locate live specimens without much trouble – after the insecticide treatments! This is really neat stuff in that bedbugs were considered a common pre–World War II pest, but became fairly rare with the widespread use of DDT and other pesticides in mainstream American homes and hotels. I had never seen a live specimen until that visit to Fort Wingate. Evidently, the environment in the high deserts of New Mexico is great for bedbugs. That, with the decreased use of certain pesticides, has allowed a resurgence of these bloodsuckers in many places in the state. In fact, hotels throughout the country are beginning to experience an inkling of the bedbug problems that were so common before the 1950s. Vertebrate pests other than rodents did not seem to be much of a problem among the Navajo schools. We did have some ravens visiting the dumpsters outside the kitchen areas. We also had a litter of puppies living in one of the dumpsters – wild dogs are common on the Reservation. Some facility maintenance men also removed several rattlesnakes from under a portable school room while I was visiting. Folks were not too excited. The Navajo will not kill these culturally important animals, and you commonly find them around buildings.

Like most Pest Control Operators, the BIA contractor was practicing as a traditional exterminator (and is the head of this family business) relying on chemical pesticide control for both prevention and remediation. However, unlike many pest control contractors, he seemed totally accepting of the IPM concept and of morphing into a diagnostician/educator. As a member of the Tribe and having a multi-decade professional relationship with the school district, he was into practicing pest management the best way possible. (Note that while his firm had practiced on the Reservation for years, he had just recently contracted for the three pilot schools we would be working in.) It was my understanding that BIA personnel had been conducting pest control prior to his contract, and that they were into the nuclear (as in "let's nuke 'em") mode of application.

According to the exterminator and the school facilities manager, all bugs were being treated at least bi-monthly with a scheduled application of a synthetic pyrethroid insecticide. I witnessed such a treatment by one of the contractor's technicians. He used approximately one gallon of solution per facility building. The insecticide was haphazardly applied to the baseboards and directly on ant mounds. In general, there was no indication that pesticides were needed for any of the standard pests we normally encountered. Thus, here was our first opportunity to demonstrate drastic pesticide use reduction. I also noted that when pesticides were indicated for very limited situations, they were not effectively used. Mattresses, walls, and baseboards were treated for the specific bedbug infestation I witnessed. However, this treatment did not include any flushing agents that would force the reclusive bloodsuckers from cracks and crevices on or around the beds, or the folds of the bedding, to expose them to the pesticide. Nor were there scheduled follow-up treatments which would break the life cycle of this pest. Timing and coverage are everything when it comes to the effective use of pesticides. Like antibiotics, if you do not use them correctly, they can do more harm than good. No monitoring traps were being used for any pests. The pest control contractor made no attempt to educate facility managers about pest prevention. Clearly, there was plenty of room for improvement in terms of sophistication and unnecessary pesticide use.

This contractor committed to upgrade his firm's methods to IPM, and I felt he would be great to work with. His firm consisted of him, his son, and nephew. The latter two were both unlicensed technicians (typical in the business). The head guy told me that the last time he saw an inspector from the New Mexico Department of Agriculture (which inspects and tests licensed Pest Control Operators) was years ago. We wanted to get them to some organized training for structural IPM. Most important, they did attend our assessment and training in Crown Point and Lake Valley and seemed to be very motivated, which was different than our experience with other programs. Kirk and Mike hoped to give them a lot of face time on their monthly visits.

The Team

Not surprisingly, when I contacted Bobby Corrigan, Jerry Jochim, Mike Lindsey, Dawn Gouge, and Carl Martin to team with me on this project, they all jumped at the chance. Mike, Carl, and I had worked on Reservations before. In fact, Carl (who had become a major risk-taker with Arizona's Structural Pest Control Board) had spent two years on the Reservation and spoke fluent Navajo. At that time Bobby and Dawn had not spent any appreciable time on an Indian Reservation. I should also mention that Bobby and Dawn were keenly interested in hantavirus management. Hell, so was I! We talked Carl Olson (who I believe is the best insect identification man in the Southwest) into joining us, after very minor arm twisting. So, it was not hard to coordinate everyone for our initial assessments and trainings. After working together for several years, we knew the drill.

We looked forward to our six-hour drives from Phoenix to our designated HQ, the El Rancho Hotel in Gallup. Normally, we would stop for one break in the desert and one in the mountains north of Strawberry, Arizona, to stretch our legs and look under rocks and logs for different arthropods. Honing our assessment skills by finding and identifying all manner of bugs is sport to folks like us. Carl Olson and Mike normally did some teaching while Jerry and Carl Martin suffered our childish delight in playing with bugs. After several years together we were really improving and integrating our skills. This all sounds like fun (and is), but it is also critical team development: A good manager recognizes when a team approach is needed, who should make up that team and how to choose the team members. Then that good manager will develop the team through briefings, debriefings, skill sharing, and social interaction: Best folks + best communication = the best result. Not only will that team produce a quality program with the best possible results, but each team member also will become their own regional team leader. It's a cascade effect. Too bad that many Extension entomologists are rarely trained to be good managers.

The local BIA team decided Chad Bourgoin would be our on-site IPM coordinator. Chad had a background in environmental management, was highly intelligent and motivated. We were very concerned he was not a professional pest manager but were pretty confident he was a quick learner. The fact is, he was the only person on-site who could do the job. It would turn out to be a problem, but one we were able to overcome. Fortunately, Chad was able to attend our staff training and midterm evaluation in Las Vegas so he could get a feel for how we conduct these types of programs. It was very good for him in terms of program initiation, pest identification and monitoring procedures. It also allowed him to see how screwed up the Clark County School District was, compared to his own school facilities. I had hoped he could stimulate BIA education facilities management on other Indian Reservations to develop the environmental specialist position. Chad's boss, Bob Villarreal, was an ex-Air Force, ex-DoD facility manager with lots of experience, and he liked what we were bringing him. He was the local boss and could make things happen. For all intents and purposes this was our school district administration, and, per our standard MOU, it looked like we had our required commitment from BIA and the Tribe to provide records, staff and take our recommended actions. The person who was really driving the BIA commitment was an environmental scientist with the BIA out of DC – Debbie McBride. She is a Navajo woman who really knew her culture firsthand. She had discovered that BIA really wanted to implement IPM. We were hopeful she could provide political clout to make this happen, and expand. Most of all, we wanted to learn from her.

We had three regulators from the Navajo EPA Pesticide Unit in Window Rock. It was my understanding that they were funded by and reported to EPA Region 9 – much like a state lead agency. The Navajo EPA folks would be partners, but mostly observers, in anticipation of the expansion of the program. The Tribal/US EPA relationship was problematic. EPA Region 9 was concerned that the Navajo pesticide staff were overloaded with current pesticide regulatory responsibilities, and did not want them to take on much more. In some ways, this relationship was much like that between US EPA Regions and most states (command and control). However, the relationship was more politically sensitive because of the sovereign status of tribes. The BIA had reasonably good control over how its schools were managed and was concerned that the Tribal grant schools would be harder to use as pilot schools. The Tribal folks felt a bit left out and it was vital to the success of this pilot project that we had their buy-in. I hoped we could develop a timeline in which we could officially bring these locally controlled grant schools on board.

How the Program Progressed

As with many pilots, this one got off to a slow start. Working with the feds is always slow. Combine that slowness with an institution that operates in the most isolated parts of our country, where time takes on a different meaning, very strange politics, and excessive travel time. I started wishing I were working with the BIA in Ohio. Of course, our white ancestors made sure there were no Reservations in Ohio. Anyway, between getting the local folks in Crown Point to organize our first training and formal assessments, the travel time to get to the Reservation, and working around the 9/11 terrorist attack travel restrictions, we did not get our training completed until the following spring. However, we did convince Chad to suspend pesticide use (other than rodent baits) in all the pilot schools, based on my initial assessment. Like I said, there just was not any need to be using pesticides when no situation indicated that pesticides would be helpful. As is typical in new programs, we had some communication problems. One of which allowed Mariano Lake to be treated for head lice just prior to our first team visit – what a waste. However, this specific incident did educate my team about the significance of "bugs" in the Navajo culture, and the pervasiveness of Madison Avenue's high-pressure sales messages.

We arrived for a briefing with Chad and Bob Villarreal at the BIA facilities building in Crown Point, the morning of our first formal assessments and staff training sessions at the pilot schools. During the briefing Chad mentioned that a classroom had been sprayed for lice. Because he had attended our Las Vegas trainings, he knew there was no such thing as a legitimate area application of pesticides for head lice, and he also knew I would get damn upset about it. In his laid-back way he related what we had heard so many times before: A teacher discovered that one of her students had head lice and demanded that the contractor spray her classroom. Because the teacher, custodian, and principal thought spraying would rid everyone of head lice, and the exterminator did not know any better (or worse, did know better but wanted the paycheck), everyone thought it was OK. Of course, this is a waste of money and dangerous to the children, but the industry promotes pesticides as a cure for any bugs, and these folks bought it.

The very interesting thing here is that culturally, the Navajo people do not hate bugs. Debbie explained that while they do not want head lice, there is no stigma regarding the presence of this human ectoparasite. In fact, the Navajo regard head lice as "the bug which brings the people together." Indeed, "preening" each other for head lice was an important social function in most cultures throughout the world. Now our "civilized cultures" have opted to apply pediculicides (many containing the same nerve agents used in pesticides) to our scalps. Unfortunately, the industry is willing to apply poison to our environment to make us believe we have no problem with head lice, and our teachers are told that no lice should be in their schools. Broadcast use of pesticides must be the answer!

The Navajo also will not step on or squash spiders. They believe spiders are symbolic of "Spider Woman" who taught them the life skill of weaving. Even so, they do not want spiders in their schools or living quarters. While the traditional Navajo will not mechanically kill a spider, there seems to be a disconnect in that it is acceptable to spray them with a pesticide to make them "go away." Is this a story of aboriginal thinking? I do not think so. This is the same logic that allows my children to eat steak, wrapped in cellophane purchased from a store, but not eat meat from an animal that I hunted, slaughtered, and butchered myself. Haven't they disconnected the wrapped meat from the flesh of a doe-eyed animal?

What caused this disconnect? Our need for convenience? Who caused this? Madison Avenue marketing, and its commercialized culture's ability to make us believe that meat comes from supermarkets, and pesticides can just make bugs disappear. Thus, we try to educate the school community, so that it can relearn the cause-and-effect relationship between community actions and the presence of pests.

Throughout 2002, Bobby, Jerry, Mike, and I conducted five one-on-one training sessions for the pest control contractor, kitchen and custodial supervisors at the pilot schools. Training dealt with identification of pests and pest-conducive situations. Working with the contractor became increasingly difficult. While he was supportive of IPM, he was unwilling to commit to training for himself and his staff so they could really implement IPM as diagnosticians/educators. The contractor was a good man, but he didn't believe that the community (school or otherwise) really wanted IPM. Further, he felt that his business did not need to go in this direction. When Chad made it a contractual obligation that the contractor use only IPM, this guy decided he had enough traditional exterminator customers to support his business, and he was probably right. In the latter part of this pilot, he declined to cooperate with our offers of training or recommendations. This actually worked out well because it forced Chad to work with his superiors to require the two BIA Pest Control Operators to take on the Eastern Navajo Agency schools and begin practicing IPM. As it turned out, these guys couldn't wait to learn more and professionalize. They even drove down to Phoenix to attend one of Dawn and Kirk's workshops on urban IPM.

Herb Holgate from Navajo EPA attended one of our training sessions. Herb really wanted us to work with the schools run by the Tribe. During this session he grabbed Mike and told him they were "going for a ride." They disappeared for about two hours. It turns out he wanted to let someone on our team know that the program should be expanded to the Tribal schools. So what he did was invite Mike, who was the eldest member of our team, to ride out to some centuries-old ruins and discuss the situation. He shared his feelings, in a very culturally significant and sacred environment, with our group's "elder." This was a communication strategy we had to respect. We realized Mike was perceived as a decision-maker. We had to be more inclusive of the "elders" in the Navajo community. Unfortunately, neither the BIA nor the EPA took this intercultural approach to communication seriously. As a result, the leader of the Navajo EPA Pesticide Program did not want to communicate with us. To date, we have not been able to integrate IPM into the Tribal schools, but because we had great folks (like Bob King at the Lake Valley School and our own Carl Martin) who helped us understand our audience, we did learn some important lessons in intercultural communication.

We did use some of those lessons learned when we presented the next two training programs for pilot school administrators, teachers, custodial and food service staff. Mostly it was a matter of learning how to show more respect and listen to this audience and how to communicate better. No doubt this is something we should apply to any audience. Most people are alike. They want to feel like what they do is important, that you respect them, and that you are not too full of your own importance. This we did well, as the team members enjoyed making fun of each other. The Navajo have a great sense of humor and found it particularly hilarious when Bobby, wearing his New York Yankees cap, would put on the accent and tell them how New Yorkers manage pests. This degree of intimacy allows us to discuss really sensitive subjects about personal hygiene, and dormitory, kitchen, and classroom-conducive conditions with any audience. In these cases, we needed always to remember that we were talking about someone's job performance and possibly their cultural values.

The only serious pests in these schools were directly related to conducive conditions. Sanitation was good and maintenance staff were "tightening up" the buildings to prevent deer mice from entering buildings. Once identified correctly, the ant problems just disappeared. Pest identification is always important, but it's critical with ants. Thanks to our Arizona entomologists, the sugar ant became the harmless pyramid ant, and with some education about conducive conditions, it became a non-pest (Gouge and Lame 2015). However, thanks to the presence of dorms in the schools, bedbugs were still a problem.

Bedbugs usually were brought into the dorms by just one or two individuals when returning from home with infested bedding or clothes. The bedbugs would feed at night, then hide and reproduce in the cracks of the beds or seams of the mattresses. On one of our drives back to Phoenix the team came up with a plan to reduce bedbugs and the pesticides that were being used against them. First, we suggested a bedbug education plan for the dorm staff, students, and their parents to provide them with the reasons why bedbugs were getting into the dorms, and how to prevent them from entering. Basically, we wanted to make sure the returning students did not bring any hitchhiking bed bugs from home into the dorms. This would be done by inspecting and laundering incoming bedding. We suggested that the BIA replace wooden bunk beds in the dorms with ones of different construction and purchase "seamless" mattresses. Apparently, the principals of these schools felt that the old army beds looked too institutional, so they purchased wooden beds homier in appearance. Unfortunately, the wooden bed construction included plenty of cracks and crevasses for bedbug harborage. The old army beds had more seamless construction. I do not know if this was by design, but I think the original military specifications for beds might just have included bedbug prevention. They probably haven't changed since World War I, when there were lots of bedbugs available for troop quarter infestation. We left it up to Chad to procure beds that prevented harborage but also were not too institutional in appearance. This falls under the heading of cultural control. We also felt we should explore the idea of using heat to kill the hidden pests. That could be done either by moving the beds into the summer sun once a year or actually increasing the temperature inside the dorms with a special heating device. Lastly, we suggested that chemical pesticides be used more effectively (very surgical applications and with a flushing agent) and noted they could be augmented with diatomaceous earth. Diatomaceous earth is abrasive to the cuticle (skin) of insects, causes their bodily fluids to leak out and desiccates them in the process. That's mechanical control. Thus, we developed an IPM program with cultural, mechanical, and chemical controls, all based on an educational approach. The start of this special IPM approach for bedbugs had just begun toward the end of our pilot and probably will take years to fully implement.

It takes time for any of our programs to fully be implemented. IPM is a process, not a miracle. In every Monroe IPM Model program that we have developed, although pesticides and pests have been drastically reduced, we have yet to get all the teachers and custodians to become partners in the IPM program, even years after the beginning of our program. Why? It is all about changing behavior! I can't think of anything more difficult or time-consuming. The only thing one can do without fear of contradiction is demonstrate success – prove it. For the Navajo pilot schools, we demonstrated immediate success by proving drastic reduction in pesticide use, and through the obvious special attention paid to the schools by the BIA facilities staff. Feds are a pain to deal with in some areas, but when it comes to procurement and contracts, once the decision has been authorized by the right folks, things can happen. This was a special lesson for dealing with this type of school system.

Toward the end of the pilot program we recommended that the Eastern Navajo Agency:

- develop a job description recognizing what a professional pest manager is and what a reasonable workload for that position should be;
- standardize BIA and Tribal pest management professional training requirements and provide joint in-service training for any pest manager;
- "professionalize" their pest management by increasing qualifications and compensation to industry standards, and consider combining and internalizing pest management responsibilities, focusing on the school as the unit of operation rather than the program area (food service, grounds, facilities); and
- develop a pesticide notification policy for all schools.

I also offered the following recommendations to the agency decisionmakers in Washington, DC:

1. US EPA and the BIA develop common goals and priorities for the expansion of IPM in American Indian and Alaska Native schools.
2. Pilot programs be demonstrated on receptive Reservations (Navajo, Hopi, Southern Pueblo, and Gila River Pima have indicated interest) as models for national expansion.
3. US EPA and BIA develop an MOU for implementing IPM in Native American schools.
4. The US EPA fund a Center for IPM in Schools in USEPA Region 9 to provide technical assistance for the expansion of this Tribal program nationwide and capitalize on this model of IPM Indian Reservations school districts throughout the Southwest.

Results

After one year in the program, we estimated there had been a 95 percent reduction of pesticide applications in the pilot schools – without additional pest occurrence or cost. Bedbugs did infest the Crown Point dorm midway through the pilot, and we had to authorize a very limited application of pesticide until we could implement alternate controls. Agency-wide expansion was implemented by the end of 2002. Because the BIA Eastern

Navajo Agency had internalized pest management, we anticipated that all 68 BIA schools (more than 14,000 children) on the Navajo, Southern Pueblo, and Hopi Reservations would be able to implement IPM within the next few years.

Perhaps the best way to suggest the program was succeeding is using a report from Chad Bourgoin during a group conference on progress in 2003: "The two PCOs are really getting into the game. Still have not used any pesticide on Eastern Navajo, and we are at 100% implementation. All schools have incorporated the IPM program into a green cleaning program, where the PCO and custodians work to solve issues related to conducive areas."

References

Abatzoglou, John T., and A. Park Williams. 2016. "Impact of Anthropogenic Climate Change on Wildfire across Western US Forests." *PNAS* 113 (42): 11770–75. https://doi.org/10.1073/pnas.1607171113.

ADEQ. 2021. "*About Us | ADEQ Arizona Department of Environmental Quality*." Phoenix, AZ: Arizona Department of Environmental Quality. www.azdeq.gov/AboutUs.

Aerts, Walter, and Denis Cormier. 2009. "Media Legitimacy and Corporate Environmental Communication." *Accounting, Organizations and Society* 34 (1): 1–27. https://doi.org/10.1016/j.aos.2008.02.005.

Agranoff, Robert J. 1986. *Intergovernmental Management: Human Services Problem-Solving in Six Metropolitan Areas*. Albany, NY: State University of New York Press.

Agyeman, Julian, David Schlosberg, Luke Craven, and Caitlin Matthews. 2016. "Trends and Directions in Environmental Justice: From Inequity to Everyday Life, Community, and Just Sustainabilities." *Annual Review of Environment and Resources* 41 (1): 321–40. https://doi.org/10.1146/annurev-environ-110615-090052.

Alexeeff, George V., John B. Faust, Laura Meehan August, Carmen Milanes, Karen Randles, Lauren Zeise, and Joan Denton. 2012. "A Screening Method for Assessing Cumulative Impacts." *International Journal of Environmental Research and Public Health* 9 (2): 648–59. https://doi.org/10.3390/ijerph9020648.

Allen, Steve, Deonie Allen, Vernon R. Phoenix, Gaël Le Roux, Pilar Durántez Jiménez, Anaëlle Simonneau, et al. 2019. "Atmospheric Transport and Deposition of Microplastics in a Remote Mountain Catchment." *Nature Geoscience*, April 1. https://doi.org/10.1038/s41561-019-0335-5.

Alves, Sarah, Joan Tilghman, Arlene Rosenbaum, and Devon C. Payne-Sturges. 2012. "U.S. EPA Authority to Use Cumulative Risk Assessments in Environmental Decision-Making." *International Journal of Environmental Research and Public Health* 9 (6): 1997–2019. https://doi.org/10.3390/ijerph9061997.

Ambec, Stefan, Mark A. Cohen, Stewart Elgie, and Paul Lanoie. 2013. "The Porter Hypothesis at 20: Can Environmental Regulation Enhance Innovation and Competitiveness?" *Review of Environmental Economics and Policy* 7 (1): 2–22. https://doi.org/10.1093/reep/res016.

Anchondo, Isela, Clair Gardner, and Sarah Hyde. 2020. "Current Trends in Environmental Civil Litigation." *American Bar Association, Section of Environment, Energy, and Resources* 34.

Andrade, José, Luís Mendes, and Luís Lourenço. 2017. "Perceived Psychological Empowerment and Total Quality Management-Based Quality Management Systems: An Exploratory Research." *Total Quality Management & Business Excellence* 28 (1–2): 76–87. https://doi.org/10.1080/14783363.2015.1050166.

Ashkenas, Ron. 2013. "How to Preserve Institutional Knowledge." *Harvard Business Review*, March 5.

Axelrod, Robert. 2009. *The Evolution of Cooperation*. Rev. ed. New York: Basic Books.

Axelrod, Robert, and Douglas Dion. 1988. "The Further Evolution of Cooperation."

Science 242 (4884): 1385–90. https://doi.org/10.1126/science.242.4884.1385.

Baliles, Gerald L. 2006. “Preserving the Chesapeake: Law, Ecology, and the Bay.” *University of Richmond Law Review* 41: 615.

Barnes, A. James, John D. Graham, and David M. Konisky. 2021. *Fifty Years at the US Environmental Protection Agency: Progress, Retrenchment, and Opportunities*. Lanham, MD: Rowman & Littlefield.

Barnes, James. 1986. “*Revised Policy Framework for State/EPA Enforcement Agreements*.” Washington, DC: US Environmental Protection Agency.

Barrett, Scott. 2003. *Environment and Statecraft: The Strategy of Environmental Treaty-Making*. Oxford: Oxford University Press.

Beck, Michael W., Iñigo J. Losada, Pelayo Menéndez, Borja G. Reguero, Pedro Díaz-Simal, and Felipe Fernández. 2018. “The Global Flood Protection Savings Provided by Coral Reefs.” *Nature Communications* 9 (1): 2186. https://doi.org/10.1038/s41467-018-04568-z.

Belcher, Oliver, Patrick Bigger, Ben Neimark, and Cara Kennelly. 2020. “Hidden Carbon Costs of the ‘Everywhere War’: Logistics, Geopolitical Ecology, and the Carbon Boot-Print of the US Military.” *Transactions of the Institute of British Geographers* 45 (1): 65–80. https://doi.org/10.1111/tran.12319.

Benotti, Mark J., Loretta A. Fernandez, Graham F. Peaslee, Gregory S. Douglas, Allen D. Uhler, and Stephen Emsbo-Mattingly. 2020. “A Forensic Approach for Distinguishing PFAS Materials.” *Environmental Forensics* 21 (3–4): 319–33. https://doi.org/10.1080/15275922.2020.1771631.

Berry, Wendell. 1971. *The Unforeseen Wilderness: Kentucky's Red River Gorge*. Red River Gorge: University of Kentucky Press.

Biesecker, Michael. 2021. “EPA-Dow Chemical Story.” *AP News*, April 27, sec. Health. https://apnews.com/article/health-chemistry-environment-politics-scott-pruitt-2350d7be5e24469ab445089bf663cdcb.

Bineham, Jeffery L. 1988. “A Historical Account of the Hypodermic Model in Mass Communication.” *Communication Monographs* 55 (3): 230–46. https://doi.org/10.1080/03637758809376169.

Booker, Brakkton. 2021. “Ex-Michigan Gov. Rick Snyder and 8 Others Criminally Charged in Flint Water Crisis.” National Public Radio, January 14, sec. National. www.npr.org/2021/01/14/956924155/ex-michigan-gov-rick-snyder-and-8-others-criminally-charged-in-flint-water-crisi.

Boyle, Kellie, and Banu Örmeci. 2020. “Microplastics and Nanoplastics in the Freshwater and Terrestrial Environment: A Review.” *Water* 12 (9): 2633. https://doi.org/10.3390/w12092633.

Braband, Lynn, Edward Horn, and Laura Sahr. 2002. “*Pest Management Practices*.” Geneva and New York: New York Symposium on Integrated Pest Management.

Brown, Juliane B., Jason M. Conder, Jennifer A. Arblaster, and Christopher P. Higgins. 2020. “Assessing Human Health Risks from Per- and Polyfluoroalkyl Substance (PFAS)-Impacted Vegetable Consumption: A Tiered Modeling Approach.” *Environmental Science & Technology* 54 (23): 15202–14. https://doi.org/10.1021/acs.est.0c03411.

Brulle, Robert J. 2018. “The Climate Lobby: A Sectoral Analysis of Lobbying Spending on Climate Change in the USA, 2000 to 2016.” *Climatic Change* 149 (3): 289–303. https://doi.org/10.1007/s10584-018-2241-z.

2020. “Denialism: Organized Opposition to Climate Change Action in the United States.” In *Handbook of US Environmental Policy*, April. www.elgaronline.com/view/edcoll/9781788972833/9781788972833.00033.xml.

Brulle, Robert J., Galen Hall, Loredana Loy, and Kennedy Schell-Smith. 2021. “Obstructing Action: Foundation Funding and US Climate Change Counter-Movement Organizations.” *Climatic Change* 166 (1): 17. https://doi.org/10.1007/s10584-021-03117-w.

Bur, Jessie. 2020. "Trump Moves to Limit Outsourcing of Government Work." *Federal Times*, August 5.

Burnett, Richard, Hong Chen, Mieczysław Szyszkowicz, Neal Fann, Bryan Hubbell, C. Arden Pope, et al. 2018. "Global Estimates of Mortality Associated with Long-Term Exposure to Outdoor Fine Particulate Matter." *PNAS* 115 (38): 9592–97. https://doi.org/10.1073/pnas.1803222115.

Cao, Hancheng, Chia-Jung Lee, Shamsi Iqbal, Mary Czerwinski, Priscilla N.Y. Wong, Sean Rintel, et al. 2021. "Large Scale Analysis of Multitasking Behavior during Remote Meetings." In *Proceedings of the 2021 CHI Conference on Human Factors in Computing Systems*, 1–13. New York: Association for Computing Machinery. https://doi.org/10.1145/3411764.3445243.

Carper, Tom. 2018. "Following Pruitt's Resignation, Carper Urges Acting Administrator Wheeler to Restore Public's Confidence in EPA's Mission, Consider Lessons of Past." Washington, DC: US Senate Committee on Environment and Public Works. www.epw.senate.gov/public/index.cfm/2018/7/following-pruitt-s-resignation-carper-urges-acting-administrator-wheeler-to-restore-public-s-confidence-in-epa-s-mission-consider-lessons-of-past.

Carson, Rachel. 1962. *Silent Spring*. Boston, MA: Houghton Mifflin Harcourt.

CG. 2021. "Strategy Consulting Services." Consultancy Global. www.consultancy.org/consulting-industry/strategy-consulting.

Charter, Martin, and Michael Jay Polonsky. 2017. *Greener Marketing: A Global Perspective on Greening Marketing Practice*. New York: Routledge.

Choy, C. Anela, Bruce H. Robison, Tyler O. Gagne, Benjamin Erwin, Evan Firl, Rolf U. Halden, et al. 2019. "The Vertical Distribution and Biological Transport of Marine Microplastics across the Epipelagic and Mesopelagic Water Column." *Scientific Reports* 9 (1): 7843. https://doi.org/10.1038/s41598-019-44117-2.

Christmann, Stefanie. 2019. "Do We Realize the Full Impact of Pollinator Loss on Other Ecosystem Services and the Challenges for Any Restoration in Terrestrial Areas?" *Restoration Ecology* 27 (4): 720–25. https://doi.org/10.1111/rec.12950.

Clark, Charles S. 2013. "Reinventing Government – Two Decades Later." *Government Executive*. https://apnews.com/article/health-chemistry-environment-politics-scott-pruitt-2350d7be5e24469ab445089bf663cdcb.

Cohen, Steve. 2018. "EPA's Collusion with the Auto Industry." New York: Columbia Climate School. https://news.climate.columbia.edu/2018/03/26/epas-collusion-auto-industry.

Cole, Luke W., and Sheila R. Foster. 2001. *From the Ground Up: Environmental Racism and the Rise of the Environmental Justice Movement*. New York: New York University Press.

Colebatch, Hal K. 2002. *Policy*. Buckingham: Open University Press.

CoP ESD. 2021a. "Water Services Environmental Services Division (ESD)." Phoenix, AZ: City of Phoenix. www.phoenix.gov:443/waterservices/envservices.

CoP OEP. 2021b. "Office of Environmental Programs – About." Phoenix, AZ: City of Phoenix. www.phoenix.gov:443/oep/about.

Corrigan, Robert M. 2001. *Rodent Control: A Practical Guide for Pest Management Professionals*. Cleveland, OH: GIE Media.

Craig, Robin Kundis, and Anna M. Roberts. 2015. "When Will Governments Regulate Nonpoint Source Pollution? A Comparative Perspective." *Boston College Environmental Affairs Law Review* 42 (1): 1–65.

Crawford, Neta. 2019. "Pentagon Fuel Use, Climate Change and the Costs of War." Providence, RI: Watson Institute for International and Public Affairs at Brown University. https://watson.brown.edu/costsofwar/files/cow/imce/papers/Pentagon%20Fuel%20Use%2C%20Climate%20Change%20and%20the%20Costs%20of%20War%20Revised%20November%202019%20Crawford.pdf.

Dechezleprêtre, Antoine, and Misato Sato. 2017. "The Impacts of Environmental Regulations on Competitiveness." *Review of Environmental Economics and Policy* 11 (2): 183–206. https://doi.org/10.1093/reep/rex013.

Dedoussi, Irene C., Sebastian D. Eastham, Erwan Monier, and Steven R. H. Barrett. 2020. "Premature Mortality Related to United States Cross-State Air Pollution." *Nature* 578 (7794): 261–65. https://doi.org/10.1038/s41586-020-1983-8.

Delk, Kayla L. 2013. "Whistleblowing – Is It Really Worth the Consequences?" *Workplace Health & Safety* 61 (2): 61–64. https://doi.org/10.1177/216507991306100203.

Deming Institute. 2021. "*Dr. Deming's 14 Points for Management*." Ketchum, ID: The W. Edwards Deming Institute. https://deming.org/explore/fourteen-points.

Dillon, Lindsey, Christopher Sellers, Vivian Underhill, Nicholas Shapiro, Jennifer Liss Ohayon, Marianne Sullivan, et al. 2018. "The Environmental Protection Agency in the Early Trump Administration: Prelude to Regulatory Capture." *American Journal of Public Health* 108 (S2): S89–94. https://doi.org/10.2105/AJPH.2018.304360.

DoD. 2021. "*Department of Defense – NEPA Review*." Washington, DC: United States Department of Defense. https://openei.org/wiki/RAPID/Roadmap/9-FD-f.

Downs, George W., and Lawrence B. Mohr. 1976. "Conceptual Issues in the Study of Innovation." *Administrative Science Quarterly* 21 (4): 700–714. https://doi.org/10.2307/2391725.

Drucker, Peter F. 1974. *Management: Tasks, Responsibilities, Practices*. New York: Harper & Row.

2009. *The Essential Drucker: The Best of Sixty Years of Peter Drucker's Essential Writings on Management*. New York: HarperCollins.

Dunn, Rob. 2012. "In Retrospect: Silent Spring." *Nature* 485 (7400): 578–79. https://doi.org/10.1038/485578a.

EIP. 2019. "*The Thin Green Line: Cuts in State Pollution Control Agencies Threaten Public Health*." Washington, DC: Environmental Integrity Project. https://environmentalintegrity.org/wp-content/uploads/2019/12/The-Thin-Green-Line-report-12.5.19.pdf.

Eisner, Alan, Gerry McNamara, and Gregory Dess. 2018. *Strategic Management: Text and Cases*. New York: McGraw-Hill Education.

Ellis, Erle C., Nicolas Gauthier, Kees Klein Goldewijk, Rebecca Bliege Bird, Nicole Boivin, Sandra Díaz, et al. 2021. "People Have Shaped Most of Terrestrial Nature for at Least 12,000 Years." *PNAS* 118 (17). https://doi.org/10.1073/pnas.2023483118.

EPI. 2018. "*Environmental Performance Index 2018*." New Haven, CT: Yale Center for Environmental Law and Policy.

ESA. 1973. "*Endangered Species Act of 1973*." Washington, DC: United States Fish and Wildlife Service.

Esty, Daniel C., and Michael E. Porter. 2002. "Ranking National Environmental Regulation and Performance: A Leading Indicator of Future Competitiveness?" In *The Global Competitiveness Report 2001–2002*, 78–101. New York: Oxford University Press.

FAI. 2021. "*Questions for New Contracting Professionals*." Washington, DC: Federal Acquisitions Institute. www.fai.gov/sites/default/files/Questions_for_new_contacting_professionals.pdf.

Farrell, Justin. 2016a. "Corporate Funding and Ideological Polarization about Climate Change." *PNAS* 113 (1): 92–7.

2016b. "Network Structure and Influence of the Climate Change Counter-Movement." *Nature Climate Change* 6 (4): 370–74. https://doi.org/10.1038/nclimate2875.

Federal Register. 2021. "*National Partnership for Reinventing Government*." Washington, DC: National Archives and Records Administration. www.federalregister.gov/agencies/national-partnership-for-reinventing-government.

Fenton, Suzanne E., Alan Ducatman, Alan Boobis, Jamie C. DeWitt, Christopher Lau, Carla Ng, et al. 2021. "Per- and Polyfluoroalkyl Substance Toxicity and Human Health Review: Current State of Knowledge and Strategies for Informing Future Research." *Environmental Toxicology and Chemistry* 40 (3): 606–30. https://doi.org/10.1002/etc.4890.

Fernandez, Sergio. 2007. "What Works Best When Contracting for Services? An Analysis of Contracting Performance at the Local Level." *Public Administration* 85: ·1119–40.

Ferrari, Cristian Alarcon, Mari Jönsson, Solomon Gebreyohannis Gebrehiwot, Linley Chiwona-Karltun, Cecilia Mark-Herbert, Daniela Manuschevich, et al. 2021. "Citizen Science As Democratic Innovation That Renews Environmental Monitoring and Assessment for the Sustainable Development Goals in Rural Areas." *Sustainability* 13 (5): 2762. https://doi.org/10.3390/su13052762.

Figura, David. 2020. "Federal Court Overturns EPA Decision Denying N.Y., N.J. Protection from Other States' Air Pollution." *New York Upstate*, July 14, sec. Outdoors.

Fournier, Al, Fred Whitford, Timothy J. Gibb, and Christian Y. Oseto. 2003. "Protecting US Schoolchildren from Pests and Pesticides." *Pesticide Outlook* 14 (1): 36–40. https://doi.org/10.1039/b300669g.

Francis, Pope. 2015. *Laudato Si: On Care for Our Common Home*. Vatican City: Our Sunday Visitor.

Freudenberg, Nicholas. 2018. "ToxicDocs: A New Resource for Assessing the Impact of Corporate Practices on Health." *Journal of Public Health Policy* 39 (1): 30–33. https://doi.org/10.1057/s41271-017-0101-0.

Friedman, Lisa. 2019. "E.P.A. Won't Ban Chlorpyrifos, Pesticide Tied to Children's Health Problems." *New York Times*, July 18, sec. Climate.

2020. "E.P.A. Rejects Its Own Findings That a Pesticide Harms Children's Brains." *New York Times*, September 23, sec. Climate.

Frydrych, Dianne. 2004. "Environmental Compliance Assurance Programs." *Journal of Environmental Management Arizona* 2 (4): 20–21.

Fuentes, Agustin. 2017. "Human Niche, Human Behaviour, Human Nature." *Interface Focus* 7 (5): 20160136. https://doi.org/10.1098/rsfs.2016.0136.

Fuller, Richard, Karti Sandilya, and David Hanrahan. 2019. "Pollution and Health Metrics." New York: Global Alliance on Health and Pollution. https://gahp.net/wp-content/uploads/2019/12/PollutionandHealthMetrics-final-12_18_2019.pdf.

GAHP. 2020. "*Air Pollution Interventions: Seeking the Intersection between Climate and Health.*" New York: Global Alliance on Health and Pollution. https://gahp.net/wp-content/uploads/2020/06/AirPollutionReport2-compressed-1.pdf.

Garner, Andra J., Michael E. Mann, Kerry A. Emanuel, Robert E. Kopp, Ning Lin, Richard B. Alley, et al. 2017. "Impact of Climate Change on New York City's Coastal Flood Hazard: Increasing Flood Heights from the Preindustrial to 2300 CE." *PNAS* 114 (45): 11861–66. https://doi.org/10.1073/pnas.1703568114.

Garnett, James L. 1992a. *Communicating for Results in Government: A Strategic Approach for Public Managers*. New York: Wiley.

Gawthrop, Louis C. 1998. *Public Service and Democracy: Ethical Imperatives for the 21st Century*. Ann Arbor, MI: University of Michigan Press.

GBC. 2015. "*Inside Federal Outsourcing: A Candid Survey of Federal Managers.*" Washington, DC: Government Business Council. http://cdn.govexec.com/media/gbc/docs/gbc_government_outsourcing_report.pdf.

Geer Wallace, M. Ariel, and James P. McCord. 2020. "High-Resolution Mass Spectrometry."

In *Breathborne Biomarkers and the Human Volatilome*. 2nd ed., edited by Jonathan Beauchamp, Cristina Davis, and Joachim Pleil, 253–70. Boston, MA: Elsevier. https://doi.org/10.1016/B978-0-12-819967-1.00016-5.

Geimer, Jennifer L., Desmond J. Leach, Justin A. DeSimone, Steven G. Rogelberg, and Peter B. Warr. 2015. "Meetings at Work: Perceived Effectiveness and Recommended Improvements." *Journal of Business Research* 68 (9): 2015–26. https://doi.org/10.1016/j.jbusres.2015.02.015.

Gillson, Lindsey, Harry Biggs, Izak P. J. Smit, Malika Virah-Sawmy, and Kevin Rogers. 2019. "Finding Common Ground between Adaptive Management and Evidence-Based Approaches to Biodiversity Conservation." *Trends in Ecology & Evolution* 34 (1): 31–44. https://doi.org/10.1016/j.tree.2018.10.003.

Gilmer, Ellen M. 2021. "*Trump Environmental Record Marked by Big Losses, Undecided Cases*." Washington, DC: Bloomberg Law. https://news.bloomberglaw.com/environment-and-energy/trump-environmental-record-marked-by-big-losses-undecided-cases.

Goldenberg, Jacob, Sangman Han, Donald R. Lehmann, and Jae Weon Hong. 2009. "The Role of Hubs in the Adoption Process." *Journal of Marketing* 73 (2): 1–13. https://doi.org/10.1509/jmkg.73.2.1.

Gouge, Dawn, and Marc Lame. 2015. "Environmental Health Professionals Work the Bugs Out: School Integrated Pest Management." *Center for Disease Control: Environmental Health Services Branch* 77 (10): 3.

Graham, Nicholas A. J., Joshua E. Cinner, Albert V. Norström, and Magnus Nyström. 2014. "Coral Reefs As Novel Ecosystems: Embracing New Futures." *Current Opinion in Environmental Sustainability* 7 (April): 9–14. https://doi.org/10.1016/j.cosust.2013.11.023.

Grant, Elaine Appleton. 2012. "*Prevailing Winds*." Cambridge, MA: Harvard T. H. Chan School of Public Health. www.hsph.harvard.edu/news/magazine/f12-six-cities-environmental-health-air-pollution.

Grullon, G. 2013. "Infographic: Pesticide Planet." *Science* 341 (6147): 730–31. https://doi.org/10.1126/science.341.6147.730.

Grim, John A. 1997. "Indigenous Traditions and Ecological Ethics in Earth's Insights." *Worldviews: Global Religions, Culture, and Ecology* 1 (1): 139–49. https://doi.org/10.1163/156853597X00281.

Gulland, Frances M. D., Ailsa J. Hall, Gina M. Ylitalo, Kathleen M. Colegrove, Tenaya Norris, Pádraig J. Duignan, et al. 2020. "Persistent Contaminants and Herpesvirus OtHV1 Are Positively Associated with Cancer in Wild California Sea Lions (*Zalophus Californianus*)." *Frontiers in Marine Science* 7. https://doi.org/10.3389/fmars.2020.602565.

Gunstone, Tari, Tara Cornelisse, Kendra Klein, Aditi Dubey, and Nathan Donley. 2021. "Pesticides and Soil Invertebrates: A Hazard Assessment." *Frontiers in Environmental Science* 9. https://doi.org/10.3389/fenvs.2021.643847.

Hadj-Hammou, Jeneen, Steven Loiselle, Daniel Ophof, and Ian Thornhill. 2017. "Getting the Full Picture: Assessing the Complementarity of Citizen Science and Agency Monitoring Data." *PLOS ONE* 12 (12): e0188507. https://doi.org/10.1371/journal.pone.0188507.

Hardin, Garrett. 1968. "The Tragedy of the Commons." *Science* 162: 1243–48.

1982. *Collective Action*. Baltimore, MD: Resources for the Future.

Hart, Stuart L. 1995. "A Natural-Resource-Based View of the Firm." *Academy of Management Review* 20 (4): 986–1014. https://doi.org/10.5465/amr.1995.9512280033.

HHS. 2015. "*About the US Department of Health and Human Services*." Washington, DC: United States Department of Health and Human Services. www.hhs.gov/about/index.html.

Hickel, Jason. 2020a. "The World's Sustainable Development Goals Aren't Sustainable." *Foreign Policy*. https://foreignpolicy.com/

2020/09/30/the-worlds-sustainable-development-goals-arent-sustainable.
2020b. "Quantifying National Responsibility for Climate Breakdown: An Equality-Based Attribution Approach for Carbon Dioxide Emissions in Excess of the Planetary Boundary." *The Lancet Planetary Health* 4 (9): e399–404. https://doi.org/10.1016/S2542-5196(20)30196-0.

Hill, Fiona. 2020. "*Public Service and the Federal Government*." Washington, DC: Brookings Institution. www.brookings.edu/policy2020/votervital/public-service-and-the-federal-government.

HUD. 2006. "*Healthy Homes Issues: Pesticides in the Home – Use, Hazards, and Integrated Pest Management*." Washington, DC: US Department of Housing and Urban Development. www.hud.gov/sites/documents/DOC_12484.PDF.

Huerta Lwanga, Esperanza, Jorge Mendoza Vega, Victor Ku Quej, Jesus de los Angeles Chi, Lucero Sanchez del Cid, Cesar Chi, et al. 2017. "Field Evidence for Transfer of Plastic Debris along a Terrestrial Food Chain." *Scientific Reports* 7 (1): 14071. https://doi.org/10.1038/s41598-017-14588-2.

Huff, James. 2007. "Industry Influence on Occupational and Environmental Public Health." *International Journal of Occupational and Environmental Health* 13 (1): 107–17. https://doi.org/10.1179/107735207800244929.

IDEM. 2021. "*About Indiana Department of Environmental Management: History*." Indianapolis: Indiana Department of Environmental Management. www.in.gov/idem/about.

IPCC. 2018. "Summary for Policy Makers." In *Global Warming of 1.5°C: An IPCC Special Report on the Impacts of Global Warming of 1.5 °C above Pre-industrial Levels and Related Global Greenhouse Gas Emission Pathways*. Geneva: World Meteorological Organization.
2019a. *Climate Change and Land: An IPCC Special Report on Climate Change, Desertification, Land Degradation, Sustainable Land Management, Food Security, and Greenhouse Gas Fluxes in Terrestrial Ecosystems*. Geneva: International Panel on Climate Change.
2019b. *IPCC Special Report on the Ocean and Cryosphere in a Changing Climate*. New York: International Panel on Climate Change.

Irwin, Aisling. 2018. "No PhDs Needed: How Citizen Science Is Transforming Research." *Nature* 562 (7728): 480–82. https://doi.org/10.1038/d41586-018-07106-5.

IUCN. 2018. "*Marine Plastics*." Gland: International Union for the Conservation of Nature. www.iucn.org/resources/issues-briefs/marine-plastics.

Jensen, Olaf P. 2019. "Pesticide Impacts through Aquatic Food Webs." *Science* 366 (6465): 566–67. https://doi.org/10.1126/science.aaz6436.

Joyce, Philip. 2013. "Outsourced Government: Have We Gone Too Far?" *Governing*, October 5.

Kammen, Daniel M., and David M. Hassenzahl. 2001. *Should We Risk It?: Exploring Environmental, Health, and Technological Problem Solving*. Princeton: Princeton University Press.

Kaufmann, Anton. 2012. "High Mass Resolution versus MS/MS." In *TOF-MS within Food and Environmental Analysis*, Comprehensive Analytical Chemistry vol. 58, edited by Amadeo R. Fernandez-Alba, 169–215. https://doi.org/10.1016/B978-0-444-53810-9.00001-8.

Kellough, J. Edward. 1998. "The Reinventing Government Movement: A Review and Critique." *Public Administration Quarterly* 22 (1): 6–20.

Kelman, Steve. 1998. "*Remaking Federal Procurement*." Cambridge, MA: John F. Kennedy School of Government at Harvard University.
2017. "'Reinventing Government,' 25 Years Later." *Federal Computer Week*. https://fcw.com/articles/2017/12/06/kelman-25-years-of-acquisition-reform.aspx.

Kennedy, Brian, Meg Hefferon, and Cary Funk. 2018. "*Students Don't Pursue STEM Because It's Too Hard, Say 52% of Americans.*" Washington, DC: Pew Research Center. www.pewresearch.org/fact-tank/2018/01/17/half-of-americans-think-young-people-dont-pursue-stem-because-it-is-too-hard.

Kettl, Donald F. 2002. "Managing Indirect Government." In *The Tools of Government: A Guide to the New Governance*, edited by Lester M. Salamon and Odus V. Elliot, 490–510. Oxford: Oxford University Press.

Khan, Jibran, Konstantinos Kakosimos, Ole Raaschou-Nielsen, Jørgen Brandt, Steen Solvang Jensen, Thomas Ellermann, and Matthias Ketzel. 2019. "Development and Performance Evaluation of New AirGIS – A GIS Based Air Pollution and Human Exposure Modelling System." *Atmospheric Environment* 198 (February): 102–21. https://doi.org/10.1016/j.atmosenv.2018.10.036.

Kiaghadi, Amin, and Hanadi S. Rifai. 2019. "Physical, Chemical, and Microbial Quality of Floodwaters in Houston Following Hurricane Harvey." *Environmental Science & Technology*, April. https://doi.org/10.1021/acs.est.9b00792.

Kingdon, John W. 2003. *Agendas, Alternatives, and Public Policies*. 2nd ed. Harlow: Longman.

2011. *Agendas, Alternatives, and Public Policies*. 2nd ed. Carmel, IN: Pearson Education Limited.

Kojola, Erik, and David N. Pellow. 2021. "New Directions in Environmental Justice Studies: Examining the State and Violence." *Environmental Politics* 30 (1–2): 100–118. https://doi.org/10.1080/09644016.2020.1836898.

Konar, Shameek, and Mark A. Cohen. 1997. "Information As Regulation: The Effect of Community Right to Know Laws on Toxic Emissions." *Journal of Environmental Economics and Management* 32 (1): 109–24. https://doi.org/10.1006/jeem.1996.0955.

Koning, John W. 1988. *The Manager Looks at Research Scientists*. New York: Science Tech Publishers.

Konisky, David M. 2015. *Failed Promises: Evaluating the Federal Government's Response to Environmental Justice*. Cambridge, MA: MIT Press.

Koontz, Tomas M., Toddi A. Steelman, JoAnn Carmin, Katrina Smith Korfmacher, Cassandra Moseley, and Craig W. Thomas. 2010. *Collaborative Environmental Management: What Roles for Government?* Abingdon, Oxon: Routledge.

Kraft, Michael E. 2017. *Environmental Policy and Politics*. New York: Taylor & Francis.

Kramar, David E., Aaron Anderson, Hayley Hilfer, Karen Branden, and John J. Gutrich. 2018. "A Spatially Informed Analysis of Environmental Justice: Analyzing the Effects of Gerrymandering and the Proximity of Minority Populations to US Superfund Sites." *Environmental Justice* 11 (1): 29–39. https://doi.org/10.1089/env.2017.0031.

Kuehn, Robert R. 1996. "The Environmental Justice Implications of Quantitative Risk Assessment." *University of Illinois Law Review* 1996: 103.

Lambur, M. T., M. E. Whalom, and F. A. Fear. 1985. "Diffusion Theory and Integrated Pest Management: Illustrations from Michigan Fruit IPM Program." *Bulletin of the Entomological Society of America* 31: 40–45.

Lame, Marc. 2005. *A Worm in the Teacher's Apple: Protecting America's School Children from Pests and Pesticides*. Bloomington, IN: AuthorHouse.

2019. "EPA Knows This Pesticide Is Dangerous, So Why Did It Reverse the Ban?" *The Hill*, February 21.

Lame, Marc, and Richard Marcantonio. 2016. "Message to Trump and State Governors: Stop the War on the Environment." *The Hill*. December 8. http://thehill.com/blogs/pundits-blog/energy-environment/309414-message-to-trump-and-state-governors-stop-the-war-on.

Landrigan, Philip J. 2001. "Children's Environmental Health: Lessons from the Past and Prospects for the Future." *Pediatric Clinics of North America* 48 (5): 1319–30. https://doi.org/10.1016/S0031-3955(05)70377-1.

Landrigan, Philip J., Anjali Garg, and Daniel B. J. Droller. 2003. "Assessing the Effects of Endocrine Disruptors in the National Children's Study." *Environmental Health Perspectives* 111 (13): 1678–82. https://doi.org/10.1289/ehp.5799.

Landrigan Phillip J., Kimmel Carole A., Correa Adolfo, and Eskenazi Brenda. 2004. "Children's Health and the Environment: Public Health Issues and Challenges for Risk Assessment." *Environmental Health Perspectives* 112 (2): 257–65. https://doi.org/10.1289/ehp.6115.

Landrigan, Philip J., Richard Fuller, Nereus J. R. Acosta, Olusoji Adeyi, Robert Arnold, Niladri (Nil) Basu, et al. 2017. "The Lancet Commission on Pollution and Health." *The Lancet* 391 (10119): 462–512. https://doi.org/10.1016/S0140-6736(17)32345-0.

Landrigan, Philip J., Richard Fuller, Howard Hu, Jack Caravanos, Maureen Cropper, David Hanrahan, et al. 2018. "Pollution and Global Health – An Agenda for Prevention." *Environmental Health Perspectives* 126 (8): 1–6. https://doi.org/10.1289/EHP3141.

Lasswell, Harold Dwight. 1948. *Power and Personality*. New York: Norton.

Latour, Bruno. 1987. *Science in Action*. Cambridge, MA: Harvard University Press.

Lazarus, Richard. 2019. "*Analytics Reveal Key Trends and Themes in Environmental Litigation*." Washington, DC: Environmental Law Institute. www.law.harvard.edu/faculty/rlazarus/docs/columns/LAZARUS_FORUM_2019_sept.pdf.

Lee, Charles. 2020. "A Game Changer in the Making? Lessons from States Advancing Environmental Justice through Mapping and Cumulative Impact Strategies." *Environmental Law Reporter* 50: 10203.

Leopold, Aldo. 1949. *A Sand County Almanac*. New York: Oxford University Press.

Li, Wenwen, Michael Batty, and Michael F. Goodchild. 2020. "Real-Time GIS for Smart Cities." *International Journal of Geographical Information Science* 34 (2): 311–24. https://doi.org/10.1080/13658816.2019.1673397.

LII. 2021. "*2 CFR § 200.24 – Cooperative Agreement*." Ithaca, NY: Cornell Law School Legal Information Institute. www.law.cornell.edu/cfr/text/2/200.24.

Livermore, Michael A. 2017a. "The Perils of Experimentation." *Yale Law Journal* 126 (3): 636–709.

2017b. "*Why Shifting Regulatory Power to the States Won't Improve the Environment*." Melbourne, Australia: The Conversation. http://theconversation.com/why-shifting-regulatory-power-to-the-states-wont-improve-the-environment-78245.

LLNL. 2020. "*Estimated US Energy Consumption in 2019: 100.2 Quads*." Lawrence Livermore National Laboratory. Livermore, CA: US Department of Energy.

Lowe, Elizabeth C., Tanya Latty, Cameron E. Webb, Mary E. A. Whitehouse, and Manu E. Saunders. 2019. "Engaging Urban Stakeholders in the Sustainable Management of Arthropod Pests." *Journal of Pest Science* 92 (3): 987–1002. https://doi.org/10.1007/s10340-019-01087-8.

Lozano, Rodrigo, Angela Carpenter, and Donald Huisingh. 2015. "A Review of 'Theories of the Firm' and Their Contributions to Corporate Sustainability." *Journal of Cleaner Production*, Bridges for a More Sustainable Future: Joining Environmental Management for Sustainable Universities (EMSU) and the European Roundtable for Sustainable Consumption and Production (ERSCP) conferences, 106 (November): 430–42. https://doi.org/10.1016/j.jclepro.2014.05.007.

Luna, Soledad, Margaret Gold, Alexandra Albert, Luigi Ceccaroni, Bernat Claramunt, Olha Danylo, et al. 2018. "Developing Mobile Applications for Environmental and Biodiversity Citizen Science: Considerations and Recommendations." In *Multimedia Tools and Applications for Environmental & Biodiversity Informatics*, edited by Alexis Joly, Stefanos Vrochidis, Kostas Karatzas, Ari Karppinen, and Pierre Bonnet, 9–30. Cham: Springer International Publishing. https://doi.org/10.1007/978-3-319-76445-0_2.

Marcantonio, Richard, and Agustin Fuentes. 2020. "A Clear Past and a Murky Future: Life in the Anthropocene on the Pampana River, Sierra Leone." *Land* 9 (3): 72. https://doi.org/10.3390/land9030072.

Marcantonio, Richard, Sean Field, and Patrick M. Regan. 2019. "Toxic Trajectories under Future Climate Conditions." *PLOS ONE* 14 (12): e0226958. https://doi.org/10.1371/journal.pone.0226958.

2020. "Toxicity Travels in a Changing Climate." *Environmental Science & Policy* 114 (December): 560–69. https://doi.org/10.1016/j.envsci.2020.09.029.

Marcantonio, Richard, Sean P. Field, Papanie Bai Sesay, and Gary A. Lamberti. 2021. "Identifying Human Health Risks from Precious Metal Mining in Sierra Leone." *Regional Environmental Change* 21 (1): 2. https://doi.org/10.1007/s10113-020-01731-5.

Marcantonio, Richard, Debra Javeline, Sean Field, and Agustin Fuentes. 2021. "Global Distribution and Coincidence of Pollution, Climate Impacts, and Health Risk in the Anthropocene." *PLOS ONE* 16 (7): e0254060. https://doi.org/10.1371/journal.pone.0254060.

Mascarenhas, Carla, João J. Ferreira, and Carla Marques. 2018. "University–Industry Cooperation: A Systematic Literature Review and Research Agenda." *Science and Public Policy* 45 (5): 708–18. https://doi.org/10.1093/scipol/scy003.

Maslow, A. H. 1943. "A Theory of Human Motivation." *Psychological Review* 50 (4): 370–96. https://doi.org/10.1037/h0054346.

Matthews, Deanna H., Gwen C. Christini, and Chris T. Hendrickson. 2004. "Five Elements for Organizational Decision-Making with an Environmental Management System." *Environmental Science & Technology* 38 (7): 1927–32. https://doi.org/10.1021/es0351239.

Mayer, Jane. 2017. *Dark Money: The Hidden History of the Billionaires behind the Rise of the Radical Right*. New York: Knopf Doubleday Publishing Group.

Mehrabian, Albert. 1971. *Silent Messages*. Belmont, CA: Wadsworth Publishing.

Mendenhall, Elizabeth. 2018. "Oceans of Plastic: A Research Agenda to Propel Policy Development." *Marine Policy* 96 (October): 291–98. https://doi.org/10.1016/j.marpol.2018.05.005.

Merriam-Webster. 2021a. "*Definition of COLLUSION*." Springfield, MA: Merriam-Webster. www.merriam-webster.com/dictionary/collusion.

2021b. "*Definition of CO-OPTATION*." Springfield, MA: Merriam-Webster. www.merriam-webster.com/dictionary/co-optation.

2021c. "*Definition of CORRUPTION*." Springfield, MA: Merriam-Webster. www.merriam-webster.com/dictionary/corruption.

2021d. "*Definition of CRISIS*." Springfield, MA: Merriam-Webster. www.merriam-webster.com/dictionary/crisis.

2021e. "*Definition of MANAGEMENT*." Springfield, MA: Merriam-Webster. www.merriam-webster.com/dictionary/management.

2021f. "*Definition of RISK*." Springfield, MA: Merriam Webster. www.merriam-webster.com/dictionary/risk.

Mitra, Amitava. 2016. *Fundamentals of Quality Control and Improvement*. New York: John Wiley & Sons.

Monk, Linda. 2013. "*Federalism | Constitution USA*." Washington, DC: Public Broadcasting Service. www.pbs.org/tpt/constitution-usa-peter-sagal/federalism.

Montrie, Chad. 2011. *A People's History of Environmentalism in the United States*. London: A&C Black.

Mora, Camilo, Bénédicte Dousset, Iain R. Caldwell, Farrah E. Powell, Rollan C. Geronimo, Coral R. Bielecki, et al. 2017.

"Global Risk of Deadly Heat." *Nature Climate Change* 7 (7): 501–506. https://doi.org/10.1038/nclimate3322.

Mora, Camilo, Daniele Spirandelli, Erik C. Franklin, John Lynham, Michael B. Kantar, Wendy Miles, Charlotte Z. Smith, et al. 2018. "Broad Threat to Humanity from Cumulative Climate Hazards Intensified by Greenhouse Gas Emissions." *Nature Climate Change* 8 (12): 1062–71. https://doi.org/10.1038/s41558-018-0315-6.

Napper, Imogen Ellen, and Richard C. Thompson. 2020. "Plastic Debris in the Marine Environment: History and Future Challenges." *Global Challenges* 4 (6): 1900081. https://doi.org/10.1002/gch2.201900081.

NASA. 2021. "*Air Quality Monitoring from Space*." Washington, DC: National Aeronautics and Space Administration. /air-quality-space.

National Archives. 2021. "*The Smith-Lever Act of 1914*." Washington, DC: National Archives Foundation. www.archivesfoundation.org/documents/smith-lever-act-1914.

ND-GAIN. 2019. "*ND-GAIN Country Index Rankings | ND-GAIN Index*." Notre Dame: Notre Dame Global Adaptation Index. http://index.gain.org/ranking.

Nevitt, Mark P., and Robert V. Percival. 2018. "Could Official Climate Denial Revive the Common Law as a Regulatory Backstop?" *Washington University Law Review* 96 (3): 54.

NFOIC. 2021. "*State Freedom of Information Laws – National Freedom of Information Coalition*." Gainesville, FL: National Freedom of Information Coalition. www.nfoic.org/state-freedom-of-information-laws.

Nguyen, Janet. 2019. "*The US Government Is Becoming More Dependent on Contract Workers*." Washington, DC: Marketplace. www.marketplace.org/2019/01/17/rise-federal-contractors.

NIAAA. 2021. "*College Drinking*." Washington, DC: National Institute on Alcohol Abuse and Alcoholism. www.niaaa.nih.gov/publications/brochures-and-fact-sheets/college-drinking.

Nisbet, Matthew C., and John E. Kotcher. 2009. "A Two-Step Flow of Influence?: Opinion-Leader Campaigns on Climate Change." *Science Communication* 30 (3): 328–54. https://doi.org/10.1177/1075547008328797.

NOAA. 2020a. "*Oil Pollution Act of 1990 Intent and Provisions*." Washington, DC: National Oceanic and Atmospheric Administration. www.gulfspillrestoration.noaa.gov/wp-content/uploads/2010/10/PPD_AP-A.pdf.

2020b. "*US Billion-Dollar Weather and Climate Disasters, 1980 – Present*." Silver Springs, MD: NOAA National Centers for Environmental Information. https://doi.org/10.25921/STKW-7W73.

2021. "*30 Years of the Oil Pollution Act: How It Helps NOAA Prepare for and Recover from Spills | NOAA Fisheries*." Washington, DC: National Oceanic and Atmospheric Administration. www.fisheries.noaa.gov/feature-story/30-years-oil-pollution-act-how-it-helps-noaa-prepare-and-recover-spills.

NRDC. 2013. "*Policy Basics: Fracking*." Washington, DC: Natural Resource Defense Council. www.nrdc.org/sites/default/files/policy-basics-fracking-FS.pdf.

NYT. 2009. "Opinion | The Halliburton Loophole." *New York Times*, November 3, sec. Opinion.

Oakland, John S., Robert J. Oakland, and Michael A. Turner. 2020. *Total Quality Management and Operational Excellence: Text with Cases*. 5th ed. London: Routledge. https://doi.org/10.4324/9781315561974.

Odum, Eugene Pleasants. 1971. *Fundamentals of Ecology*. 3rd ed. Berkeley: University of California Press.

OIG. 2007. "*Strategic Agricultural Initiative Needs Revisions to Demonstrate Results*." Washington, DC: US Environmental Protection Agency Office of the Inspector General.

2018. "*Management Weaknesses Delayed Response to Flint Water Crisis*." Washington, DC: US Environmental Protection Agency Office of the Inspector General.

O'Leary, Rosemary. 2020. *The Ethics of Dissent: Managing Guerrilla Government*. 3rd ed. Washington, DC: CQ Press.

O'Leary, Rosemary, Robert F. Durant, Daniel J. Fiorino, and Paul S. Wieland. 1999. *Managing for the Environment: Understanding the Legal, Organizational, and Policy Challenges*. San Francisco: Jossey-Bass.

O'Neill, Daniel W., Andrew L. Fanning, William F. Lamb, and Julia K. Steinberger. 2018. "A Good Life for All within Planetary Boundaries." *Nature Sustainability* 1 (2): 88–95. https://doi.org/10.1038/s41893-018-0021-4.

OP NCEE. 2014. "Regulatory and Non-regulatory Approaches to Pollution Control." In *Guidelines for Preparing Economic Analyses*. Washington, DC: US Environmental Protection Agency National Center for Environmental Economics in the Office of Policy. www.epa.gov/sites/production/files/2017-09/documents/ee-0568-04.pdf.

Osborne, David, and Ted Gaebler. 1993. *Reinventing Government: How the Entrepreneurial Spirit Is Transforming the Public Sector*. New York: Plume.

Ostrom, Elinor. 1990. *Governing the Commons: The Evolution of Institutions for Collective Action*. Cambridge: Cambridge University Press.

OTA. 1995. *Environmental Policy Tools: A User's Guide*. OTA-ENV-634. Washington, DC: US Congress, Office of Technology Assessment.

Oxford. 2021a. "Definition of SCROUNGING." In *Oxford English Dictionary*. Oxford: Oxford University Press.

2021b. "Definition of POLITICAL WILL." In *Oxford English Dictionary*. Oxford: Oxford University Press.

Pearlstein, Steven. 2014. "The Federal Outsourcing Boom and Why It's Failing Americans." *Washington Post*, January 31, sec. Business. www.washingtonpost.com/business/the-federal-outsourcing-boom-and-why-its-failing-americans/2014/01/31/21d03c40-8914-11e3-833c-33098f9e5267_story.html.

Pelch, Katherine E., Anna Reade, Taylor A. M. Wolffe, and Carol F. Kwiatkowski. 2019. "PFAS Health Effects Database: Protocol for a Systematic Evidence Map." *Environment International* 130 (September): 104851. https://doi.org/10.1016/j.envint.2019.05.045.

Perlow, Leslie A., Constance Noonan Hadley, and Eunice Eun. 2017. "Stop the Meeting Madness." *Harvard Business Review*, July 1.

Persson, Linn, Bethanie M. Carney Almroth, Christopher D. Collins, Sarah Cornell, Cynthia A. de Wit, Miriam L. Diamond, et al. 2022. "Outside the Safe Operating Space of the Planetary Boundary for Novel Entities." *Environmental Science & Technology*, January. https://doi.org/10.1021/acs.est.1c04158.

Pimentel, David, ed. 1991. *Handbook of Pest Management in Agriculture*. 2nd ed. Boca Raton, FL: CRC Press.

Pinker, Steven. 2017. "The Second Law of Thermodynamics." *Edge*. www.edge.org/response-detail/27023.

Polk, Emily, and Sibyl Diver. 2020. "Situating the Scientist: Creating Inclusive Science Communication through Equity Framing and Environmental Justice." *Frontiers in Communication* 5. https://doi.org/10.3389/fcomm.2020.00006.

Pope, Arden, Jacob S. Lefler, Ezzati Majid, Joshua D. Higbee, Julian D. Marshall, Kim Sun-Young, et al. 2020. "Mortality Risk and Fine Particulate Air Pollution in a Large, Representative Cohort of US Adults." *Environmental Health Perspectives* 127 (7): 077007. https://doi.org/10.1289/EHP4438.

Popovich, Nadja, Livia Albeck-Ripka, and Kendra Pierre-Louis. 2020. "The Trump Administration Rolled Back More Than 100 Environmental Rules. Here's the Full List." *New York Times*, October 16, sec. Climate. www.nytimes.com/interactive/2020/climate/trump-environment-rollbacks-list.html.

Porter, Michael E. 1991. "America's Green Strategy." *Scientific American* 264 (4): 168.
2011. *Competitive Advantage of Nations: Creating and Sustaining Superior Performance*. New York: Simon and Schuster.

Porter, Michael E., and Claas van der Linde. 1995. "Toward a New Conception of the Environment-Competitiveness Relationship." *Journal of Economic Perspectives* 9 (4): 97–118. https://doi.org/10.1257/jep.9.4.97.

Potoski, Matthew, and Aseem Prakash. 2004. "The Regulation Dilemma: Cooperation and Conflict in Environmental Governance." *Public Administration Review* 64 (2): 152–63. https://doi.org/10.1111/j.1540-6210.2004.00357.x.

Powell, Colin. 2012. *It Worked for Me: In Life and Leadership*. New York: HarperCollins.

Pralle, Sarah. 2019. "Drawing Lines: FEMA and the Politics of Mapping Flood Zones." *Climatic Change* 152 (2): 227–37. https://doi.org/10.1007/s10584-018-2287-y.

Ranco, Darren J., Catherine A. O'Neill, Jamie Donatuto, and Barbara L. Harper. 2011. "Environmental Justice, American Indians and the Cultural Dilemma: Developing Environmental Management for Tribal Health and Well-being." *Environmental Justice* 4 (4): 221–30. https://doi.org/10.1089/env.2010.0036.

Ratto, Fabrizia, Benno I. Simmons, Rebecca Spake, Veronica Zamora-Gutierrez, Michael A. MacDonald, Jennifer C. Merriman, et al. 2018. "Global Importance of Vertebrate Pollinators for Plant Reproductive Success: A Meta-Analysis." *Frontiers in Ecology and the Environment* 16 (2): 82–90. https://doi.org/10.1002/fee.1763.

Raymond, Colin, Radley M. Horton, Jakob Zscheischler, Olivia Martius, Amir AghaKouchak, Jennifer Balch, et al. 2020. "Understanding and Managing Connected Extreme Events." *Nature Climate Change* 10 (7): 611–21. https://doi.org/10.1038/s41558-020-0790-4.

Raymond, Colin, Tom Matthews, and Radley M. Horton. 2020. "The Emergence of Heat and Humidity Too Severe for Human Tolerance." *Science Advances* 6 (19): eaaw1838. https://doi.org/10.1126/sciadv.aaw1838.

Reade, Anna, Tracy Quinn, and Judith S. Schreiber. 2019. "*PFAS in Drinking Water 2019*." Washington, DC: Natural Resource Defense Council.

Rein, Lisa. 2017. "Federal Employees Are Suspicious of Trump's New Government Efficiency Initiative." *Washington Post*, March 27, sec. PowerPost.

Rice, Ronald E., and Everett M. Rogers. 1980. "Reinvention in the Innovation Process." *Knowledge* 1 (4): 499–514. https://doi.org/10.1177/107554708000100402.

Rich, Nathaniel. 2018. "Losing Earth: The Decade We Almost Stopped Climate Change." *New York Times*, August 1, sec. Magazine. www.nytimes.com/interactive/2018/08/01/magazine/climate-change-losing-earth.html.

Roberts, James R., and Catherine J. Karr. 2012. "Pesticide Exposure in Children." *Pediatrics* 130 (6): e1765–88. https://doi.org/10.1542/peds.2012-2758.

Robinson, John P. 1976. "Interpersonal Influence in Election Campaigns: Two Step-Flow Hypotheses." *Public Opinion Quarterly* 40 (3): 304–19. https://doi.org/10.1086/268307.

Rockström, Johan, Will Steffen, Kevin Noone, Åsa Persson, F. Stuart Chapin, Eric Lambin, et al. 2009. "Planetary Boundaries: Exploring the Safe Operating Space for Humanity." *Ecology and Society* 14 (2). www.jstor.org/stable/26268316.

Rogelberg, Steven G., Joseph A. Allen, Linda Shanock, Cliff Scott, and Marissa Shuffler. 2010. "Employee Satisfaction with Meetings: A Contemporary Facet of Job Satisfaction." *Human Resource Management* 49 (2): 149–72. https://doi.org/10.1002/hrm.20339.

Rogers, Everett M. 1995. *Diffusion of Innovations*. 5th ed. New York: Free Press.

Rogers, Everett, and D. Lawrence Kincaid. 1981. *Communication Networks: Toward* a *New Paradigm for Research*. London: Free Press.

Rogers, Everett M., and Thomas M. Steinfatt. 1999. *Intercultural Communication*. Long Grove, IL: Waveland Press.

Romm, Joseph. 2011. "The Next Dust Bowl." *Nature* 478 (7370): 450–51. https://doi.org/10.1038/478450a.

Romney, Alexander C., Isaac H. Smith, and Gerardo A. Okhuysen. 2019. "In the Trenches: Making Your Work Meetings a Success." *Business Horizons* 62 (4): 459–71. https://doi.org/10.1016/j.bushor.2019.02.003.

Ross, Rachel. 2019. "What Are PFAS?" *Live Science*. www.livescience.com/65364-pfas.html.

Rossetto, Rudy, Giovanna De Filippis, Iacopo Borsi, Laura Foglia, Massimiliano Cannata, Rotman Criollo, and Enric Vázquez-Suñé. 2018. "Integrating Free and Open Source Tools and Distributed Modelling Codes in GIS Environment for Data-Based Groundwater Management." *Environmental Modelling & Software* 107 (September): 210–30. https://doi.org/10.1016/j.envsoft.2018.06.007.

Royse, David, Bruce A. Thyer, and Deborah K. Padgett. 2015. *Program Evaluation: An Introduction to an Evidence-Based Approach*. Boston, MA: Cengage Learning.

SAA. 2021. "*Institutional Memory*." Washington, DC: Society of American Archivists. https://dictionary.archivists.org/entry/institutional-memory.html.

Salop, Steven C., and David T. Scheffman. 1983. "Raising Rivals' Costs." *American Economic Review* 73 (2): 267–71.

Sandman, Peter M. 1993. *Responding to Community Outrage: Strategies for Effective Risk Communication*. Fairfax, VA: American Industrial Hygiene Association.

Schaider, Laurel A., Simona A. Balan, Arlene Blum, David Q. Andrews, Mark J. Strynar, Margaret E. Dickinson, et al. 2017. "Fluorinated Compounds in US Fast Food Packaging." *Environmental Science & Technology Letters* 4 (3): 105–11. https://doi.org/10.1021/acs.estlett.6b00435.

Schiermeier, Quirin. 2015. "The Science behind the Volkswagen Emissions Scandal." *Nature*, September. https://doi.org/10.1038/nature.2015.18426.

Schmidheny, Stephan, and Federico J. L. Zorraquin. 1996. *Financing Change: The Financial Community, Eco-efficiency, and Sustainable Development*. Cambridge, MA: MIT Press. https://ideas.repec.org/a/eee/soceco/v26y1997i4p459-463.html.

Shelley, Donna, Gbenga Ogedegbe, and Brian Elbel. 2014. "Same Strategy Different Industry: Corporate Influence on Public Policy." *American Journal of Public Health* 104 (4): e9–11. https://doi.org/10.2105/AJPH.2013.301832.

Shrader-Frechette, Kristin. 2002. *Environmental Justice: Creating Equality, Reclaiming Democracy*. Oxford: Oxford University Press.

2007. *Taking Action, Saving Lives: Our Duties to Protect Environmental and Public Health*. New York: Oxford University Press.

Smith, Adam, and Jessica L. Matthews. 2015. "Quantifying Uncertainty and Variable Sensitivity within the US Billion-Dollar Weather and Climate Disaster Cost Estimates." *Natural Hazards* 77 (3): 1829–51. https://doi.org/10.1007/s11069-015-1678-x.

Smith, Madeleine, David C. Love, Chelsea M. Rochman, and Roni A. Neff. 2018. "Microplastics in Seafood and the Implications for Human Health." *Current Environmental Health Reports* 5 (3): 375–86. https://doi.org/10.1007/s40572-018-0206-z.

Snow, J. 1856. "On the Mode of Communication of Cholera." *Edinburgh Medical Journal* 1 (7): 668–70.

Solomon, Gina M., Rachel Morello-Frosch, Lauren Zeise, and John B. Faust. 2016. "Cumulative Environmental Impacts: Science and Policy to Protect Communities." *Annual Review of Public Health* 37 (1): 83–96. https://

doi.org/10.1146/annurev-publhealth-032315-021807.

SPEA. 2014. *Government Outsourcing: A Practical Guide for State and Local Governments*. Indiana University School of Public and Environmental Affairs.

Sridhar, R., V. Sachithanandam, T. Mageswaran, R. Purvaja, R. Ramesh, A. Senthil Vel, and E. Thirunavukkarasu. 2016. "A Political, Economic, Social, Technological, Legal and Environmental (PESTLE) Approach for Assessment of Coastal Zone Management Practice in India." *International Review of Public Administration* 21 (3): 216–32. https://doi.org/10.1080/12294659.2016.1237091.

Stankey, George H., Roger N. Clark, and Bernard T. Bormann. 2005. *Adaptive Management of Natural Resources: Theory, Concepts, and Management Institutions*. Corvallis, OR: US Department of Agriculture, Forest Service, Pacific Northwest Research Station.

Stapleton, Philip J., Margaret A. Glover, and S. Petie Davis. 2001. *Environmental Management Systems: An Implementation Guide for Small and Medium-Sized Organizations*. Electronic resource. 2nd ed. Washington, DC: US Environmental Protection Agency.

Starling, Grover. 2005. *Managing the Public Sector*. 9th ed. Belmont, CA: Wadsworth Publishing.

Steffen, Will, Wendy Broadgate, Lisa Deutsch, Owen Gaffney, and Cornelia Ludwig. 2015. "The Trajectory of the Anthropocene: The Great Acceleration." *The Anthropocene Review* 2 (1): 81–98.

Steffen, Will, Katherine Richardson, Johan Rockström, Sarah E. Cornell, Ingo Fetzer, Elena M. Bennett, et al. 2015. "Planetary Boundaries: Guiding Human Development on a Changing Planet." *Science* 347 (6223): 1259855. https://doi.org/10.1126/science.1259855.

Steffen, Will, Johan Rockström, Katherine Richardson, Timothy M. Lenton, Carl Folke, Diana Liverman, et al. 2018. "Trajectories of the Earth System in the Anthropocene." *PNAS*, August: 201810141. https://doi.org/10.1073/pnas.1810141115.

Stone, Richard. 2002. "Counting the Cost of London's Killer Smog." *Science* 298 (5601): 2106–107. https://doi.org/10.1126/science.298.5601.2106b.

Sunderland, Elsie M., Xindi C. Hu, Clifton Dassuncao, Andrea K. Tokranov, Charlotte C. Wagner, and Joseph G. Allen. 2019. "A Review of the Pathways of Human Exposure to Poly- and Perfluoroalkyl Substances (PFASs) and Present Understanding of Health Effects." *Journal of Exposure Science & Environmental Epidemiology* 29 (2): 131–47. https://doi.org/10.1038/s41370-018-0094-1.

Sutter, Paul S. 2013. "The World with Us: The State of American Environmental History." *Journal of American History* 100 (1): 94–119. https://doi.org/10.1093/jahist/jat095.

Taherkhani, Mohsen, Sean Vitousek, Patrick L. Barnard, Neil Frazer, Tiffany R. Anderson, and Charles H. Fletcher. 2020. "Sea-Level Rise Exponentially Increases Coastal Flood Frequency." *Scientific Reports* 10 (1): 1–17. https://doi.org/10.1038/s41598-020-62188-4.

TEF. 2020. "*Hazard vs Risk*." Cincinnati, OH: Toxicology Education Foundation. https://toxedfoundation.org/hazard-vs-risk.

TI. 2020. "*The Leadership Training Market*." Raleigh, NC: Training Industry, Inc. https://trainingindustry.com/wiki/leadership/the-leadership-training-market.

Tinline, Phil. 2016. "Too Good to Be Forgotten – Why Institutional Memory Matters." *BBC News*, March 21, sec. Business. www.bbc.com/news/business-35821782.

TLB. 2021. "*You Get What You Pay For: New Final Rule Limits Use of LPTA*." Chicago: Taft Law Bulletin. https://1npdf11.onenorth.com/pdfrenderer.svc/v1/abcpdf11/GetRenderedPdfByUrl/you-get-what-you-pay-for-new-final-rule-limits-use-of-lpta.pdf/?

url=https%3A%2F%2Fwww.taftlaw.com%2Fpdf%2Fnews-events%2Flaw-bulletins%2Fyou-get-what-you-pay-for-new-final-rule-limits-use-of-lpta.

Tollefson, Jeff. 2018. "US Environmental Group Wins Millions to Develop Methane-Monitoring Satellite." *Nature* 556 (7701): 283. https://doi.org/10.1038/d41586-018-04478-6.

TPC. 2015. "Building Trust in an Age of Rage: Communicating and Collaborating with the Public and Other Stakeholders to Get Things Done." Scottsdale, AZ: The Participation Company. www.epa.gov/sites/production/files/2015-09/documents/tues_atlanta_9_10_245_godec.pdf.

Tropman, John E. 2003. *Making Meetings Work: Achieving High Quality Group Decisions*. London: Sage.

UN. 2021. "The 17 Goals | Sustainable Development." United Nations. https://sdgs.un.org/goals.

UN FAO. 2017. "Watershed Management." Rome: United Nations Food and Agriculture Organization. https://elearning.fao.org/course/view.php?id=649.

UNEP. 2020. "*The Emissions Gap Report 2020*." Nairobi, Kenya: United Nations Environmental Program. www.unep.org/emissions-gap-report-2020.

UNESCO. 2017. "*Wastewater: The Untapped Resource*." Paris: United Nations Educational, Scientific, and Cultural Organization. www.unep.org/resources/publication/2017-un-world-water-development-report-wastewater-untapped-resource.

US CDC. 2019. "*Per- and Polyfluorinated Substances (PFAS) Factsheet | National Biomonitoring Program | CDC*." Atlanta, GA: US Centers for Disease Control and Prevention. www.cdc.gov/biomonitoring/PFAS_FactSheet.html.

US DoJ. 2021. "*Freedom of Information Act*." Washington, DC: US Department of Justice. www.foia.gov.

US EPA. 1969. "*National Environmental Policy Act*." Overviews and Factsheets. Washington, DC: United States Environmental Protection Agency. www.epa.gov/nepa.

1991. "*Contract Administration*." Washington, DC: US Environmental Protection Agency. https://nepis.epa.gov/Exe/ZyNET.exe/9101P0MW.TXT?ZyActionD=ZyDocument&Client=EPA&Index=1986+Thru+1990&Docs=&Query=&Time=&EndTime=&SearchMethod=1&TocRestrict=n&Toc=&TocEntry=&QField=&QFieldYear=&QFieldMonth=&QFieldDay=&IntQFieldOp=0&ExtQFieldOp=0&XmlQuery=&File=D%3A%5Czyfiles%5CIndex%20Data%5C86thru90%5CTxt%5C00000030%5C9101P0MW.txt&User=ANONYMOUS&Password=anonymous&SortMethod=h%7C-&MaximumDocuments=1&FuzzyDegree=0&ImageQuality=r75g8/r75g8/x150y150g16/i425&Display=hpfr&DefSeekPage=x&SearchBack=ZyActionL&Back=ZyActionS&BackDesc=Results%20page&MaximumPages=1&ZyEntry=1&SeekPage=x&ZyPURL.

1995. "*Environmental Policy Tools: A User's Guide*." Washington, DC: US Environmental Protection Agency.

1996. "*RCRA Public Participation Manual, 1992 Edition*." Washington, DC: US Environmental Protection Agency. www.epa.gov/sites/default/files/2017-01/documents/rcra_pub_participtn_man.pdf.

1999. "Consideration of Cumulative Impacts." US Environmental Protection Agency. www.epa.gov/sites/production/files/2014-08/documents/cumulative.pdf.

2001. "*US EPA Risk Assessment Guidance for Superfund: Volume III – Part A, Process for Conducting Probabilistic Risk Assessment*." EPA 540-R-02-002. Washington, DC: US Environmental Protection Agency. www.epa.gov/sites/production/files/2015-09/documents/rags3adt_complete.pdf.

2010. "*2010–2014 Pollution Prevention (P2) Program Strategic Plan February 2010*." Washington, DC: US Environmental Protection Agency. www.epa.gov/p2/2010-

2014-pollution-prevention-program-strategic-plan.
2011. "*Environmental Justice*." Washington, DC: US Environmental Protection Agency. www.epa.gov/environmentaljustice.
2013a. "*National Strategy for Improving Oversight of State Enforcement Performance*." Washington, DC: US Environmental Protection Agency. www.epa.gov/compliance/national-strategy-improving-oversight-state-enforcement-performance.
2013b. "*Our Mission and What We Do.*" Overviews and Factsheets. Washington, DC: US Environmental Protection Agency. www.epa.gov/aboutepa/our-mission-and-what-we-do.
2013c. "*Public Participation in the NPDES Permit Issuance Process*." Washington, DC: US Environmental Protection Agency. www3.epa.gov/npdes/pubs/publicparticipation.pdf.
2013d. "*The Origins of EPA*." Collections and Lists. Washington, DC: US Environmental Protection Agency. www.epa.gov/history/origins-epa.
2013e. "*Summary of Executive Order 12898 – Federal Actions to Address Environmental Justice in Minority Populations and Low-Income Populations*." Overviews and Factsheets. Washington, DC: US Environmental Protection Agency. www.epa.gov/laws-regulations/summary-executive-order-12898-federal-actions-address-environmental-justice.
2013f. "*Summary of the Clean Water Act*." Overviews and Factsheets. Washington, DC: US Environmental Protection Agency. www.epa.gov/laws-regulations/summary-clean-water-act.
2013g. "*Consent Decree Agreements*." Overviews and Factsheets. Washington, DC: US Environmental Protection Agency. www.epa.gov/enforcement/negotiating-superfund-settlements.
2014a. "*Framework for Human Health Risk Assessment to Inform Decision Making*." Washington, DC: US Environmental Protection Agency. www.epa.gov/risk/framework-human-health-risk-assessment-inform-decision-making.
2014b. "*Introduction to the Public Participation Toolkit*." Washington, DC: US Environmental Protection Agency. www.epa.gov/sites/production/files/2014-05/documents/ppg_english_full-2.pdf.
2014c. "*Overview of EPA's Brownfields Program*." Overviews and Factsheets. Washington, DC: US Environmental Protection Agency. www.epa.gov/brownfields/overview-epas-brownfields-program.
2014d. "*Learn about Dioxin*." Overviews and Factsheets. Washington, DC: US Environmental Protection Agency. www.epa.gov/dioxin/learn-about-dioxin.
2014e. "*Contacts in the Office of Pesticide Programs, Science Division (BEAD, EFED, HED)*." Overviews and Factsheets. Washington, DC: US Environmental Protection Agency. www.epa.gov/pesticide-contacts/contacts-office-pesticide-programs-science-division-bead-efed-hed.
2014f. "*National Pollutant Discharge Elimination System (NPDES)*." Collections and Lists. Washington, DC: US Environmental Protection Agency. www.epa.gov/npdes.
2014g. "*Learn about Sustainability*." Overviews and Factsheets. Washington, DC: US Environmental Protection Agency. www.epa.gov/sustainability/learn-about-sustainability.
2014h. "*Assessing Pesticides under the Endangered Species Act.*" Other Policies and Guidance. Washington, DC: US Environmental Protection Agency. www.epa.gov/endangered-species/assessing-pesticides-under-endangered-species-act.
2014i. "*Federal Inventories Reform Act*." Washington, DC: US Environmental Protection Agency. www.epa.gov/contracts/federal-activities-inventory-reform-act.
2015a. "*Understanding the Safe Drinking Water Act*." Washington, DC: US

Environmental Protection Agency. www.epa .gov/sdwa/overview-safe-drinking-water-act.

2015b. "*Overview of the Clean Air Act and Air Pollution.*" Collections and Lists. Washington, DC: US Environmental Protection Agency. www.epa.gov/clean-air-act-overview.

2015c. "*About the TSCA Chemical Substance Inventory.*" Overviews and Factsheets. Washington, DC: US Environmental Protection Agency. www.epa.gov/tsca-inventory/about-tsca-chemical-substance-inventory.

2015d. "*Types of Nonpoint Source Pollution.*" Overviews and Factsheets. Washington, DC: US Environmental Protection Agency. www .epa.gov/nps/types-nonpoint-source-pollution.

2015e. "Superfund: CERCLA Overview." Overviews and Factsheets. US EPA. September 9, 2015. www.epa.gov/superfund/superfund-cercla-overview.

2015f. "*Basic Information about Nonpoint Source (NPS) Pollution.*" Overviews and Factsheets. Washington, DC: US Environmental Protection Agency. www.epa .gov/nps/basic-information-about-nonpoint-source-nps-pollution.

2016a. "*Overview of the Pesticide Environmental Stewardship Program.*" Overviews and Factsheets. Washington, DC: US Environmental Protection Agency. www .epa.gov/pesp/overview-pesticide-environmental-stewardship-program.

2016b. "*Basic Information on PFAS.*" Overviews and Factsheets. Washington, DC: US Environmental Protection Agency. www.epa.gov/pfas/basic-information-pfas.

2017. "*Population Surrounding US EPA Superfund, RCRA CA, and Brownfield Sites.*" Washington, DC: US Environmental Protection Agency. www.epa.gov/sites/production/files/2015-09/documents/weballsites9.28.15.pdf.

2020a. "*Toxic Release Inventory Data and Tools.*" Data and Tools. Washington, DC: US Environmental Protection Agency. www .epa.gov/toxics-release-inventory-tri-program/tri-data-and-tools.

2020b. "*TRI National Analysis Executive Summary 2019.*" Washington, DC: US Environmental Protection Agency. https://gispub.epa.gov/trina2019/execsum/#sectors.

2020c. "*Stop Sale, Use, or Removal Orders Issued to Amazon.com Services LLC.*" Overviews and Factsheets. Washington, DC: US Environmental Protection Agency. www .epa.gov/enforcement/stop-sale-use-or-removal-orders-issued-amazoncom-services-llc.

2020d. "*Regional and Geographic Offices.*" Collections and Lists. Washington, DC: US Environmental Protection Agency. www.epa .gov/aboutepa/regional-and-geographic-offices.

2021a. "*Basic Information on PFAS.*" Overviews and Factsheets. Washington, DC: US Environmental Protection Agency. www .epa.gov/pfas/basic-information-pfas.

2021b. "*Cleanups in My Community: Cleaning Up Our Land, Water and Air Digital Database.*" Washington, DC: US Environmental Protection Agency. https://ofmpub.epa.gov/apex/cimc/f?p=cimc:map:0:::71.

2021c. "*Code of Federal Regulations 1508.1 Definitions of the National Environmental Policy Act.*" Text. Washington, DC: US Environmental Protection Agency. www.ecfr .gov.

2021d. "*Criteria Air Pollutants.*" Other Policies and Guidance. Washington, DC: US Environmental Protection Agency. www.epa .gov/criteria-air-pollutants.

2021e. "*EPA's Budget and Spending 1970–2020.*" Overviews and Factsheets. Washington, DC: US Environmental Protection Agency. www.epa.gov/planandbudget/budget.

2021f. "*Flint Drinking Water Response.*" Overviews and Factsheets. Washington, DC: US Environmental Protection Agency. www .epa.gov/flint.

2021g. "*Hazardous Air Pollutants.*" Collections and Lists. Washington, DC: US Environmental Protection Agency. www.epa .gov/haps.

2021h. "*History of the Resource Conservation and Recovery Act (RCRA) and the Hazardous and Solid Waste Amendments of 1984.*" Other Policies and Guidance. Washington, DC: US Environmental Protection Agency. www.epa.gov/rcra/ history-resource-conservation-and-recovery-act-rcra.

2021i. "*National Priorities List (NPL) Sites as of May 2021.*" Data and Tools. Washington, DC: US Environmental Protection Agency. www.epa.gov/superfund/national-priorities-list-npl-sites-state.

2021j. "*Summary of the Resource Conservation and Recovery Act.*" Overviews and Factsheets. Washington, DC: US Environmental Protection Agency. www.epa .gov/laws-regulations/summary-resource-conservation-and-recovery-act.

2021k. "*US Environmental Protection Agency Laws and Regulations.*" Collections and Lists. Washington, DC: US Environmental Protection Agency. www.epa.gov/laws-regulations/regulations.

2021l. "*Watershed Academy.*" Overviews and Factsheets. Washington, DC: US Environmental Protection Agency. www.epa .gov/watershedacademy.

US EPA. 2021a. "*About EPA Organizational Structure.*" Collections and Lists. Washington, DC: US Environmental Protection Agency. www.epa.gov/aboutepa.

US EPA. 2021b. "*Learn about Sustainability.*" Overviews and Factsheets. Washington, DC: US Environmental Protection Agency. www .epa.gov/sustainability/learn-about-sustainability.

USA. 2020. "*Commanders Guide to Environmental Requirements 2019.*" San Antonio, Texas: US Army Environmental Command. https://aec.army.mil/application/ files/6015/5016/3759/Commanders_Guide__ FINAL_online_version.pdf.

USCG. 2021. "*Oil Pollution Act of 1990.*" Washington, DC: US Coast Guard. www .uscg.mil/Mariners/National-Pollution-Funds-Center/About_NPFC/opa.

USDA. 2021. "*About the US Department of Agriculture.*" Washington, DC: United States Department of Agriculture. www.usda.gov/ our-agency/about-usda.

USFR. 2021. "Climate Change and Financial Stability." US Federal Reserve. www .federalreserve.gov/econres/notes/feds-notes/ climate-change-and-financial-stability-20210319.htm.

USGCRP. 2018. "*Impacts, Risks, and Adaptation in the United States: Fourth National Climate Assessment, Volume II: Report-in-Brief.*" Washington, DC: United States Global Change Research Program.

USGS, ed. 2006. *Pesticides in the Nation's Streams and Ground Water, 1992–2001.* The Quality of Our Nation's Waters 1291. Reston, VA: US Geological Survey.

USMC. 2018. *MCDP 1 Warfighting*. United States Marine Corps. Quantico, VA: Independently published.

Vermeulen, Roel, Emma L. Schymanski, Albert-László Barabási, and Gary W. Miller. 2020. "The Exposome and Health: Where Chemistry Meets Biology." *Science* 367 (6476): 392–96. https://doi.org/10.1126/ science.aay3164.

Waldo, Dwight. 1980. *The Enterprise of Public Administration: A Summary View*. Novato, CA: Chandler & Sharp Publishers.

Watts, Nick, Neil Adger, Qiang Zhang, Peng Gong, Hugh Montgomery, and Anthony Costello. 2015. "Health and Climate Change: Policy Responses to Protect Public Health." *The Lancet* 386: 1861–914.

Watts, Nick, Markus Amann, Sonja Ayeb-Karlsson, Kristine Belesova, Timothy

Bouley, Maxwell Boykoff, et al. 2017. "The Lancet Countdown on Health and Climate Change: From 25 Years of Inaction to a Global Transformation for Public Health." *The Lancet* S0140–6736 (17): 32464–69. https://doi.org/10.1016/S0140-6736(17)32464-9.

Watts, Nick, Markus Amann, Nigel Arnell, Sonja Ayeb-Karlsson, Kristine Belesova, Maxwell Boykoff, et al. 2019. "The 2019 Report of The Lancet Countdown on Health and Climate Change: Ensuring That the Health of a Child Born Today Is Not Defined by a Changing Climate." *The Lancet* 394 (10211): 1836–78. https://doi.org/10.1016/S0140-6736(19)32596-6.

Webb, J. Angus, Robyn J. Watts, Catherine Allan, and John C. Conallin. 2018. "Adaptive Management of Environmental Flows." *Environmental Management* 61 (3): 339–46. https://doi.org/10.1007/s00267-017-0981-6.

WEF. 2020. "*Public Trust Doctrine*." Sacramento, CA: Water Education Foundation. www.watereducation.org/aquapedia/public-trust-doctrine.

Weiland, Paul S. 1999. "Preemption of Local Efforts to Protect the Environment: Implications for Local Government Officials." *Virginia Environmental Law Journal* 18 (4): 467–506.

Westover, Robert Hudson. 2016. "Conservation versus Preservation? | US Forest Service." www.fs.usda.gov/features/conservation-versus-preservation.

Whitehead, Heather D., Marta Venier, Yan Wu, Emi Eastman, Shannon Urbanik, Miriam L. Diamond, et al. 2021. "Fluorinated Compounds in North American Cosmetics." *Environmental Science & Technology Letters* 8 (7): 538–44. https://doi.org/10.1021/acs.estlett.1c00240.

Wilson, Edward O. 1987. "The Little Things That Run the World (the Importance and Conservation of Invertebrates)." *Conservation Biology* 1 (4): 344–46.

2010. *The Creation: An Appeal to Save Life on Earth*. New York: W. W. Norton & Company.

Wolanski, Eric, Robert Richmond, Laurence McCook, and Hugh Sweatman. 2003. "Mud, Marine Snow and Coral Reefs: The Survival of Coral Reefs Requires Integrated Watershed-Based Management Activities and Marine Conservation." *American Scientist* 91 (1): 44–51.

Woodrow, K. D. 1994. "The Proposed Federal Environmental Sentencing Guidelines: A Model for Corporate Environmental Compliance Programs." *BNA Environment Reporter* 25: 325.

Woods, Neal D. 2020. "An Environmental Race to the Bottom? 'No More Stringent' Laws in the American States." *Publius: The Journal of Federalism*, no. pjaa031 (October). https://doi.org/10.1093/publius/pjaa031.

Woolhandler, Steffie, David U. Himmelstein, Sameer Ahmed, Zinzi Bailey, Mary T. Bassett, Michael Bird, et al. 2021. "Public Policy and Health in the Trump Era." *The Lancet* 397 (10275): 705–53. https://doi.org/10.1016/S0140-6736(20)32545-9.

Worland, Justin. 2018. "The EPA Head's Controversial Room Rental Is Not His Only Tie to Natural Gas." *Time*. https://time.com/5222286/scott-pruitt-apartment-natural-gas.

WRCOG. 2021. "*What Are Councils of Governments? | WRCOG, CA*." Riverside, CA: Western Riverside Council of Governments. https://wrcog.us/246/What-are-Councils-of-Governments.

Wright, Stephanie L., and Frank J. Kelly. 2017. "Plastic and Human Health: A Micro Issue?" *Environmental Science & Technology* 51 (12): 6634–47. https://doi.org/10.1021/acs.est.7b00423.

Wu, Jin, Derek Watkins, Josh Williams, Shalini Venugopal Bhagat, Hari Kumar, Jeffrey

Gettleman, et al. 2020. "Who Gets to Breathe Clean Air in New Delhi?" *New York Times*, December 17, sec. World. www.nytimes.com/interactive/2020/12/17/world/asia/india-pollution-inequality.html.

Yamamuro, Masumi, Takashi Komuro, Hiroshi Kamiya, Toshikuni Kato, Hitomi Hasegawa, and Yutaka Kameda. 2019. "Neonicotinoids Disrupt Aquatic Food Webs and Decrease Fishery Yields." *Science* 366 (6465): 620–23. https://doi.org/10.1126/science.aax3442.

Ziman, John. 1996. "Is Science Losing Its Objectivity?" *Nature* 382 (August): 751–54. https://doi.org/10.1038/382751a0.

Index